Environmental Footprints and Eco-design of Products and Processes

Series Editor

Subramanian Senthilkannan Muthu, Green Story Inc.
Toronto, Canada

Indexed by Scopus

This series aims to broadly cover all the aspects related to environmental assessment of products, development of environmental and ecological indicators and eco-design of various products and processes. Below are the areas fall under the aims and scope of this series, but not limited to: Environmental Life Cycle Assessment; Social Life Cycle Assessment; Organizational and Product Carbon Footprints; Ecological, Energy and Water Footprints; Life cycle costing; Environmental and sustainable indicators; Environmental impact assessment methods and tools; Eco-design (sustainable design) aspects and tools; Biodegradation studies; Recycling; Solid waste management; Environmental and social audits; Green Purchasing and tools; Product environmental footprints; Environmental management standards and regulations; Eco-labels; Green Claims and green washing; Assessment of sustainability aspects.

Rubee Singh • Wasswa Shafik • Vikas Kumar
Editors

Blockchain Technologies for Smart Circular Economy and Organisational Sustainability

Springer

Editors
Rubee Singh
Institute of Business Management
GLA University
Mathura, Uttar Pradesh, India

Wasswa Shafik
School of Digital Science
Universiti Brunei Darussalam
Bandar Seri Begawan, Brunei Darussalam

Vikas Kumar
University of Portsmouth
Portsmouth, UK

ISSN 2345-7651 ISSN 2345-766X (electronic)
Environmental Footprints and Eco-design of Products and Processes
ISBN 978-3-031-89182-3 ISBN 978-3-031-89183-0 (eBook)
https://doi.org/10.1007/978-3-031-89183-0

This Springer imprint is published by the registered company Springer Nature Switzerland AG
The registered company address is: Gewerbestrasse 11, 6330 Cham, Switzerland

If disposing of this product, please recycle the paper.

Preface

Depletion of natural resources points to the necessity for alternative mechanisms in the form of a sustainable living strategy. In this realm, both a smart circular economy and organizational sustainability are prime candidates. The smart or digital circular economy combines the principles of circularity with digital technologies like blockchain, AI, IoT, and big data. Digital circular economy goes beyond the traditional circular economy by leveraging digitalization to optimize resource use, enhance product lifecycle management, and enable real-time decision-making. It holds immense promise in accelerating the transition to sustainable economic systems by promoting transparency, traceability, and efficiency across supply chains. However, while the digital circular economy offers transformative potential, its adoption requires overcoming significant technical, regulatory, and organizational barriers. Exploring the interplay between blockchain technologies and the digital circular economy reveals new opportunities for innovation, creating systems that are not only more sustainable but also more adaptive and resilient in addressing environmental and economic challenges.

Blockchain technologies are envisioned to coexist with the principles of the smart circular economy and address the global issues of sustainable development. Given the potential of blockchain systems to consistently work with the principles of a smart circular economy, there are still distinct aspects that require further research and assessment. Current perspectives on addressing these drawbacks regarding research in the field of the smart circular economy. Globally, adding blockchain to the domain of sustainability has received considerable attention from scholars. Exploring the properties of blockchain in economic-related systems can lead to insights into how blockchain can help sustainability efforts. Overall, and given the paramount implications for a vast range of economic-related sectors, this topic of research seems very ripe for investigation. An understanding of current breakthroughs in the marriage of blockchain and the domain-specific smart circular economy is key. However, a rigorous overview of the current breakthroughs could be helpful for scholars, practitioners, and stakeholders in economic-related sectors. Several sectors are likely to benefit from an overview of recent developments that could have transformative impacts on their systems.

The proponents of blockchain technologies indeed envision a revolution for industries and societal systems that would provide a new approach to value creation, the fourth revolution or Industry 4.0. While much of this is yet to be realized, it is fine for dreamers to dream, and there has been intriguing confluence in the literature in both the industry, broader business literature, and academic literature in envisioning blockchain's role in a more responsible world... what we call contributing to sustainable development goal (SDG) 12 and even beyond. Simultaneously, however, naysayers have questioned the potential of blockchain to influence very much at all, as its phrases hyping smokes and mirrors might have become parts of our future potential. Thus, while scholars are currently 'baptizing' or 'rearing' this field, vast expectations from public and private entities have been brewed, thus cultivating societal hopes and dreams. We strongly agree that the official adoption and embrace of blockchain can be much more than simply a new technological inflection but, in fact, reflects societal innovation and inclusion.

It was argued that one of the most enticing implications of blockchain technology can be found in supply chain management. It is important to account for the role of blockchain technology in promoting trust and transparency within supply chains. Participants in this study suggested that blockchain technology can enhance accountability within supply chains. A range of approaches to supply chain transparency and traceability systems were presented, and the importance of connecting distributed nodes and interoperability between different systems was also highlighted. According to the community of practice, blockchain technology is also suitable to accelerate and automate data collection in distributed locations. Measuring and reporting sustainability outcomes were identified as good indicators of best practice. However, it was also noted that simply having access to the technology does not automatically deliver sustainability. A shift first needs to occur in terms of promoting an understanding of the relationship between sustainability measures in blockchain technology and improving the methods to document later. Barriers to the implementation of blockchain technology may occur at the organization, adoption, legal, and funding phases.

Where thought and practice are aligned, opportunities for research into blockchain's wider contributions to society will appear. An inescapable reality bounding such future work is that SDGs are too broad and complex to be solved by one corporation, one sector, or one form of research. Thus, given that research is becoming indelibly transdisciplinary, concertation and dialogue in this tri-locational space provide a meeting place for three important, and often prismatic, facets of scholars: those about theoretical insights about technology and the society it affects; those from industry continuing to innovate and apply new technologies; and a new set of interdisciplinary scholars thinking critically across traditional academic borders, working in close collaboration with industry on innovative ideas. Theory, research, and applications of the new, such as blockchain, can contribute to all ten of the SDGs that are pinned to sustainability. Definition of all 17 SDG's. Thus, there is yet much to discuss and talk about in this space, and we suggest that the main goal, which is to understand research, industry implementation, and practical policy, can be greatly assisted by understanding both what the industry is currently thinking and how these thoughts are changing as reality meets the dream.

At the same time, silver bullets often attract irrational levels of hope, and we need to maintain a clear-eyed understanding of the ongoing and enduring challenges and limits to implementing blockchain strategies. This series provides one such dialogue. We hope that this book will help elevate shared knowledge and interdisciplinary scholarship efforts in contributing to a more sustainable world, forming more future-defining practices, and more rounded and careful future research on this technology's potential and limitations. Showing through a dialogue between scholars currently baptizing a field what they think of and how their thoughts have more practically actualized through time is helpful for future research, discourse, and business strategies. Thus, we hope for more than a grounded look at a wider phenomenon in this book; we hope the provided insights help in any sustainable future as it is suitable for students, researchers, and professionals.

Mathura, Uttar Pradesh, India — Rubee Singh
Bandar Seri Begawan, Brunei Darussalam — Wasswa Shafik
Portsmouth, UK — Vikas Kumar

Acknowledgements

We would like to express our gratitude to all those who have contributed to the creation of this book. Firstly, we thank the researchers and experts who have shared their knowledge and insights with us, providing valuable information on the current state of 'Blockchain Technologies for Smart Circular Economy and Organisational Sustainability'. We also extend our appreciation to the organizations and institutions that have supported our work, including government agencies, non-governmental organizations, and academic institutions.

Finally, we would like to acknowledge the invaluable contributions of our colleagues and editors, who have worked tirelessly to ensure that this book is informative, engaging, and accessible to a wide range of readers. We hope that this book will inspire further research and action towards a smart circular economy and blockchain technologies for organizations to promote sustainability as per United Nations agenda. We would like to express our sincere gratitude to the experts and stakeholders who contributed their valuable insights and perspectives to this book. Their contributions have been instrumental in shaping our understanding of the challenges and opportunities involved in achieving smart circular economy with the use of blockchain technologies globally. We hope that this book will serve as a useful resource for policymakers, practitioners, and researchers working towards a circular economy and a sustainability future for the continent.

We are immensely thankful to **Series Editor, Prof. Subramanian Senthikannan Muthu, Publishing-Executive Editor Mr. Zachary Romano**, and the entire team at Springer Nature for bringing out this publication in a very professional and timely manner.

Rubee Singh
Wasswa Shafik
Vikas Kumar

About the Book

The increasing need for a sustainable world and the integration of digital technology have helped redirect attention towards a circular economy. This raises the question of how blockchain can work within the circular economy. Research on the integration of blockchain with the circular economy is in its early stages. Decentralized systems and platforms from blockchain technology can change supply chain mechanisms or structural functions to achieve a circular economy. In a circular economy system, resources are traced through a virtual cycle to support future reproduction and create as little waste as possible. In a general sense, blockchain enhances transparency, sustainability, and security through cryptographic, social, and technical mechanisms. Blockchain's intrinsic and enabling features position it as a tool for updating traditional business models. This shift will affect different processes, including production, waste management, inventory management, and supply chain management. Blockchain can be used to underpin functions by which stakeholders have access to necessary yet trustworthy information concerning product life cycles and processes. More specifically, researchers proposed cases for the use of blockchain to track the origin and journey of products. The real-time sharing of certified data from IoT devices and along digital and physical supply chain nodes can give rise to more efficient, shared, and connected supply and recycling chains. Blockchain could establish a dynamic, real-time, secure, certifiable mechanism for shifting goods and products and information concerning them from one process to another based on the demand for action, the availability or need of resources, and the commercial viability of the act. Circular capabilities for a product and its components could be established via blockchain in a way that provides trustworthy information, making this book:

- Dissect how blockchain enhances transparency, traceability, and accountability in supply chains to drive sustainable and waste-free systems.
- Present practical frameworks and strategies to integrate blockchain into organizational sustainability practices for long-term benefits.

- Validate blockchain's role in optimizing resources, reducing waste, and increasing energy efficiency within a circular economy model.
- Feature global case studies that highlight successful blockchain applications across industries, such as manufacturing, energy, and agriculture.
- Examine emerging trends, ethical considerations, and future opportunities for blockchain in advancing sustainable development goals.

Contents

About the Editors

Rubee Singh, Ph.D. is an Assistant Professor of Human Resources at the Institute of Business Management, GLA University, India. She is also a Postdoctoral (D. Litt) Researcher at the Department of Management Studies, Kumaun University, Nainital, India. She holds Ph.D. in Management from Noida International University, India, and an MBA in HR from Dr. A.P.J. Abdul Kalam Technical University, India. She has more than 11 years of experience in research and teaching. Apart from teaching and research, she has handled many administrative positions in premier institutes affiliated with A.K.T.U. She has conducted a number of FDPs, conferences, workshops, and HR conclaves in the area of sustainability, CSR, AI, HR, human factors, women empowerment, and human rights. She is a United Nations Volunteer and a contributing author for Voices of Youth (UNICEF). She is an honorary life member of the Women Economic Forum and National Secretary of IATM (Youth Wing), International Anti-Terrorism Movement in India. Dr. Singh is the Managing Editor of IJARSH India. She serves on the editorial boards of more than 30 international journals. Additionally, she is an Advisory Board Member of the Heliyon Journal, Elsevier, and Research in Globalization and a reviewer for many reputed journals, including the *European Journal of Innovation Management* (EJIM), the *British Food Journal*, *Sustainability*, and the *Social Responsibility Journal*. She is an Associate Editor of *Circular Economy and Sustainability Journal (Scopus-Q1), Springer*.

She is a life member of the International Association of Engineers, Thailand, IMRTC USA, ISRD UK, among others. She is a bestselling author with significant experience as an excellent editor, eminent researcher, great academician, global ambassador (children's), and social activist. She has presented many papers at national and international conferences and has published 25+ academic books with Springer, Elsevier, De Gruyter, and Emerald Publishing. Additionally, she has authored more than 100+ research papers in SCI, ABDC-A, Scopus, and refereed/UGC-indexed journals. She has an excellent research track record with 1100+ citations on Google Scholar and H-index-21. Her areas of research include sustainability, digital HRM, sustainable HRM, circular economy, digital CE, net zero emissions, climate change, Industry 4.0, artificial intelligence, blockchain technology, and corporate social responsibility. Being a thought leader, her books are honoured by being added to the library, Ohio Christian University, University of California, Corporacion Universitaria Unisabaneta, UMass Chan Medical School, Tunxis Community College, USA, Aston University, UK, and the University of Port Harcourt, Nigeria. Her book, titled *Training and Development*, was added to the MBA syllabus of the 'Institute of Aeronautical Engineering' (Jawaharlal Nehru Technological University, Hyderabad). Dr. Singh has been honoured with many prestigious awards, including Most Influential Woman Leaders in 2022, consecutive Best Researcher Awards of the Year 2022–2023 and 2023–2024 by GLA University, India, Child Ambassador Award of the Year 2022, Iconic Faculty Award 2019, International Young Researchers Award 2021, Young Scientist Award 2020, and Nari Ratan Award 2023. The list of awards and appreciation is long, but Dr. Singh has always prioritized her work over recognition, making her unique and commendable in her field.

Wasswa Shafik, Ph.D. a remarkable computer scientist, information technologist, and educator, hails from the vibrant capital city of Uganda, Kampala. His passion extends beyond technology, encompassing people's health, justice, and the responsible use of technological advancements, regardless of race, religion, or sex. With a strong educational background, he holds a bachelor's degree in information technology engineering with a minor in mathematics from Ndejje University, Uganda. He further pursued a master of engineering degree in information technology engineering (MIT) with a computer and communication network option from Yazd University, Iran, and a PhD in computer science from Universiti Brunei Darussalam, Brunei Darussalam. As a member of the IEEE, Shafik has made significant contributions in various domains, such as smart agriculture, health informatics, computer vision, network science, and security and privacy. His passion for sustainable solutions and his deep understanding of the health and environmental challenges we face today have propelled him to the forefront of sustainable measures. Shafik's interest extends to Smart Cities, where he investigates the intersection of big data, IoT, and artificial intelligence. His work delves into energy management, privacy, and performance evaluation in the context of Internet of Things (IoT) applications within smart urban environments. Additionally, Shafik explores fog computing architectures, privacy, and security solutions. His work also encompasses the evaluation and analysis of fog-mobile edge performance in IoT scenarios. Wasswa's expertise extends to artificial intelligence and its application in cybersecurity and cyberdefense. He reviews the theoretical understanding of deep learning in the context of analysis of UAV biomedical engineering technologies. He has published more than 80 research articles, book sections, and conferences in many reputed journals, including Scopus Q1, Q2, and SCI in different publishers. He has also served as departmental support for mathematics for data science, advanced topics in computing, and advanced algorithms at the School of Digital Science, Universiti Brunei Darussalam. He has served in various capacities, including as a community data officer at Pace-Uganda, a research associate at

TechnoServe, a research assistant at PSI-Uganda, a research lead at the Socio-economic Data Center (SEDC-Uganda), the head of the Dig Connectivity Research Laboratory (DCRLab), Kampala, Uganda, and acting managing director at the Asmaah Charity Organisation. Shafik's journey is one of curiosity, innovation, and a deep-rooted desire to positively impact the world through technology, education, and sustainability.

Vikas Kumar, Ph.D. is a Deputy Vice-Chancellor Research and Innovation at University of Portsmouth, UK. Previously, he was an Associate Dean for Research Innovation and Enterprise at the Faculty of Business, Law and Social Sciences, Birmingham City University, UK. He is also a Professor of Operations and Supply Chain Management. Prof. Kumar has over a decade of teaching and research experience. Prior to joining BCU, he was Director of Research and Scholarship at Bristol Business School, University of the West of England. Prof. Kumar has also worked in India, Ireland, and Hong Kong. He earned his Ph.D. in Management from the University of Exeter, United Kingdom. He did his graduation (B. Tech) at the National Institute of Advanced Manufacturing Technology, Ranchi (formerly NIFFT), India. He has published over 300 journal articles and conference papers, including A*/A as per ABDC and 4*/3* as per ABS-indexed journals. He has an excellent research track record, with 20,500 citations on Google Scholar and an h-index of 77. He serves on the editorial boards of several international journals and has guest-edited several special issues in high-impact journals. He is also a reviewer for more than 15 international journals, including ABS 4/4* and ABS 3* journals. Prof. Kumar works very closely with industries and has secured research funding in excess of £1 million from various research agencies, including Innovate UK, EPSRC, the British Council, the British Academy, the Newton Fund, and the Science Foundation of Ireland. He has worked on several international collaborative projects with researchers from several UK universities and researchers from Brazil, Vietnam, Thailand, Indonesia, India, Mexico, Taiwan, Columbia, Peru, Costa Rica, and Turkey. His research interests include net zero emissions, circular economy, sustainability, Industry 4.0, digital supply chains, sustainable supply chains, and operational excellence.

Part I
Foundations of Blockchain and Smart Circular Economy

Chapter 1
Integrating Sustainable HRM Practices to Enhance Environmental Performance in the Digital Circular Economy

Rubee Singh

Abstract Globalization and unexpected technological upheavals have compelled manufacturers to seek new strategies to maintain the sustainability of their businesses. This research seeks to present a conceptual model that illustrates the significance of human resource management in promoting the adoption of a digital circular economy, which can subsequently impact an organization's environmental performance. This study examines existing literature concerning the role of Industry 4.0 in circular economy practices and the influence of human resources on the adoption of circular economy principles in business. Following an extensive review, the literature is integrated into a conceptual model. The literature research indicated that only a limited number of frameworks incorporate Industry 4.0 technologies for circular economy activities. The function of HRM is overlooked in the move from traditional business management systems to the establishment of a digital circular economy. Integrating HRM practices that merge circular methodologies with the new technologies of Industry 4.0 facilitates the circular advancement of an organization while enhancing its environmental performance. The approach presented in this study integrates conventional HRM practices with the innovative concept of leveraging Industry 4.0 technology into the adoption of a circular economy to improve an organization's environmental performance. The conceptual model provides theoretical and practical implications for optimizing human resources to improve organizational environmental performance via digital circular economy activities.

Keywords HRM Practices · Digital Circular Economy · Environmental Performance · Industry 4.0 · Organizational Performance · Sustainability

---

R. Singh (✉)
Institute of Business Management, GLA University, Mathura, India

Department of Management Studies, Kumaun University, Nainital, India

R. Singh et al. (eds.), *Blockchain Technologies for Smart Circular Economy and Organisational Sustainability*, Environmental Footprints and Eco-design of Products and Processes, https://doi.org/10.1007/978-3-031-89183-0_1

1.1 Introduction

The circular economy (CE) framework is utilized by corporations to manage the after-use phase of not only material but human resources as well, with the goal of creating a sustainable organization (Kirchherr et al., 2017). D'Adamo (2019) emphasized the significance of HR strategies in guaranteeing a consistent and enduring adoption of CE in the organization. Prior research has emphasized that green HRM encompasses various practices such as hiring, skill enhancement, management of performance and compensation of employees, and leadership (Singh et al., 2022). These practices contribute to the generation of green innovation, which manifests as advancements in eco-innovation for the development of an environmental management culture (Renwick et al., 2016). Furthermore, existing research has demonstrated that big data analytics (BDA) is considered a crucial component of CE implementation in the context of HRM for enhanced environmental performance. BDA encompasses integration of information, data visualization techniques, and adapting to an evolving data system for decision-making (Jabbour et al., 2019a). This information is utilized to achieve desired environmental performance within the context of CE.

Prior research primarily focused on a direct or immediate association between HR practices, sustainability, and circular economy performance (CEP), while also pointing out the limited amount of specific information available on how HRM can be a crucial factor leading to CEP, particularly the green HRM (Jabbour et al., 2019b; Dibia et al., 2020). Employee wellbeing is viewed as an important aspect that fosters sustainability in the workplace (Zeshan et al., 2024). Marrucci et al. (2021) found a direct correlation between the needs of institutions and stakeholders and employing a green approach to manage HR. The results highlight the potential for green HRM to become an exclusive practice in top-performing companies. Furthermore, prior studies have refrained from thoroughly examining the association of human resource management (HRM) and CE in light of Industry 4.0 (I4.0). Additionally, digitalization is a recent trend in companies that lack a strong emphasis on improving CEP. Furthermore, the implementation of CE is required for industries that prioritize the improvement of sustainable environmental development (Pagoropoulos et al., 2017).

Digitalization is now viewed as a means to facilitate and expedite the shift towards a CE due to its innovative technologies. The digital technologies (DTs) of Industry 4.0 (I4.0), such as blockchain, artificial intelligence, BDA, and the Internet of Things (IoT), are anticipated to improve product quality through innovative production practices and circular business models. Additionally, these DTs will facilitate knowledge exchange and foster connections among the different parties involved in the value chain (Pagoropoulos et al., 2017). Furthermore, DTs are expected to enhance the role of citizens and consumers by providing them with information and education and empowering them to actively participate in developing a CE. The existing studies on DCE have primarily concentrated on examining how digital technologies facilitate circular strategies (Cagno et al., 2021), identifying instances of its use (Ranta et al., 2021; Singh et al., 2023) and the difficulties encountered during implementation. However, Khan et al. (2024) considered factors like top management involvement and supportive policies as a part of the framework needed for implementing a technology-based CE, but the authors have not specifically addressed the basic functions of HRM.

Piyathanavong et al. (2024) explicitly focused on the organizational culture as a facilitator for transitioning into a DCE. Hence, the current body of research, however, lacks a thorough examination of the potential impact that HRM practices can exert on DCE. In fact, numerous authors are advocating for further investigation into the effects of digital technology applications in the context of the CE (Cagno et al., 2021; Singh et al., 2024; Ranta et al., 2021). This paper addresses the limitations of existing research by providing a conceptual model of the potential impact of HRM on the DCE and the impact of DCE on environmental performance (EP). The paper offers implications related to role of HRM in implementing a DCE enabled by data and DTs in resource-intensive sectors and different geographical locations. This study seeks to contribute to existing theory by proposing a novel link between HRM and DCE and their mutually advantageous impacts on the EP. Ultimately, our research suggests that organizations should foster a culture that emphasizes the adoption of DCE in order to enhance environmental performance and customer experience simultaneously.

The paper has the following structure. After the introduction section, there is a literature review section. These sections broaden our understanding regarding the basic concept of CE and the role of I4.0 in the facilitation of CE. Further, the role of HRM in the adoption of DCE is discussed, and then the conceptual model of the study is presented. The next section discusses the research methodology of the paper. Subsequently, implications of the study are provided both for theory and practice. Finally, the conclusion of the paper is given along with limitations and areas that demand further attention from the researchers.

1.2 Literature Review

1.2.1 Circular Economy

CE solutions aim to maximize the longevity of product value by minimizing waste. They sustain resources within the value chain beyond the point where a product has become no longer useful, allowing it to be reused, consistently creating and preserving value for users across multiple life stages (Singh et al., 2023). Nevertheless, transitioning to a CE requires modifying the entire value chain. To achieve success, there is a need to have technology and processes that support this goal. This includes finding innovative methods to transform waste into valuable products and influencing the consumers' behavior (Pagoropoulos et al., 2017). In the CE, resources are optimally utilized in loops for their prolonged lifespan (Kirchherr et al., 2017).

However, Jabbour et al. (2019a) argued that recycling might be a less feasible approach as it diminishes the value derived from material processing to produce a new item. The purpose behind such processes is the minimization of resource usage, enabling the system to function at its optimal efficiency and averting inefficiency. Remanufacturers in a circular economy should utilize state-of-the-art technology and provide a warranty for their remanufactured items (Khan et al., 2022). Products should be redesigned to facilitate their prolonged usage, repairability, upgradability, remanufacturing, and eventual recycling rather than being discarded and replaced, often ending up in landfills.

1.2.2 Digital Circular Economy: The Integration Between CE and Industry 4.0

The requirements of I4.0 were driven by the necessity for interaction between humans and computers, live supervision, data analytics, and improved product management. Industry 4.0 aims to improve the interaction between machines and between humans and machines, with the goal of integrating social and technical components to create a prosperous human society (Sony & Naik, 2020). Digitalization is crucial for industries to successfully implement I4.0. These technologies offer advantages such as enhanced flexibility, reduced costs, improved service, increased customization of products, and more (Singh et al., 2024). An emerging concern is that the transition to the digital era could have a detrimental effect on environmental sustainability (Nwokolo et al., 2024). The pandemic has significantly affected operations management in every sector, highlighting the unsustainability and lack of resilience in current approaches (Belhadi et al., 2021). Therefore, there is a requirement to transition towards sustainable operations management through CE adoption. The integration of CE and I4.0 will ensure a continuous movement of resources for establishing a sustainable system of distribution (Belhadi et al., 2021), thereby enhancing the efficiency of operations management. Strategically using the convergence of I4.0 and CE can assist industries in attaining operational excellence. Therefore, it is crucial to integrate technology with CE to advance towards a sustainable future.

1.2.3 Role of HRM in the Adoption of Digital CE

Given the essential elements of the CE literature, it is clear that it also possesses notable characteristics of sustainability. Hence, there is a requirement for the human aspect within the CE for attaining sustainability. One way to promote a sustainable system of management, including circular initiatives, as well as other components of organizational sustainability, is by emphasizing the organizational characteristics of the CE (Jabbour et al., 2013; Singh et al., 2022). In order to support CE aims, it is highly vital to provide green training and hire green staff (Jabbour & Jabbour, 2014). Empowering employees, eco-focused hiring and selecting, environment training, environmentally conscious performance evaluation and awards, the creation of sustainable organizational cultures, and green teams all contribute to achieving sustainability (Renwick et al., 2016; Obeidat et al., 2023). In the HR context, it encompasses aspects such as collaboration, organizational culture, the process of hiring and selecting employees, training programs, evaluating performance, and providing incentives (Singh et al., 2022; Subramanian & Suresh, 2022). The involvement of top management is crucial to drive the employee motivation towards CE (Khan et al., 2024).

It is apparent from the above discussion that the role of HRM inside a CE organization is still in its first phases and lacks clarity. Zeshan et al. (2024) found that most authors do not adequately address the impact of social wellbeing on sustainability. Furthermore, Kirchherr et al. (2017) argue that after analyzing more than a hundred scientific papers on CE, the existing knowledge and understanding of CE mostly overlooks the social dimensions. To address the gap in research on this phenomenon, Jabbour et al. (2019b) used an environmentally friendly approach to exploring the link between HR and the role of HRM in CE organizations. They developed a theory base for exemplifying the potential connections between HRM and CE practices at the firm level. Jabbour et al. (2019b) proposed a conceptual framework that suggests CE-enabling HRM is linked to GHRM practices and dimensions. Jabbour et al. (2019b) suggested that the commitment of top-level managers and leadership may support the practices and dimensions of HR and HRM that are hypothesized to enable organizational CE practices. Marrucci et al. (2021) specifically considered the role of HRM in implementing CE and its impact on the environmental performance of an organization. Table 1.1 presents an overview of the most relevant studies.

While several studies examine HRM's role in CE (e.g., Jabbour et al., 2019b; Subramanian & Suresh, 2022), they primarily focus on green HRM without exploring how digital technologies facilitate CE transitions. On the other hand, studies in the manufacturing sector are also scarce as this sector faces unique challenges in adopting CE principles. Additionally, the influence of HRM is mainly associated with sustainability, but DCE and environmental performance outcomes are often understudied. This gap suggests a need for research that establishes how specific HRM practices can improve environmental metrics in a manufacturing sector and how HRM can directly support DCE adoption. We consider four aspects in this chapter to develop a model regarding the impact of HR practices on the adoption of DCE.

Table 1.1 Articles explicitly examined both HRM and CE

	Authors	Objective	Findings
1	Jabbour et al. (2019b)	To explore the role of HR managers in advancing CE, emphasizing the 'human side' of CE	Identifies HR managers as central to CE implementation through leadership, skill-building, and aligning human resources with CE goals
2	Dibia et al. (2020)	To critically assess the role of HRM in CE across six case studies in various organizations	HRM influences CE by fostering social outcomes like job creation and wellbeing, but practical applications remain underdeveloped
3	Marrucci et al. (2021)	To examine the impact of green HRM on CE and performance in certified organizations	Finds that green HRM practices positively influence CE by improving environmental performance in certified organizations
4	Subramanian and Suresh (2022)	To develop a model linking organizational learning and green HRM for CE transition	Organizational learning, paired with green HRM, promotes circular transformation, especially in small- to medium-sized firms
5	Obeidat et al. (2023)	To analyze how green HRM influences CE in manufacturing industries	Green HRM practices significantly enhance CE in manufacturing, aiding in resource efficiency and environmental sustainability
6	Khan et al. (2024)	To propose a framework for integrating Smart CE practices in manufacturing	Smart CE framework aids in the digital transformation of CE in manufacturing
7	Piyathanavong et al. (2024)	To examine industry 4.0 and CE through a developmental culture lens	Developmental culture promotes industry 4.0 and CE integration, offering a competitive edge
8	Zeshan et al. (2024)	To explore sustainable workplaces focusing on employee wellbeing through the job demand-resource model	Sustainable workplaces enhance employee wellbeing, but HRM'S role in CE is not extensively covered

Source: Author's compilation

1.2.4 Organizational Culture

The influence of organizational culture is not just on the management of a business but it can also be seen in its objectives and strategies. It embodies the ingrained principles and convictions that are collectively held by employees inside an organization. For the construction of organizational culture, the management of work practices and their change throughout time inside organizations is crucial (Renwick et al., 2016). It is an important component of the general operations of a business as it plays a crucial role in facilitating clarification, communication, and objectivity regarding various job tasks and work procedures. This, in turn, has a significant impact on the organizational effectiveness and innovative practices for managing day-to-day functions of the business (McLaughlin et al., 2008). Within the organization, culture serves as a means to connect strategy and its implementation by utilizing resources to facilitate and oversee change (Anh Vu et al., 2022). Leadership has been viewed as a prominent factor in establishing a suitable and favorable organizational culture (Sarros et al., 2008). Recent research highlighted the importance of organizational culture in fostering HR orientation for increased performance and improved the overall adoption of the evolved ways of managing processes and business operations. Moreover, the organization's culture is a strategic approach toward DCE adoption (Piyathanavong et al., 2024). In light of this context, we can propose that:

Proposition 1 The organization's culture has a significant impact on the adoption of DCE.

1.2.5 Employees' Skills and Abilities

The skills and abilities possessed by employees in a company are seen as valuable tangible resources that contribute to enhancing business efficiency and establishing a sustained competitive edge (Mousavi et al., 2018). These characteristics enable organizations to effectively apply innovation-driven sustainable methods, such as CE (Khan et al., 2020). Thus, the presence of competent human resources is crucial for the development, modification, adjustment, and dissemination of eco-friendly practices within commercial organizations. It is important to find a middle ground between having knowledge unique to design and having diverse skills applicable across disciplines for participating in and carrying out CE initiatives in an effective way (Charnley et al., 2011). A study examined the importance of specific competencies and skills in organizations for enhancing internal capabilities and achieving optimization of operations and resources (De los Rios & Charnley, 2017). The corporations engaged in remanufacturing aim to optimize product design to increase product lifespan and improve business operations through resource and process optimization and innovation; such companies will need to provide special training for building skills, enhancing capabilities, and establishing partnerships among different groups to promote the adoption of CE (Bourguignon, 2016). Given these viewpoints, we can infer that:

Proposition 2 The skills and abilities of employees have a significant impact on DCE adoption.

1.2.6 Leadership

An organizational leader is accountable for overseeing a certain team. It is essential to define the specific traits of leaders in circular businesses within this particular setting. In the CE context, leaders are obligated to possess knowledge about societal issues and play their role in promoting and advancing positive environmental and social practices, making a significant contribution to the widespread CE adoption (Blass & Hayward, 2015). Consequently, leaders in firms that embrace circular notions play a crucial role in promoting environmentally friendly activities as they guide individuals toward the intended goals. Belaud et al. (2019) assert that visionary leadership is a compelling characteristic for companies, as it serves as a motivator and advocate for the pursuit of growth by presenting a chance to enhance the effectiveness of the CE outcomes. However, according to Dimante et al. (2016), factors such as competencies in sustainable development, approaching work with a global mindset, thinking long-term, and having the ability to measure results and share lessons learned are crucial. Ambekar et al. (2019) theoretically constructed a model for the implementation of low-carbon practices within organizations. This methodology has the ability to assess specific attributes, such as leadership, as seen by the firm. Khan et al. (2024) specifically examined how top management involvement is instrumental in driving DCE initiatives along with supportive policies in this direction. So, it can be concluded that:

Proposition 3 Leadership has a significant impact on the adoption of DCE.

1.2.7 Team Development

The process of imparting skills and knowledge to a group of individuals while enhancing collaboration among them to achieve a common goal is referred to as team development. Lee and Chang (2008) have demonstrated that teams are essential in the integration of the CE inside a company and have a beneficial influence on employee job satisfaction. Prajogo et al. (2014) suggested that cross-functional teams can be employed in the design of environmentally friendly products. These teams should include individuals from several departments within a firm, such as engineers and marketing specialists, among others. Jabbour et al. (2013) refer to this team as the green team since it encompasses many mindsets that equally embrace green thinking, resulting in an improved and expedited process of designing. To broaden the scope of involvement and communication across teams, extending it beyond the boundaries of the corporation to include the entire supply chain (Prajogo et al., 2014). Renwick et al. (2013) suggest the adoption of a compensation system to incentivize environmental performance, including recognizing and rewarding personnel who make the greatest contributions to sustainability. Companies that prioritize sustainability in their operations actively seek out individuals with exceptional expertise in environmental matters during the recruitment process (Bhushan & Singh, 2024). So, it is proposed that:

Proposition 4 Team development has a significant impact on the DCE adoption.

Historically, performance has been associated with financial indicators. Nevertheless, organizations have progressively begun to reconcile economic indicators with the environment and social performance in order to acknowledge the

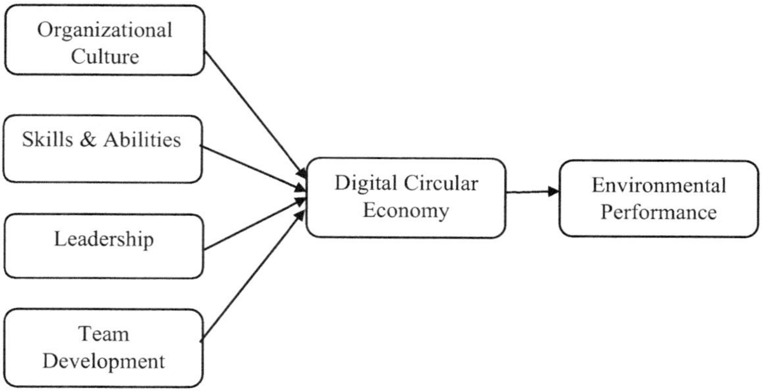

Fig. 1.1 Conceptual model

diverse advantages across various aspects (Epstein & Roy, 2003). To achieve genuine changes, there is a necessity to use eco-friendly approaches in the form of CE initiatives for enhancing organizational performance in terms of sustainability, which is significant in facilitating and directing the transition liner systems of working in circular systems. The sustainable practices of a business yield advantages in the form of high-level productivity and enhanced corporate reputation, which can serve as incentives for organizations to invest in circular practices (Dey et al., 2019). Research has examined the interconnection of CE practices and environmental; however, the existing literature on this topic has not yet reached a definitive conclusion (van Loon et al., 2021). This is further significant in offering valuable guidelines for organizations involved in small or medium-scale operations, particularly given their reluctance to take risks and the constraints on their investment resources. Yang et al. examine how specific CE practices can aid in enhancing the overall environmental performance through a synergistic impact. Consequently, we establish the following:

Proposition 5 The implementation of DCE has a significant impact on environmental performance.

On the basis of all the propositions, we present a conceptual model showcasing the relationship among HRM, DCE, and EP, as shown in Fig. 1.1.

1.3 Research Methodology

This paper is based on secondary data including journal articles and book chapters downloaded from different databases. Initially the author focused on the quality of papers, and hence, articles were downloaded only from Scopus and Web of Science databases. However, due to limited research on DCE, Google Scholar is also considered to increase the number of sources included. First of all, to offer a thorough understanding of the basic nature of CE, the articles defining CE are considered and then articles on the application of digital technology of Industry 4.0 for CE, which is known as DCE, are searched to get a clear picture on this aspect. Then the articles detailing the linkage between HRM and CE ARE considered to deal with the four areas: organizational culture, leadership, employees' skill, and team development. The impact of circular economy on the environmental performance of the organization is also discussed in the paper. Finally, on the basis of existing literature a conceptual model is developed to synthesize the findings of literature.

1.4 Implications

1.4.1 Theoretical Implications

This work provides substantial contributions to the limited theoretical literature in the field of digital circular economy. This study also adds to the research conducted on the role of HRM in adopting CE practices. The main contribution is to demonstrate that by using HRM and DCE, organizations may fully harness the inherent capabilities of HR to enhance their environmental performance. Furthermore, we propose that the human resources of businesses have an effect on the firm's ability to absorb new knowledge, which in turn affects their environmental performance. Furthermore, the study has established a clear correlation between HRM and DCE. The deployment of DCE can result in a competitive advantage for firms in terms of sustainable development. Furthermore, this study highlights the emerging role of Industry 4.0 as a facilitator between HRM and environmental performance through circular economy practices. Furthermore, the study emphasized the theoretical role of DCE between HRM and EP. This role is crucial in facilitating or advancing the sustainable development of the business. DCE is a necessary component of the digitalization period in the 21st for improving EP. Essentially, the elements discussed in this article are strategically important in the modern business environment and provide novel guidance for additional research.

1.4.2 Practical Implications

In addition to contributions to theory, implications of this research for practice are also furnished that are beneficial for contemporary organizations, as it focuses on the success of implementing DCE for meeting sustainability goals. Managers should take into account the impact of HRM on the adoption of DCE while implementing circular practices. Managers must recognize that altering consumption patterns and implementing practices such as reuse, repair, recycling, refurbishment, and multiple uses involves more than just utilizing digital technologies. These actions bring about a fundamental transformation through HRM. Our study suggests that practitioners should carefully consider the impact of HRM in technically advanced environments in order to attain better environmental consequences. The study would provide useful insights for framing policies related to the creation of a strategic operation system, which aims at promoting sustainable HR. It would involve extensive data analysis and contribute to the improvement of the EP. Therefore, the model offers various suggestions to senior executives of organizations to deviate from their conventional human resource practices and prioritize HRM as a means of making a positive impact on society and the environment. Therefore, this study will provide guidance to managers regarding the significance of the CE and its direct connection with environmental preservation. Policymakers in the government should create an organized system and initialization plan to assist executives and managers in organizations in gaining a deeper understanding of the strategies and practices of CE. This will enable them to develop the necessary capabilities and skills for managing the implementation and transformation in an organization effectively. Prior studies have indicated that inadequate leadership and insufficient dedication from top-level executives can limit the smooth implementation of CE. Therefore, visionary leadership must be involved in influencing the factors supporting the effective incorporation of CE practices.

1.5 Conclusion

This paper examines the adoption of DCE, primarily focusing on the HRM activities of firms for devising a conceptual framework with the objective of filling the existing gap related to CE dynamics by conducting a thorough literature assessment of the relationship between human resources and the function of HRM in firm-level CE adoption. This is accomplished by analyzing four aspects of HR including organizational culture, skills and abilities of employees, leadership, and team development for adopting DCE operations. According to the research, as compared to other areas of an organization, like the designing phase and production process, marketing and selling activities, and supply chain management, the examination shows that HRM is not currently viewed as a priority in the business strategy of an organization. In the CE context analyzed for this research, the crucial organizational function has not received attention. The implications on managerial and critical perspectives of HR and HRM within CE have the capacity to significantly contribute to the existing body of scholarly literature on the role of HRM in firm-level CE pursuits.

In addition to implications for theory and practice, there are a few limitations as well, indicating potential future directions through many recommendations. The first component is the research process, which entails the utilization of secondary sources of data to draw conclusions. Furthermore, the paper's implications exclusively pertain to a mere four HR practices. Consequently, conducting a more comprehensive examination of all the practices could potentially offer alternate perspectives. The current study proposed only the conceptual framework of HRM in relation to the DCE by specifically examining the intermediary function of DCE for environmental performance just on the basis of theoretical propositions. Therefore, future research is much needed by employing empirical testing of the conceptual model proposed. Upcoming research should focus on activities related to knowledge management, including organizational learning, knowledge generation, and knowledge sharing for building DCE. Building on Piyathanavong et al. (2024), researchers could explore the role of HRM in aligning DCE initiatives to drive environmental performance. In the future, longitudinal studies may be conducted to track how HRM practices can be transformed for adopting DCE over time, offering insights into the effectiveness of different HRM strategies in driving sustainable change as Jabbour et al. (2019b) indicated. Although Marrucci et al. (2021) discuss the influence of green HRM on environmental outcomes, the relationship has not been thoroughly explored. Future research could develop frameworks or metrics to quantify HRM's contribution to environmental performance within CE.

References

Ambekar, S., Prakash, A., & Patyal, V. S. (2019). Role of culture in low carbon supply chain capabilities. *Journal of Manufacturing Technology Management, 30*(1), 146–179. https://doi.org/10.1108/JMTM-01-2018-0024

Anh Vu, T., Plimmer, G., Berman, E., & Ha, P. N. (2022). Performance management in the Vietnam public sector: The role of institution, traditional culture and leadership. *International Journal of Public Administration, 45*(1), 49–63. https://doi.org/10.1080/01900692.2021.1912018

Belaud, J. P., Prioux, N., Vialle, C., & Sablayrolles, C. (2019). Big data for agri-food 4.0: Application to sustainability management for by-products supply chain. *Computers in Industry, 111*(1), 41–50. https://doi.org/10.1016/j.compind.2019.06.006

Belhadi, A., Kamble, S., Fosso Wamba, S., & Queiroz, M. M. (2021). Building supply-chain resilience: An artificial intelligence-based technique and decision-making framework. *International Journal of Production Research*, 1–21. https://doi.org/10.1080/00207543.2021.1940669

Bhushan, M., & Singh, R. (2024). Talent management facilitates net-zero transition through employee green behavior. In R. Singh, S. Khan, A. Kumar, S. Luthra, & H. Chokshi (Eds.), *Net zero economy, corporate social responsibility and sustainable value creation. CSR, sustainability, ethics & governance*. (pp. [specific pages]). Springer. https://doi.org/10.1007/978-3-031-55779-8_7

Blass, E., & Hayward, P. (2015). Developing globally responsible leaders: What role for business schools in the future? *Futures, 66*(1), 35–44. https://doi.org/10.1016/j.futures.2014.12.008

Bourguignon, D. (2016). Closing the loop. New circular economy package. *European Parliamentary Research Service*, Briefing January 2016. Retrieved June 12, 2021, from https://tinyurl.com/2p9hcknw.

Cagno, E., Neri, A., Negri, M., Bassani, C. A., & Lampertico, T. (2021). The role of digital technologies in operationalizing the circular economy transition: A systematic literature review. *Applied Sciences, 11*, 3328. https://doi.org/10.3390/app11083328

Charnley, F., Lemon, M., & Evans, S. (2011). Exploring the process of whole system design. *Design Studies, 32*(2), 156–179. https://doi.org/10.1016/j.destud.2010.08.002

D'Adamo, I. (2019). Adopting a circular economy: Current practices and future perspectives. *Social Sciences, 8*(12), 328. https://doi.org/10.3390/socsci8120328

De los Rios, I. C., & Charnley, F. J. (2017). Skills and capabilities for a sustainable and circular economy: The changing role of design. *Journal of Cleaner Production, 160*, 109–122. https://doi.org/10.1016/j.jclepro.2016.10.096

Dey, P. K., Malesios, C., De, D., Chowdhury, S., & Abdelaziz, F. B. (2019). Could lean practices and process innovation enhance supply chain sustainability of small and medium-sized enterprises? *Business Strategy and the Environment, 28*(4), 582–598. https://doi.org/10.1002/bse.2269

Dibia, C., Oruh, E., Anderson, M., & Dirpal, G. (2020). Human resource management and circular economy: A critical perspective. In *British Academy of Management 2020 Conference: Innovation for a Sustainable Future* (pp. 1–20). Retrieved from https://virtual.oxfordabstracts.com/#/event/1520/submission/477

Dimante, D., et al. (2016). Development of business education for circular economy in Latvia. In *International Conference New Challenges of Economic and Business Development–2016: Society, Innovations and Collaborative Economy, Latvia, Riga* (pp. 12–14).

Epstein, M. J., & Roy, M. J. (2003). Making the business case for sustainability: Linking social and environmental actions to financial performance. *Journal of Corporate Citizenship, 9*, 79–96.

Jabbour, C. J. C., & Jabbour, A. B. L. D. S. (2014). Low-carbon operations and production: Putting training in perspective. *Industrial and Commercial Training, 46*(6), 327–331. https://doi.org/10.1108/ICT-12-2013-0082

Jabbour, C. J. C., Santos, F. C. A., Fonseca, S. A., & Nagano, M. S. (2013). Green teams: Understanding their roles in the environmental management of companies located in Brazil. *Journal of Cleaner Production, 46*, 58–66. https://doi.org/10.1016/j.jclepro.2012.09.018

Jabbour, C. J. C., de Sousa, L., Jabbour, A. B., Sarkis, J., & Filho, M. G. (2019a). Unlocking the circular economy through new business models based on large-scale data: An integrative framework and research agenda. *Technological Forecasting and Social Change, 144*, 546–552. https://doi.org/10.1016/j.techfore.2017.09.010

Jabbour, C. J. C., Sarkis, J., de Sousa, L., Jabbour, A. B., Renwick, D. W. S., Singh, S. K., Grebinevych, O., Kruglianskas, I., & Filho, M. G. (2019b). Who is in charge? A review and a research agenda on the 'human side' of the circular economy. *Journal of Cleaner Production, 222*, 793–801. https://doi.org/10.1016/j.jclepro.2019.03.038

Khan, O., Daddi, T., & Iraldo, F. (2020). Microfoundations of dynamic capabilities: Insights from circular economy business cases. *Business Strategy and the Environment, 29*(3), 1479–1493. https://doi.org/10.1002/bse.2447

Khan, S., Ali, S. S., & Singh, R. (2022). Determinants of remanufacturing adoption for circular economy: A causal relationship evaluation framework. *Applied System Innovation, 5*(4), 62. https://doi.org/10.3390/asi5040062

Khan, S., Singh, R., Alnahas, J., Abbate, S., & Centobelli, P. (2024). Navigating the smart circular economy: A framework for manufacturing firms. *Journal of Cleaner Production, 144007*.

Kirchherr, J., Reike, D., & Hekkert, M. (2017). Conceptualizing the circular economy: An analysis of 114 definitions. *Resources, Conservation and Recycling, 127*, 221–232. https://doi.org/10.1016/j.resconrec.2017.09.005

Lee, Y. D., & Chang, H. M. (2008). Relations between teamwork and innovation in organizations and the job satisfaction of employees: A factor analytic study. *International Journal of Management, 25*(1), 732–739.

Marrucci, L., Daddi, T., & Iraldo, F. (2021). The contribution of green human resource management to the circular economy and performance of environmental certified organisations. *Journal of Cleaner Production, 319*, 128859.

McLaughlin, P., Bessant, J., & Smart, P. (2008). Developing an organisation culture to facilitate radical innovation. *International Journal of Technology Management, 44*(3–4), 298–323. https://doi.org/10.1504/IJTM.2008.020699

Mousavi, S., Bossink, B., & van Vliet, M. (2018). Dynamic capabilities and organizational routines for managing innovation towards sustainability. *Journal of Cleaner Production, 203*, 224–239. https://doi.org/10.1016/j.jclepro.2018.08.215

Nwokolo, S. C., Singh, R., Khan, S., & Kumar, A. (2024). Threats to the rapidity of sustainability transitions posed by technological changes. In *Technological pathways for Africa's net-zero economy* (pp. 39–75). Elsevier. https://doi.org/10.1016/B978-0-443-31486-5.00002-1

Obeidat, S. M., Abdalla, S., & Al Bakri, A. A. K. (2023). Integrating green human resource management and circular economy to enhance sustainable performance: An empirical study from the Qatari service sector. *Employee Relations: The International Journal, 45*(2), 535–563.

Pagoropoulos, A., Pigosso, D. C., & McAloone, T. C. (2017). The emergent role of digital technologies in the circular economy: A review. *Procedia CIRP, 64*, 19–24.

Piyathanavong, V., Olapiriyakul, S., Garza-Reyes, J. A., Kumar, V., Huynh, V. N., & Karnjana, J. (2024). Implementing industry 4.0 and circular economy through the developmental culture perspective—Driving a competitive advantage in the manufacturing industry. *Business Strategy and the Environment.*

Prajogo, D. I., Tang, A. K. Y., & Lai, K. H. (2014). The diffusion of environmental management system and its effect on environmental management practices. *International Journal of Operations & Production Management, 34*(1), 565–585. https://doi.org/10.1108/IJOPM-10-2012-0448

Ranta, V., Aarikka-Stenroos, L., & Väisänen, J. M. (2021). Digital technologies catalyzing business model innovation for circular economy: Multiple case study. *Resources, Conservation & Recycling, 164*, 105155. https://doi.org/10.1016/j.resconrec.2020.105155

Renwick, D. W. S., Redman, T., & Maguire, S. (2013). Green human resource management: A review and research agenda. *International Journal of Management Reviews, 15*(1), 1–14. https://doi.org/10.1111/j.1468-2370.2011.00328.x

Renwick, D. W. S., Jabbour, C. J. C., Muller-Camen, M., Redman, T., & Wilkinson, A. (2016). Introduction: Contemporary developments in green (environmental) HRM scholarship. *International Journal of Human Resource Management, 27*(2), 1–16. https://doi.org/10.1080/09585192.2015.1069254

Sarros, J. C., Cooper, B. K., & Santora, J. C. (2008). Building a climate for innovation through transformational leadership and organizational culture. *Journal of Leadership & Organizational Studies, 15*(2), 145–158. https://doi.org/10.1177/1548051808324100

Singh, R., Khan, S., & Centobelli, P. (2022). Investigating the interplay between social performance and organisational factors supporting circular economy practices. *Sustainability, 14*(24), 16781.

Singh, R., Khan, S., & Dsilva, J. (2023). A framework for assessment of critical factor for circular economy practice implementation. *Journal of Modelling in Management, 18*(5), 1476–1497.

Singh, R., Gupta, A., & Kandpal, V. (2024). ESG excellence in the digital world: Circular economy E-business solutions. In E. D. R. Santibanez Gonzalez, V. Kandpal, P. K. Ozili, & P. Chatterjee (Eds.), *Fintech for ESG and the circular economy* (pp. 193–204). Wiley. https://doi.org/10.1002/9781394234035.ch10

Sony, M., & Naik, S. (2020). Industry 4.0 integration with socio-technical systems theory: A systematic review and proposed theoretical model. *Technology in Society, 61*, 101248. https://doi.org/10.1016/j.techsoc.2020.101248

Subramanian, N., & Suresh, M. (2022). The contribution of organizational learning and green human resource management practices to the circular economy: A relational analysis—Part I. *The Learning Organization, 29*(5), 428–442. https://doi.org/10.1108/TLO-04-2022-0089

van Loon, P., Diener, D., & Harris, S. (2021). Circular products and business models and environmental impact reductions: Current knowledge and knowledge gaps. *Journal of Cleaner Production, 288*, 125627. https://doi.org/10.1016/j.jclepro.2020.125627

Zeshan, M., Morelli, M., Rasool, S., Centobelli, P., & Cerchione, R. (2024). Empowering sustainable workplaces: A perspective on employee Well-being in the light of the job demand resource model. *Sustainable Development.*

Chapter 2
Sustainable Finance and Green Investments: The Role of Blockchain Circular Economy

Daravath Vikas Nayak, A. Arun Kumar, Gowri B. Chanal, and Srinivas Bolagani

Abstract This study examines how blockchain technology can improve accountability and transparency in sustainable finance and use, emphasizing how it affects investor confidence and environmental results. Although funding environmentally sustainable projects requires sustainable finance, issues like data transparency, fund misallocation, and environmental outcome verification still exist. These issues can be resolved by blockchain, which ensures transparent, real-time tracking of fund allocations and environmental impacts through its decentralized and immutable ledger system. The study examines how blockchain can increase trust and transparency in sustainable finance, evaluates how well it tracks and validates environmental outcomes, and highlights the opportunities and problems of integrating blockchain into sustainable finance. Insights into how blockchain-based systems can promote greater trust and boost market participation in sustainable finance are provided by this further investigation into the effect of blockchain on investor participation. In addition to addressing obstacles like regulatory uncertainty, technological difficulties, and data quality concerns, this study demonstrates the revolutionary potential of blockchain technology for improving the effectiveness, legitimacy, and worldwide reach of sustainable finance. According to the findings, if suitable legal frameworks, technical standards, and international cooperation are created to facilitate its broad adoption, blockchain technology may be essential to expanding sustainable finance.

Keywords Blockchain · Transparency · Trust · Efficiency · Sustainable Finance

D. V. Nayak · S. Bolagani
Loyola Academy (Degree & PG College), Secunderabad, Telangana, India

A. A. Kumar (✉) · G. B. Chanal
ICFAI Law School, ICFAI Foundation for Higher Education (Deemed to be University Under Sec 3 of UGC Act,1956), Hyderabad, Telangana, India
e-mail: arunkumar@ifheindia.org

R. Singh et al. (eds.), *Blockchain Technologies for Smart Circular Economy and Organisational Sustainability*, Environmental Footprints and Eco-design of Products and Processes, https://doi.org/10.1007/978-3-031-89183-0_2

2.1 Introduction

Innovative forms of sustainable finance, such as blended finance and climate funds, have emerged due to the global need to address environmental sustainability and climate change. According to blockchain for environmentally sustainable economies (2022), these sustainable finances are fixed-income securities intended to fund initiatives with a positive environmental impact, like clean transportation, sustainable agriculture, and renewable energy. Even though sustainable finance is becoming increasingly popular, there are still issues with trust, accountability, and transparency, especially regarding funding distribution and oversight. The market's scalability has been hampered by these problems and the possibility of greenwashing, which misrepresents environmental benefits (Mulligan et al., 2024). A decentralized digital ledger called blockchain technology provides revolutionary answers to these problems.

Additionally, blockchain makes it possible to track and report environmental metrics precisely, guaranteeing that creative financial investments produce measurable sustainability results (Schulz & Feist, 2021). Blockchain integration in sustainable finance has enormous potential, but obstacles remain to overcome, including high implementation costs, scalability issues, and regulatory uncertainty (Mohan Kumar, 2024). Technology companies, financial institutions, and legislators must work together to remove these obstacles. This study aims to identify blockchain technology's role in bringing transparency and accountability to sustainable finance.

Sustainable finance, which aims to facilitate environmentally sustainable economic activities, is essential to combat climate change and accomplish global sustainability goals. Despite its significance, sustainable finance faces several significant obstacles. One of the main obstacle is the absence of accountability and transparency, especially regarding the issuance and use of green bonds. Environmental friendly projects can be financed through sustainable finance, but the lack of uniform reporting standards raises questions about the distribution and administration of funds. Issuers may engage in greenwashing, which is the practice of misrepresenting projects as environmentally sustainable if there are unclear disclosure requirements (Abou Jaoude & George Saade, 2019).

Conventional reporting systems depend on third-party certifications and recurring, manual audits, which are frequently costly, time-consuming, and prone to errors (Kumar et al., 2022). These inefficiencies cause a discrepancy between the desired and actual results, making it difficult to gauge the full impact of sustainable finance initiatives. Projects that promise to lower carbon emissions, for example, might not offer trustworthy data to support their claims, which would further erode stakeholder confidence. There are notable regional differences in the still-evolving regulatory environment for sustainable finance (Kumar et al., 2024).

Sustainable finance instruments' development, marketing, and oversight are inconsistent due to a lack of standardized standards and policies (Mulligan et al., 2024). These disparities make it more challenging to match sustainable finance programs with international sustainability objectives, like those outlined in the Paris

Agreement. Furthermore, because issuers and investors are still unsure of the risks and compliance requirements, regulatory uncertainty may deter private sector involvement. Sustainable finance frequently entails more extraordinary upfront expenses than conventional financing methods. To confirm their environmental credentials, sustainable finance requires a lot of paperwork and certification procedures (Ozili, 2023).

Smaller issuers and developing nations may find these extra expenses unaffordable, restricting the availability and expandability of sustainable finance options. Additionally, mainstream investors more accustomed to traditional financial products may be put off by the perceived complexity of sustainable finance instruments. Although blockchain technology can potentially improve accountability and transparency in sustainable finance, the adoption of this technology is still relatively low (Mohan Kumar, 2024). By combining these technologies, reporting could be made more efficient, expenses could be decreased, and data accuracy could be increased.

However, obstacles like initial setup costs, technical complexity, and a lack of experience hinder widespread adoption, making traditional systems susceptible to errors and inefficiencies. The difficulties sustainable finance faces, from investor trust and transparency to technological and regulatory barriers, highlight the necessity of all-encompassing solutions to realize its full potential (Bhushan & Singh, 2024). Sustainable finance has the potential to significantly contribute to the global shift towards sustainability by surmounting these obstacles. There are currently few empirical studies on the use of blockchain in sustainable finance, despite the literature's emphasis on the technology's potential to improve accountability and transparency in this area. More research is needed to determine how blockchain affects investor confidence and whether it can produce quantifiable environmental results. By investigating how blockchain is changing sustainable finance, this study seeks to close these gaps.

2.2 Objectives of the Study

1. To research how blockchain technology can improve accountability and transparency in issuing and using sustainable finance.
2. To comprehend how blockchain affects investor participation and trust in sustainable finance.
3. To investigate how well blockchain works to monitor and confirm sustainable finance projects' environmental results.
4. To determine the opportunities and difficulties of incorporating blockchain technology into sustainable finance.

2.3 Blockchain Technology's Potential to Improve Accountability and Transparency in the Issuance and Use of Sustainable Finance

Distributed ledger technology for recording transactions across various systems with greater transparency is blockchain technology. Each block or unit relates to the previous transaction access time stamp, making it hard to tamper. One of the core features of blockchain technology is its decentralization; it operates without a central authority, meaning that no single entity controls the data. Blockchain is a powerful tool for enhancing transparency, particularly in financial markets (Khan et al., 2024). Blockchain's decentralized nature is beneficial in sustainable finance, as it can ensure that all parties' investors, issuers, and auditors have access to the same accurate and real-time information.

By facilitating increased transparency and traceability across the sustainable finance life cycle, blockchain technology can help address these problems (Shahbaz Khan et al., 2024). The transaction details, including the amount of money raised, the project's environmental objectives, and the planned use of the proceeds, can be documented on a blockchain when sustainable finance is issued. All stakeholders, including investors, regulators, and auditors, have access to this ledger, guaranteeing no room for doubt about the distribution of funds (Singh et al., 2023). The transparency of the blockchain also makes it easier to guarantee that money is spent in a way that aligns with sustainable finance goals. Each disbursement of funds to projects and information about the project's status can be documented on the blockchain. The real-time tracking, investors and other stakeholders can confirm that the funds are being used for their intended purposes, which boosts system trust and lowers the possibility of fraud or misallocation (Chenguel & Mansour, 2024).

Blockchain has the potential to improve significantly post-issuance reporting in addition to increasing transparency during the issuance phase. Investors in sustainable finance frequently worry about the actual environmental impact of the projects they are funding and how their money is spent (Singh et al., 2022). The issuer usually handles traditional post-issuance reporting, which can lead to delays, inconsistent data, and incomplete information. Blockchain can monitor how money is allocated to various project phases, such as acquisition, building, and operation. Every transaction is documented on the blockchain as funds are distributed and milestones are reached, creating an unchangeable and transparent audit trail (Singh et al., 2024a, b).

In addition to expediting the reporting process, this automation guarantees that the data is correct, current, and verifiable by all stakeholders (Mishra & Kaushik, 2023). One of the main issues with sustainable finance is the possibility of greenwashing, in which environmentally friendly projects are promoted as such but fall short of the environmental standards that investors were promised. By offering a transparent and verifiable record of how money is being spent, blockchain technology can help reduce this risk. Blockchain reduces the possibility of greenwashing by ensuring that the flow of funds can be independently verified by recording each transaction on an immutable ledger (Khan et al., 2023).

Blockchain technology can also improve accountability in the market for sustainable finance. The obligation of issuers to utilize the money raised through sustainable finance in line with the declared environmental goals is known as accountability. Blockchains are an effective tool for ensuring that issuers are held responsible for their actions because they can produce an unchangeable record of transactions (Khan et al., 2022). The transparent and traceable records on the blockchain make it simple to identify instances in which an issuer deviates from the original goal of sustainable finance, such as using funds for a project that is not environmental friendly (Singh et al., 2024a, b).

Real-time fund tracking enables stakeholders, including investors, to hold issuers responsible for any misappropriation or misallocation of funds (Rathnasiri et al., 2024). Furthermore, the decentralized structure of blockchains guarantees that no one party controls the data, lowering the possibility of record manipulation or falsification (Di Vaio & Varriale, 2020). Blockchain can improve accountability by making it easier for independent auditors to participate. Conventional auditing procedures may rely on erroneous or incomplete data and can be expensive and time-consuming. The auditing process is made more dependable and efficient by blockchain, which gives auditors direct access to an unchangeable and transparent record of every transaction.

2.4 The Impact of Blockchain on Investor Trust and Participation in Sustainable Finance

Sustainable finance is one of the many industries in which blockchain technology, which is renowned for its transparency, decentralization, and immutability, has the potential to transform. In the battle against climate change, sustainable finance has become a crucial tool for funding initiatives that benefit the environment. However, issues with sustainable finance include inconsistent reporting, lack of transparency in fund utilization, and greenwashing. Blockchain-based solutions promise to solve these problems by offering an open and verifiable way to monitor investments (Arun Kumar et al., 2024).

To improve the overall efficacy of sustainable finance markets, this objective investigates how investor trust and involvement in sustainable finance are impacted by incorporating blockchain technology into sustainable finance systems. Blockchain's primary characteristic is its capacity to offer a shared, transparent, and impenetrable ledger, which makes it especially appropriate for applications where accountability, trust, and verification are essential (Mhlanga, 2023).

Debt instruments known as sustainable finance are offered to fund initiatives with a positive environmental impact, like energy efficiency, renewable energy, and sustainable infrastructure. Despite its quick growth, the sustainable finance industry has encountered issues with accountability and transparency. Investors frequently question the accuracy of post-issuance reports, and there is still a serious risk of greenwashing, in which projects' environmental benefits are exaggerated. By offering transparent, verifiable, and real-time information, blockchain technology has the potential to address these issues and may be crucial in increasing investor confidence and involvement in sustainable finance. The growth of any financial market, particularly in sustainable finance, depends on investor trust (Nayak et al., 2024).

From the issuance of sustainable finance to the distribution of funds to projects, blockchain technology can offer a transparent record of every transaction. Investor confidence that their money is going to real green projects is increased by this transparency, which also lowers the chance of fund misallocation (Islam et al., 2021).

Blockchain technology can facilitate ongoing, real-time tracking of sustainable finance-funded projects. Smart contracts, which automatically carry out predetermined actions when specific conditions are met, like the distribution of funds to project stages, are used to accomplish this. Furthermore, real-time environmental data from the projects can be obtained by integrating blockchain with other technologies, like the Internet of Things (IoT). For instance, sensors could monitor how well renewable energy installations are performing or how energy-efficient buildings are, with the blockchain automatically recording the data (Khan et al., 2022).

Investors are reassured by this degree of openness, which also enables them to confirm that environmental goals are being fulfilled (Chen & Volz, 2022). Investors can use the unchangeable audit trail this produces to confirm that money is being spent wisely and that the stated results of green projects are accurate. Investor mistrust may result from post-issuance reporting prone to mistakes or even deliberate manipulation without blockchain. Because no one party can change the records, the decentralized nature of blockchain also lowers the risk of fraud (Chukwunweike, 2024). As a result, there is more trust in sustainable finance in general and sustainable finance in particular.

Trust is essential to expanding sustainable finance, as is greater involvement from institutional and individual investors. More investors may become involved in sustainable finance due to blockchain technology's ability to resolve transparency concerns, particularly those who have been apprehensive in the past because they doubt the veracity and credibility of green claims. The following elements demonstrate how blockchain technology may affect investors' involvement in sustainable finance. To make well-informed investment decisions, investors are increasingly searching for real-time data. Investors can closely monitor the performance of their investments thanks to blockchain's ability to provide ongoing updates on the use of proceeds and the status of green projects (Khan et al., 2022).

Greenwashing is less likely thanks to blockchain technology's capacity to track project environmental impact and validate the actual use of funds. Investors can use blockchain technology to independently confirm that sustainable finance proceeds are going toward appropriate environmental uses. The sustainable finance space will be drawn, especially institutional investors with a fiduciary duty to guarantee the integrity and accuracy of their investments (Ghosh, 2020). Reducing greenwashing contributes to developing a more reliable and resilient sustainable finance market.

Because blockchain technology is decentralized, cross-border investment is easier. Blockchain-enabled sustainable finance can draw international investors by allowing for transparent, real-time reporting and monitoring across all geographies (Khan et al., 2022). This global accessibility increases participation in the global sustainable finance community by expanding the pool of possible investors. The ability of blockchain technology to facilitate safe, verifiable transactions also guarantees that investors from various jurisdictions can trust that their investments are secure. Therefore, blockchain can increase the sustainability finance industry's global reach and inclusivity, drawing in institutional and individual investors (Zhao, 2024). Blockchain can also increase market efficiency and liquidity by expediting the issuance and trading of sustainable finance.

Blockchain can reduce the time and expense associated with issuing and trading sustainable finance by doing away with middlemen like banks and facilitating peer-to-peer transactions directly. This efficiency is incredibly alluring for investors seeking quick and economical ways to enter and exit the market. Furthermore, retail investors can participate in the market without making significant capital commitments thanks to the ease with which blockchain-based sustainable finance can be separated into smaller, tradeable units. More involvement in sustainable finance may result from enhanced market accessibility and liquidity (Oluwafemi Elias et al., 2024). Blockchain technology can significantly increase investor participation and trust in sustainable finance. Blockchain tackles important issues linked to accountability and the possibility of greenwashing by offering transparency, real-time reporting, and an unchangeable audit trail.

Furthermore, the overall appeal of sustainable finance is increased by blockchain's capacity to lower transaction costs, boost market efficiency, and ease cross-border investment. Increased participation in sustainable finance is anticipated as more institutional and retail investors learn about the benefits of blockchain technology in improving transparency and lowering risks. This will ultimately propel the sector's expansion. In this sense, blockchain technology can be a game-changer in fostering trust and increasing investor involvement in sustainable finance, which will aid in meeting global sustainability goals (Arun Kumar et al., 2024).

2.5 The Challenges and Opportunities Associated with Integrating Blockchain Technology in Sustainable Finance

The regulatory ambiguity surrounding blockchain and sustainable finance is one of the biggest obstacles to adopting blockchain technology in these markets. Blockchain regulations are still evolving, and legal frameworks for issuing sustainable finance vary by nation and region. It is challenging for issuers, investors, and other stakeholders to confidently adopt blockchain solutions due to the absence of a unified global regulatory framework. Some jurisdictions may have stringent data privacy laws that contradict the transparency features of blockchain technology, or they may not acknowledge blockchain records as legally binding. These legal nuances may make it difficult to issue sustainable finance across borders and could prevent blockchain-based sustainable finance systems from scaling globally (Kouam, 2024).

Complex environmental data, including waste diversion, energy generation, and carbon emissions reductions, are frequently included in sustainable finance projects. These data can be challenging to quantify and validate. Third-party certifications or audits, which can be expensive and time-consuming, may be necessary to confirm emission reductions (Varma, 2019). Additionally, it can be logistically challenging to integrate blockchain with current environmental monitoring systems, mainly if these systems are not standardized across projects or jurisdictions. Investor confidence and involvement in sustainable finance may be impacted by low-quality data, which can compromise blockchain's ability to track environmental outcomes (Kölbel et al., 2020).

Blockchain has many advantages, but technical issues hamper its widespread use in sustainable finance. The main problem is scalability. The environmental advantages of sustainable finance may be outweighed by the slow transaction speeds and high energy consumption of current blockchain networks, especially those that employ Proof-of-Work (PoW) consensus mechanisms (Lăzăroiu et al., 2023). The blockchain platform must be able to manage a growing volume of transactions without sacrificing its sustainability or efficiency as sustainable finance expands. Furthermore, the development and application of blockchain technology necessitates a substantial amount of processing power and specific technical expertise.

The absence of standardization in the issuance of sustainable finance and the application of blockchain technology presents another difficulty in incorporating blockchain technology into sustainable finance. Since different issuers use different frameworks to measure environmental outcomes and ensure compliance with sustainability criteria, sustainable finance is still in its infancy. Stakeholders may encounter challenges guaranteeing that blockchain solutions are compatible across markets and jurisdictions without standardized guidelines for blockchain integration (Schletz et al., 2020). The best ways to use blockchain to monitor environmental outcomes, report impact, and validate data are not universally agreed upon. Investors may find it more challenging to evaluate the caliber of sustainable finance and the veracity of reported results if there is a lack of standardization, which could result in dispersed blockchain implementations.

This fragmentation might be lessened and increased market participation and transparency could be promoted by standardizing blockchain systems for sustainable finance. Significant financial and technological resources may be needed to develop and deploy blockchain technology for sustainable finance (Arun Kumar et al., 2024). Blockchain platforms' development, integration with current systems, and adherence to legal and regulatory requirements require investment from issuers and stakeholders. These expenses may limit the scalability of blockchain solutions in emerging markets and may be unaffordable for smaller issuers. Furthermore, ongoing maintenance, upgrades, and operational expenses may be necessary for blockchain solutions used in sustainable finance, which could strain issuers and eventually lower the financial viability of blockchain-based sustainable finance (Naderi & Tian, 2022).

Blockchain technology can help allay worries about "greenwashing," the practice of issuers making exaggerated claims about the environmental benefits of their projects. Blockchain helps to guarantee that sustainable finance keeps its environmental commitments by facilitating ongoing monitoring of environmental performance and offering unchangeable evidence of project results (Adedoyin Tolulope Oyewole et al., 2024). A wider range of environmentally conscious investors may be drawn in, and participation in sustainable finance may rise because of this enhanced transparency. By automating crucial procedures, blockchain technology can significantly increase the effectiveness of sustainable finance issuance and management.

Smart contracts, self-executing agreements written on the blockchain, can release payments when certain conditions are met, initiate environmental reporting, and automate the distribution of funds (Afua Addy et al., 2024). Issuers and investors may experience less administrative strain because of this automation, which could result in quicker transactions and cheaper expenses. Furthermore, blockchain can simplify procedures and lower transaction costs by doing away with middlemen like banks and clearinghouses, opening sustainable finance to a larger pool of investors. Along with enabling real-time updates and cutting down on the time and expense involved with conventional monitoring techniques, blockchain also offers an effective platform for tracking and reporting environmental outcomes (Nagarajan & Afjal, 2023).

Innovative financial products may be made possible by incorporating blockchain technology into sustainable finance. Blockchain technology could improve liquidity and reduce transaction costs in sustainable finance by representing ownership and rights to future payments. Furthermore, blockchain may make it easier to create new financial products like impact investment funds or green derivatives that combine the advantages of sustainable finance with other sustainability-focused goods (Schulz & Feist, 2021). These developments might increase the number of options accessible to investors and support the development of sustainable finance.

Blockchain technology integration offers opportunities as well as challenges for sustainable finance. Blockchain offers many advantages, such as increased transparency, efficiency, and global integration, but regulatory uncertainty, data quality concerns, technological constraints, and costs significantly hamper adoption. Sustainable finance can become more dependable and efficient for funding sustainable development by tackling these issues and utilizing blockchain's advantages (Arun Kumar et al., 2024). Blockchain has the potential to revolutionize sustainable finance, boosting investor confidence and propelling global advancements in environmental sustainability with the correct infrastructure, legal frameworks, and innovative investment.

2.6 Managerial Implications

For issuers, investors, and regulators, incorporating blockchain technology into sustainable finance has some managerial ramifications. Blockchain provides issuers with real-time reporting, improved transparency in project execution, and a simplified fund allocation process that can lower operating costs and foster investor trust. To increase the effectiveness of reporting systems and automate contract execution, managers should make an investment in learning about blockchain technology. By guaranteeing the traceability of funds and confirming the environmental impact of financed projects, blockchain lowers the risk of greenwashing and increases investor confidence in sustainable finance.

To make better investment decisions, investors should concentrate on comprehending how blockchain technology helps to reduce these risks. To guarantee compliance, data privacy, and cross-border interoperability, regulators must establish clear guidelines for the broad adoption of blockchain technology. To establish a uniform framework for blockchain in sustainable finance, regulators must encourage cooperation.

2.7 Scope for Further Study and Limitations of the Study

Regulations, are very important for integrating blockchains into sustainable finance and should be the focus of future research. Future research may also examine the scalability of various blockchain consensus mechanisms to satisfy the expanding demand for sustainable finance. To fully grasp the extent of blockchain's contribution to sustainable finance, evaluating the long-term financial and environmental effects of blockchain-based sustainable finance on sustainable projects will also be essential. In addition to highlighting the various issues encountered worldwide, comparative studies of blockchain adoption across various sustainable finance and geographical areas will assist in identifying best practices.

There are several restrictions on this study. First, it mainly uses theoretical approaches, which might not adequately represent the practical difficulties of implementing blockchain in sustainable finance. Second, the analysis's applicability may change over time due to the dynamic nature of blockchain technology and regulatory frameworks. Third, the findings' generalizability was limited by the lack of a thorough investigation into regional variations in blockchain implementation and sustainable finance practices. Finally, the study lacks stakeholder interviews and empirical testing, which could offer a more in-depth understanding of the real-world opportunities and difficulties associated with combining blockchain technology with sustainable finance.

2.8 Conclusion

Transparency, accountability, and efficiency in sustainable finance could all be significantly improved by incorporating blockchain technology. Blockchain can solve major issues like data verification, fund misallocation, and greenwashing by offering real-time, unchangeable records of transactions and environmental impacts. This will increase investor confidence. Its value for issuers and participants in sustainable finance is further highlighted by its potential to simplify procedures, cut expenses, and allow global scalability. However, issues like data quality, technological barriers, and regulatory obstacles must be resolved to fully realize blockchain's potential in this field.

References

Abou Jaoude, J., & George Saade, R. (2019). Blockchain applications—Usage in different domains. *IEEE Access, 7*, 45360–45381. https://doi.org/10.1109/ACCESS.2019.2902501

Adedoyin Tolulope Oyewole, Adeoye, O. B., Addy, W. A., Okoye, C. C., & Ofodile, O. C. (2024). Enhancing global competitiveness of U.S. SMES through sustainable finance: A review and future directions. *International Journal of Management & Entrepreneurship Research, 6*(3), 634–647. https://doi.org/10.51594/ijmer.v6i3.876

Afua Addy, W., Chrisanctus Ofodile, O., Bukola Adeoye, O., Tolulope Oyewole, A., Chinazo Okoye, C., Odeyemi, O., & James Ololade, Y. (2024). Data-driven sustainability: How Fintech innovations are supporting green finance. *Engineering Science & Technology Journal, 5*(3), 760–773. https://doi.org/10.51594/estj/v5i3.871

Arun Kumar, A., Sowmyya, T., Sambrani, S., & Shekhar, S. (2024). Behavioral changes and public engagement in net-zero policies: Issues and challenges. In R. Singh & D. Crowther (Eds.), *Transition towards a sustainable future . Approaches to global sustainability, markets, and governance*. Springer. https://doi.org/10.1007/978-981-97-5756-5_14

Bhushan, M., & Singh, R. (2024). Talent management facilitates net-zero transition through employee green behavior. In R. Singh, S. Khan, A. Kumar, S. Luthra, & H. Chokshi (Eds.), *Net zero economy, corporate social responsibility and sustainable value creation . CSR, sustainability, ethics & governance*. Springer. https://doi.org/10.1007/978-3-031-55779-8_7

Blockchain for Environmentally Sustainable Economies. (2022). https://doi.org/10.37602/IJSSMR.2022.52049.

Chen, Y., & Volz, U. (2022). Scaling up sustainable investment through blockchain-based project bonds. *Development Policy Review, 40*, 3. https://doi.org/10.1111/dpr.12582

Chenguel, M. B., & Mansour, N. (2024). Green finance: Between commitment and illusion. *Competitiveness Review: An International Business Journal, 34*(1), 179–192.

Chukwunweike, J. N., Yussuf, M., Okusi, O., Bakare, T. O., & Abisola, A. J. (2024). The role of deep learning in ensuring privacy integrity and security: Applications in AI-drivencybersecurity solutions. *World Journal of Advanced Research and Reviews, 23*(02), 1778–1790. https://doi.org/10.30574/wjarr.2024.23.2.2550

Di Vaio, A., & Varriale, L. (2020). Blockchain technology in supply chain management for sustainable performance: Evidence from the airport industry. *International Journal of Information Management, 52*. https://doi.org/10.1016/j.ijinfomgt.2019.09.010

Ghosh, K. (2020). Blockchain technology: A catalyst for change in finance. In *Drishtikon: A Management Journal* (Vol. 12, Issue 1). http://www.publishingindia.com/drishtikon/ Electroniccopyavailableat, https://ssrn.com/abstract=3906573.

Islam, N., Marinakis, Y., Olson, S., White, R., & Walsh, S. (2021). *ORE Open Research Exeter title is Blockchain mining profitable in the long run? a note on versions is Blockchain mining profitable in the long run?* http://hdl.handle.net/10871/124303.

Khan, S., Ali, S. S., & Singh, R. (2022). Determinants of remanufacturing adoption for circular economy: A causal relationship evaluation framework. *Applied System Innovation, 5*(4), 62. https://doi.org/10.3390/asi5040062

Khan, S., Khan, M. I., & Singh, R. (2023). Modeling the barriers of Blockchain technology implementation in supply chain. *Journal of Industrial Integration and Management, 08*(03), 367-385. https://doi.org/10.1142/S2424862223500070

Khan, S., Ullah, A., Liu, Y., & Kashif, M. (2024). Examining the blockchain and green innovation technologies on sustainability (ESG): the moderating role of global financial integration. In *Journal of Sustainable Finance and Investment*. Taylor and Francis Ltd. https://doi.org/10.1080/20430795.2024.2441204.

Kölbel, J. F., Heeb, F., Paetzold, F., & Busch, T. (2020). Can sustainable investing save the world? Reviewing the mechanisms of investor impact. *Organization & Environment, 33*(4), 554–574. https://doi.org/10.1177/1086026620919202

Kouam, A. W. F. (2024). *Blockchain technology and sustainable finance: Exploring the potential for climate change mitigation and adaptation*. https://doi.org/10.21203/rs.3.rs-5508660/v1.

Kumar, S., Sharma, D., Rao, S., Lim, W. M., & Mangla, S. K. (2022). Past, present, and future of sustainable finance: Insights from big data analytics through machine learning of scholarly research. *Annals of Operations Research*. https://doi.org/10.1007/s10479-021-04410-8

Kumar, K., Singh, R., Sharma, A., & Lehri, B. P. S. (2024). Climate change adaptation strategies for achieving net-zero economy. In R. Singh & D. Crowther (Eds.), *Transition towards a sustainable future. Approaches to global sustainability, markets, and governance*. Springer. https://doi.org/10.1007/978-981-97-5756-5_8

Lăzăroiu, G., Bogdan, M., Geamănu, M., Hurloiu, L., Ionescu, L., & Ştefănescu, R. (2023). Artificial intelligence algorithms and cloud computing technologies in blockchain-based fintech management. *Oeconomia Copernicana, 14*(3), 707–730. https://doi.org/10.24136/oc.2023.021

Mhlanga, D. (2023). Block chain technology for digital financial inclusion in the industry 4.0, towards sustainable development? In *Frontiers in blockchain* (Vol. 6). Frontiers Media SA. https://doi.org/10.3389/fbloc.2023.1035405

Mishra, L., & Kaushik, V. (2023). Application of blockchain in dealing with sustainability issues and challenges of financial sector. *Journal of Sustainable Finance and Investment, 13*(3), 1318–1333. https://doi.org/10.1080/20430795.2021.1940805

Mohan Kumar, B. (2024). Using Blockchain technology for sustainable finance reporting. *Educational Administration Theory and Practice*, 1182–1187. https://doi.org/10.53555/kuey.v30i6.5462

Mulligan, C., Morsfield, S., & Cheikosman, E. (2024). Blockchain for sustainability: A systematic literature review for policy impact. *Telecommunications Policy, 48*, 2. https://doi.org/10.1016/j.telpol.2023.102676

Naderi, N., & Tian, Y. (2022). Leveraging Blockchain technology and tokenizing green assets to fill the green finance gap. *Energy Research Letters, 3*, 3. https://doi.org/10.46557/001c.33907

Nagarajan, C. D., & Afjal, M. (2023). Innovative applications and implementation challenges of blockchain technology in the financial sector. In *Handbook of research on AI-based technologies and applications in the era of the Metaverse* (pp. 279–297). IGI Global. https://doi.org/10.4018/978-1-6684-8851-5.ch014

Nayak, D. V., Arun Kumar, A., Suryadevara, R., & Khan, S. (2024). Financing the transformation to net zero goals: IREDA investment for sustainable initiatives. In R. Singh & D. Crowther (Eds.), *Transition towards a sustainable future. Approaches to global sustainability, markets, and governance*. Springer. https://doi.org/10.1007/978-981-97-5756-5_4

Oluwafemi Elias, Awotunde, O. J., Oladepo, O. I., Azuikpe, P. F., Samson, O. A., Oladele, O. R., & Ogunruku, O. O. (2024). The evolution of green fintech: Leveraging AI and IoT for sustainable financial services and smart contract implementation. *World Journal of Advanced Research and Reviews, 23*(1), 2710–2723. https://doi.org/10.30574/wjarr.2024.23.1.2272

Ozili, P. K. (2023). Assessing global interest in decentralized finance, embedded finance, open finance, ocean finance and sustainable finance. *Asian Journal of Economics and Banking, 7*(2), 197–216. https://doi.org/10.1108/ajeb-03-2022-0029

Rathnasiri, M. S. H., Dewasiri, N. J., & Kumar, A. A. (2024). Policy framework and implementation strategies for Sri Lanka's transition to a net-zero economy. In R. Singh, S. Khan, A. Kumar, S. Luthra, & H. Chokshi (Eds.), *Net zero economy, corporate social responsibility and sustainable value creation. CSR, sustainability, ethics & governance*. Springer. https://doi.org/10.1007/978-3-031-55779-8_3

Schletz, M., Nassiry, D., Lee, M., & Kyoon. (2020). *Blockchain and tokenized securities: The potential for green finance*. https://www.adb.org/publications/blockchain-tokenized-securities.

Schulz, K., & Feist, M. (2021). Leveraging blockchain technology for innovative climate finance under the Green Climate Fund. *Earth System Governance, 7*. https://doi.org/10.1016/j.esg.2020.100084

Shahbaz Khan, Singh, R., Alnahas, J., Abbate, S., & Centobelli, P. (2024). Navigating the smart circular economy: A framework for manufacturing firms. *Journal of Cleaner Production, 480*, 144007. https://doi.org/10.1016/j.jclepro.2024.144007

Singh, R., Khan, S., & Centobelli, P. (2022). Investigating the interplay between social performance and organisational factors supporting circular economy practices. *Sustainability, 14*(24), 16781. https://doi.org/10.3390/su14241678

Singh, R., Khan, S., & Dsilva, J. (2023). A framework for assessment of critical factor for circular economy practice implementation. *Journal of Modelling in Management, 18*(5), 476–1497. https://doi.org/10.1108/JM2-06-2021-0145

Singh, R., Gupta, A., Bajpai, A., & Kandpal, V. (2024a). Exploring challenges of circular economy initiatives for smart cities. In V. Kandpal, E. D. Santibanez-Gonzalez, P. Chatterjee, & M. K. Nallapaneni (Eds.), *Smart cities and circular economy* (pp. 59–69). Emerald Publishing Limited, Leeds. https://doi.org/10.1108/978-1-83797-957-820241005

Singh, R., Gupta, A., Bajpai, A., & Kandpal, V. (2024b). *Exploring challenges of circular economy initiatives for smart cities* (pp. 59–69). Emerald Publishing Limited eBooks. https://doi.org/10.1108/978-1-83797-957-820241005

Varma, J. R. (2019). Blockchain in finance. *Vikalpa, 44*(1), 1–11. https://doi.org/10.1177/0256090919839897

Zhao, Y. (2024). Empowering sustainable finance: The convergence of AI, Blockchain, and big data analytics. *Advances in Economics, Management, and Political Sciences, 85*(1), 267–273. https://doi.org/10.54254/2754-1169/85/20240925

Chapter 3
Blockchain-Enabled Strategies for Sustainable Business and Organisational Growth

Gazi Kurnaz and Cumali Kılıç

Abstract This chapter examines the transformative role of blockchain technology in fostering sustainable business and organisational growth. It highlights blockchain's capacity to enhance supply chain transparency, promote innovative business models and embed ethical practices into corporate strategies. Blockchain's decentralized and immutable features enable organisations to address challenges such as resource inefficiency, unethical practices and stakeholder mistrust. The integration of blockchain with the principles of the smart circular economy offers a pathway for redefining linear economic models, emphasizing reuse, remanufacturing and recycling to achieve sustainability goals. Furthermore, blockchain-enabled tokenization and decentralized decision-making frameworks are explored as tools for incentivizing sustainability and improving resource management. Despite its potential, blockchain adoption faces barriers, including regulatory uncertainties and technological standardization issues. This chapter underscores the necessity for strategic frameworks, stakeholder collaboration and a culture of continuous learning to fully leverage blockchain's potential for sustainable practices. By addressing these challenges, blockchain technology can become a cornerstone for achieving long-term sustainability and resilience in businesses and organisations.

Keywords Blockchain Technology · Sustainable Business Practices · Smart Circular Economy · Supply Chain Transparency · Decentralized Decision-Making

G. Kurnaz · C. Kılıç (✉)
Department of Business Administration, Faculty of Economics and Administrative Sciences, Artuklu University, Mardin, Turkey
e-mail: gazi@artuklu.edu.tr; cumalikilic@artuklu.edu.tr

R. Singh et al. (eds.), *Blockchain Technologies for Smart Circular Economy and Organisational Sustainability*, Environmental Footprints and Eco-design of Products and Processes, https://doi.org/10.1007/978-3-031-89183-0_3

3.1 Introduction

Blockchain technology is increasingly recognized as a transformative force in promoting sustainable business practices and organisational growth (Khan et al., 2023). Its decentralized nature and inherent transparency can significantly enhance supply chain management, ethical business practices and overall sustainability performance (Saberi et al., 2019). This synthesis explores how blockchain-enabled strategies can foster sustainable business and organisational growth through various mechanisms.

One of the primary advantages of blockchain technology is its ability to enhance transparency and traceability within supply chains (Centobelli et al., 2022). This is particularly crucial in sectors such as food and manufacturing, where sustainability is paramount (Singh et al., 2024a). For instance, blockchain can facilitate better sustainability communication by providing immutable records of product origins and processing methods, thereby allowing consumers to make informed choices (Cao et al., 2023). Furthermore, the integration of blockchain in supply chains can enhance trust among stakeholders, including producers, distributors and consumers, which is essential for fostering collaborative efforts towards sustainability (Kumar et al., 2024; Joo & Han, 2021). The ability to track products throughout their lifecycle not only helps in adhering to regulatory standards but also in mitigating risks associated with fraud and unethical practices (Mercuri et al., 2021).

Moreover, blockchain technology can serve as a catalyst for new business models that prioritize sustainability. Organisations leveraging blockchain can create value through innovative business models that emphasize resource efficiency and circular economy principles (Singh et al., 2024b; Ghobakhloo, 2023) and companies can utilize blockchain to develop models that support remanufacturing and recycling, thereby contributing to sustainable development goals (Govindan, 2022). The flexibility of blockchain allows businesses to adapt their operational frameworks to incorporate sustainability as a core component, thus driving long-term organisational growth (Moiana, 2023).

In addition to operational enhancements, blockchain can also influence corporate governance and ethical practices. The technology's transparency can promote accountability and ethical behaviour among businesses, which is increasingly demanded by consumers and stakeholders (Ronaghi & Mosakhani, 2021). By embedding ethical considerations into blockchain systems, companies can enhance their social sustainability efforts, thereby improving their public image and stakeholder trust (Aljumah, 2023). This alignment of business ethics with technological innovation is crucial for fostering a sustainable business ecosystem (Umar, 2024).

However, the adoption of blockchain technology is not without challenges. Many organisations face barriers such as a lack of standardization and understanding of blockchain's potential benefits (Govindan, 2022). To overcome these hurdles, businesses must develop clear strategies and frameworks that outline the integration of blockchain into their operations. This includes identifying key success factors and understanding the specific characteristics of blockchain-inspired business models

(Lam, 2023). Additionally, organisations must engage in continuous learning and adaptation to fully leverage the capabilities of blockchain for sustainable growth (Yu, 2024).

In conclusion, blockchain technology presents a significant opportunity for businesses to enhance their sustainability practices and drive organisational growth. By improving transparency, fostering innovative business models and promoting ethical practices, blockchain can serve as a powerful enabler of sustainable business strategies. However, successful implementation requires overcoming existing barriers and developing a comprehensive understanding of blockchain's potential within the context of sustainability.

3.2 Understanding Blockchain Technology

Understanding blockchain technology requires a comprehensive exploration of its fundamental principles, architecture and applications across various sectors (Wang et al., 2021). Blockchain, at its core, is a decentralized digital ledger that records transactions across many computers in such a way that the registered transactions cannot be altered retroactively (Dong et al., 2023; Leema et al., 2021). This immutability is a key feature that enhances trust and security in digital transactions, making blockchain a transformative technology in numerous fields, including finance, supply chain management and healthcare.

The architecture of blockchain consists of several layers, including the data layer, network layer, consensus layer and application layer. Each layer plays a crucial role in the functionality of blockchain systems. The data layer is responsible for the structure of the blockchain, where transactions are grouped into blocks. The network layer facilitates communication between nodes, while the consensus layer ensures agreement among distributed nodes regarding the validity of transactions. Finally, the application layer encompasses the various applications that utilize blockchain technology, such as smart contracts and decentralized applications (dApps) (Cao et al., 2023; Paik et al., 2019).

Smart contracts, which are self-executing contracts with the terms of the agreement directly written into code, represent a significant advancement enabled by blockchain technology (Mik, 2017). They automate processes and reduce the need for intermediaries, thereby increasing efficiency and reducing costs (Bartoletti & Pompianu, 2017). The potential applications of smart contracts span various industries, including finance, real estate and healthcare, where they can streamline operations and enhance transparency (Mohite & Kale, 2023).

In addition, blockchain technology is characterized by its consensus mechanisms, which are protocols that ensure all nodes in the network agree on the validity of transactions. Common consensus algorithms include Proof of Work (PoW) and Proof of Stake (PoS), each with its advantages and challenges regarding scalability, energy consumption and security (Touloupou et al., 2022; Zhang et al., 2021).

Understanding these mechanisms is crucial for evaluating the performance and reliability of different blockchain implementations.

The implications of blockchain extend beyond technical specifications; they also encompass economic and social dimensions. For instance, blockchain can enhance supply chain transparency by providing immutable records of product provenance, which is essential for industries like food and pharmaceuticals (Juma et al., 2019). Also, the technology fosters trust in transactions between parties that do not necessarily trust each other, thereby enabling new business models and collaborative frameworks (Massaro et al., 2020; Kolb et al., 2019).

Despite its potential, the adoption of blockchain technology faces several challenges, including regulatory uncertainties, interoperability issues and the need for standardization across platforms (Reshi, 2023). Addressing these challenges is vital for the broader acceptance and integration of blockchain into existing systems. In conclusion, blockchain technology represents a paradigm shift in how transactions are conducted and recorded. Its decentralized nature, combined with features like immutability and transparency, positions it as a powerful tool for enhancing trust and efficiency across various sectors. As research and development continue, understanding the intricacies of blockchain will be essential for leveraging its full potential.

3.3 Smart Circular Economy and Organisational Sustainability

The concept of a smart circular economy (SCE) integrates digital technologies with circular economy principles to enhance organisational sustainability (Singh et al., 2022). This approach emphasizes the importance of resource efficiency, waste reduction and the sustainable management of materials throughout their lifecycle. By leveraging digital technologies, organisations can optimize their operations, foster innovation and contribute to broader sustainability goals (Khan et al., 2024).

At its core, the SCE seeks to redefine traditional linear economic models, which typically follow a “take-make-dispose” pattern, into a circular framework that emphasizes reuse, remanufacturing and recycling (Pirrone, 2023; Lobo et al., 2021). Digital technologies such as the Internet of Things (IoT), artificial intelligence (AI) and blockchain play a crucial role in facilitating this transition. For instance, IoT devices can monitor resource usage in real time, enabling organisations to optimize their processes and reduce waste (Karunarathna et al., 2024; Kristoffersen et al., 2020). Similarly, blockchain technology can enhance transparency and traceability in supply chains, ensuring that materials are sourced sustainably and reused effectively (Bressanelli et al., 2022).

The integration of smart technologies into circular economy practices can significantly improve organisational sustainability (Singh et al., 2023). Research indicates that organisations adopting smart circular strategies can achieve better sustainability outcomes, particularly in terms of resource efficiency and waste management (Bressanelli et al., 2022; Maria et al., 2022). For example, digital platforms can facilitate the sharing of resources among organisations, thereby reducing the overall demand for new materials and minimizing waste (Pirrone, 2023). Furthermore, the use of predictive analytics can help organisations anticipate resource needs and optimize inventory management, further enhancing sustainability (Kristoffersen et al., 2020).

However, transitioning to a SCE is not without challenges. Organisations often face barriers such as a lack of awareness, insufficient technological infrastructure and resistance to change within organisational cultures (Lobo et al., 2021). To overcome these obstacles, it is essential for organisations to foster a culture of sustainability that encourages innovation and collaboration among employees (Bhushan & Singh, 2024; Wijethilake et al., 2021; Lozano, 2018). Also, leadership plays a critical role in this process, as effective leaders can inspire and mobilize their teams towards sustainable practices (Klapper et al., 2020).

Additionally, the successful implementation of smart circular strategies requires a holistic approach that considers the entire ecosystem of stakeholders involved in the value chain (Khan et al., 2022a). This includes engaging suppliers, customers and regulatory bodies to create a collaborative environment that supports sustainability initiatives (Fobbe, 2020). By aligning the values and goals of all stakeholders, organisations can enhance their sustainability performance and contribute to the achievement of the United Nations Sustainable Development Goals (SDGs) (Singh et al., 2024c; Kayıkçı et al., 2021).

At last, the SCE represents a promising paradigm for enhancing organisational sustainability. By integrating digital technologies with circular economy principles, organisations can optimize resource use, reduce waste and foster innovation. However, realizing the full potential of this approach requires overcoming existing barriers and fostering a culture of sustainability that engages all stakeholders in the value chain.

3.4 Strategies for Implementing Blockchain in Sustainable Practices

3.4.1 Supply Chain Transparency and Traceability

Supply chain transparency and traceability are critical components of modern supply chain management, particularly in the context of sustainability and ethical practices (Baklaga, 2024; Böhmecke-Schwafert et al., 2022). Transparency refers to the visibility of information and processes throughout the supply chain, allowing stakeholders to track the movement of goods, monitor production processes and ensure compliance with ethical and regulatory standards (Kashmanian, 2017; Huzaifa, 2024). Traceability, on the other hand, is the ability to identify and verify the components and chronology of events in all steps of a process chain, which is essential for ensuring product integrity and safety (Francisco & Swanson, 2018; Park & Li, 2021).

One of the primary benefits of enhanced supply chain transparency is the ability to mitigate risks associated with unethical practices, such as labour exploitation and environmental degradation (Khan et al., 2022b). As consumers increasingly demand information about the origins of the products they purchase, companies are compelled to disclose their supply chain practices to maintain credibility and brand value (Munir et al., 2022). Regulatory initiatives like the California Transparency in Supply Chains Act require companies to disclose their efforts to combat human trafficking and slavery, thereby promoting ethical practices within supply chains (Jin et al., 2015). This increased scrutiny not only helps consumers make informed choices but also incentivizes companies to adopt more sustainable practices.

The integration of technologies such as blockchain and the IoT significantly enhances supply chain transparency and traceability (Kılıç, 2024; Bendarag et al., 2022; Parmentola et al., 2021). Blockchain technology provides a decentralized and immutable ledger of transactions, ensuring that all parties have access to accurate and tamper-proof information regarding product origins and movements (Oriekhoe, 2024). This capability is particularly valuable in industries where traceability is crucial, such as food and pharmaceuticals, as it enables rapid recall capabilities for unsafe goods and enhances overall product safety (Huzaifa, 2024). Additionally, IoT devices can collect real-time data on product conditions and locations, further improving visibility across the supply chain (Rejeb et al., 2019; Hellani et al., 2021).

Moreover, transparency in supply chains can lead to improved operational efficiency and decision-making (Udeh et al., 2024). With access to comprehensive data about their supply chains, organisations can identify inefficiencies, reduce waste and optimize resource allocation (Xu et al., 2021). This not only contributes to cost savings but also supports sustainability goals by minimizing environmental impact (Kashmanian, 2017; Park & Li, 2021). Furthermore, transparent supply chains facilitate better collaboration among partners, as stakeholders can share information and work together to address challenges and improve overall performance (Brun et al., 2020; Holloway, 2024).

However, while achieving supply chain transparency, organisations often face barriers such as data silos, lack of standardization and resistance to change from stakeholders (Sodhi & Tang, 2019). To overcome these barriers, companies must invest in technology and infrastructure that support data sharing and collaboration. Additionally, fostering a culture of transparency within organisations and among supply chain partners is essential for encouraging open communication and trust (Ko et al., 2022; Liu, 2023).

Supply chain transparency and traceability are vital for promoting sustainability and ethical practices in today's global economy. By leveraging technologies like blockchain and IoT, organisations can enhance visibility across their supply chains, mitigate risks and improve operational efficiency. As consumer demand for transparency continues to grow, companies that prioritize these aspects will be better positioned to succeed in an increasingly competitive marketplace.

3.4.2 *Tokenization and Incentivizing Sustainability*

Tokenization, facilitated by blockchain technology, is emerging as a powerful mechanism for incentivizing sustainability across various sectors (Harish et al., 2023). By converting tangible and intangible assets into digital tokens, organisations can create new economic models that promote sustainable practices and enhance stakeholder engagement (Akella et al., 2023). This synthesis explores how tokenization can drive sustainability initiatives, focusing on its applications in sustainable finance, behaviour change and collaborative platforms.

One of the most significant applications of tokenization in promoting sustainability is in the realm of sustainable finance (Luzzani et al., 2021). Tokenization allows for fractional ownership of sustainable assets, enabling a broader range of investors to participate in funding projects with positive environmental or social impacts (Zhang et al., 2024). For instance, through the use of smart contracts, organisations can create tokens that represent shares in renewable energy projects or sustainable agriculture initiatives (Örnek & Kabak, 2023). This approach not only democratizes access to investment opportunities but also aligns financial resources with sustainability goals, thereby unlocking new funding avenues for green projects (Kim & Kim, 2024; Kontopanou, 2023). The ability to trade these tokens on blockchain platforms enhances liquidity and provides transparency, further encouraging investment in sustainable initiatives.

Tokenization can also serve as an effective tool for incentivizing behaviour change among individuals and organisations (Meneguzzo et al., 2024). By implementing token economies, organisations can reward sustainable behaviors, such as recycling, energy conservation, or participation in sustainability programs (Bułkowska, 2023). For example, tokens can be awarded to consumers for purchasing eco-friendly products or participating in community clean-up events. These tokens can then be redeemed for discounts, products, or services, creating a tangible incentive for individuals to engage in sustainable practices (Barclay et al., 2022). This mechanism not only encourages positive behaviour but also fosters a culture of sustainability within communities.

In the context of collaborative platforms, tokenization can facilitate partnerships between various stakeholders, including businesses, researchers and consumers (Narayan & Tidström, 2020). By utilizing blockchain technology, organisations can create verifiable and transparent systems that track contributions and rewards within collaborative projects (Leible et al., 2019). For instance, a platform designed for small and medium-sized enterprises (SMEs) could use tokens to incentivize collaboration with research institutions, allowing SMEs to access valuable resources and expertise without the burden of high costs (Udokwu, 2024). This collaborative approach not only enhances innovation but also promotes sustainable practices by aligning the interests of diverse stakeholders towards common sustainability goals (Wang & Zhong, 2022).

The successful implementation of tokenization for sustainability requires careful design and consideration of the underlying economic models (Sunyaev et al., 2021).

Organisations must ensure that their token economies are structured to provide real value to participants while promoting sustainable outcomes (Kim & Chung, 2018). This involves creating mechanisms that balance incentives with the long-term sustainability of the initiatives, avoiding potential pitfalls such as token inflation or misalignment of interests among stakeholders (Kim & Kim, 2024; Wang & Zhong, 2022). Additionally, fostering trust and transparency within tokenized systems is crucial for encouraging participation and ensuring the integrity of sustainability initiatives (Tomlinson et al., 2020; Schmidt, 2023).

Eventually, tokenization represents a transformative opportunity for incentivizing sustainability across various sectors. By leveraging blockchain technology, organisations can create new economic models that promote sustainable practices, encourage behaviour change and facilitate collaboration among stakeholders. As the adoption of tokenization continues to grow, it will be essential for organisations to design effective token economies that align incentives with sustainability goals, ultimately contributing to a more sustainable future.

3.4.3 Decentralized Decision-Making for Sustainable Operations

Decentralized decision-making is increasingly recognized as a vital approach for fostering sustainable operations across various sectors (Tišma & Škrtić, 2023). This governance model empowers local stakeholders, enhances responsiveness to community needs and promotes innovation through collaborative efforts. By leveraging decentralized frameworks, organisations can align their operational strategies with sustainability goals, ultimately contributing to more resilient and adaptive systems (Nurgazina et al., 2021).

One of the primary advantages of decentralized decision-making is the empowerment of local communities (Ahmad & Abu Talib, 2015). When decision-making authority is distributed among local actors, it allows for a more nuanced understanding of local contexts and challenges (Richardson et al., 2019). This is particularly relevant in sustainable development initiatives, where local knowledge and participation are crucial for effective resource management (Khlaifat et al., 2018). Local governments can tailor sustainability initiatives to meet the specific needs of their communities, ensuring that interventions are relevant and effective (Cho, 2018). This localized approach not only enhances accountability but also fosters a sense of ownership among community members, which is essential for the long-term success of sustainability initiatives (Atisa et al., 2020; Elliott, 2023).

Furthermore, decentralized decision-making can facilitate innovation and adaptability in organisations. By breaking down hierarchical structures, organisations can encourage diverse perspectives and ideas to emerge, leading to more creative solutions to sustainability challenges (Karim et al., 2023; Vendette, 2023). For example, blockchain technology exemplifies this shift by enabling decentralized autonomous organisations (DAOs), where governance rules are encoded in smart contracts, allowing stakeholders to participate in decision-making processes without centralized control (Beck et al., 2018; Arruñada & Garicano, 2018). This model not only enhances transparency but also allows for more agile responses to changing environmental conditions and stakeholder expectations (Chen et al., 2020; Kondova & Barba, 2019).

In addition to fostering innovation, decentralized decision-making can improve the efficiency of resource allocation. When local entities have the authority to make decisions, they can respond more quickly to emerging needs and allocate resources more effectively (Cho, 2018; Wang et al., 2023). This is particularly important in the context of environmental management, where timely interventions can significantly impact sustainability outcomes (Atisa et al., 2020; Malawi, 2024). Also, local governments can implement targeted initiatives to promote renewable energy adoption or waste reduction, aligning their strategies with broader sustainability goals while addressing local priorities (Smoke, 2018).

However, the successful implementation of decentralized decision-making requires careful consideration of governance structures and processes. Organisations must establish clear frameworks that define roles, responsibilities and accountability mechanisms to ensure effective collaboration among stakeholders (Chen et al., 2020; Schädler, 2023). Additionally, capacity building is essential to equip local actors with the skills and resources needed to engage in decision-making processes effectively (Hamal, 2018). This includes providing training and support to enhance local governance capabilities, ensuring that communities can manage resources sustainably and equitably (Atisa et al., 2020; Alsagheer, 2023).

In conclusion, decentralized decision-making represents a promising approach for enhancing sustainable operations. By empowering local stakeholders, fostering innovation and improving resource allocation, this governance model can contribute to more resilient and adaptive systems. However, organisations must invest in establishing effective governance frameworks and capacity-building initiatives to realize the full potential of decentralized decision-making in promoting sustainability.

3.5 Discussions

Blockchain technology has emerged as a transformative tool for addressing critical sustainability challenges in modern organisations. This research has demonstrated its potential to enhance transparency, foster innovative business models, and integrate ethical practices into organisational operations and suggests that blockchain's decentralized and immutable features enable organisations to improve trust among stakeholders, streamline supply chains, and promote sustainability-driven strategies. These advancements can directly impact organisational growth and long-term resilience by creating a framework for more accountable and efficient business practices.

A key theme that emerged is the role of blockchain in advancing the principles of the smart circular economy. By enabling real-time traceability and promoting resource efficiency, blockchain allows organisations to transition from traditional linear economic models to sustainable circular systems. This aligns with global sustainability goals, particularly in addressing waste reduction and material reuse challenges. However, the successful implementation of these systems requires comprehensive strategies that address technological and organisational barriers.

The integration of blockchain-enabled tokenization and decentralized decision-making frameworks highlights new avenues for incentivizing sustainable practices and enhancing stakeholder engagement. These mechanisms not only promote environmental and social sustainability but also encourage innovation by enabling collaborative ecosystems. However, the scalability of such systems and the alignment of stakeholder interests remain pressing challenges that require further exploration.

Despite its transformative potential, the adoption of blockchain technology is hindered by several challenges, including regulatory uncertainties, lack of standardization, and the need for robust technological infrastructure. Addressing these challenges will require multi-stakeholder collaboration, investments in technological education, and the establishment of clear governance frameworks. Moreover, fostering a culture of sustainability within organisations is crucial for ensuring the effective adoption of blockchain-based solutions.

Future research in this area should focus on evaluating the long-term impacts of blockchain on sustainability outcomes across diverse industries. Thus, empirical studies examining the scalability of blockchain solutions, their cost-effectiveness, and their ability to adapt to varying regulatory environments will be critical. Additionally, exploring the interplay between blockchain technology and other emerging digital technologies, such as AI and IoT, could provide deeper insights into creating holistic solutions for sustainability challenges.

Likewise, this research underscores the significant potential of blockchain technology in driving sustainable organisational growth. While barriers to adoption exist, strategic interventions and a commitment to innovation can help organisations unlock blockchain's full potential, contributing to a more sustainable and resilient global economy.

3.6 Concluding Remarks

Blockchain technology presents a transformative opportunity for advancing sustainability in businesses and organisations. Its inherent characteristics—decentralization, transparency, and immutability—offer innovative solutions to pressing challenges in supply chain management, ethical governance, and resource efficiency. By integrating blockchain into operational and strategic frameworks, organisations can achieve enhanced accountability, foster stakeholder trust, and contribute to global sustainability goals.

This research has highlighted how blockchain-enabled strategies, such as tokenization, decentralized decision-making, and principles of the smart circular economy, can drive innovation and sustainability. These approaches redefine traditional business models by promoting recycling, remanufacturing and resource optimization. They also encourage ethical practices and provide mechanisms for rewarding sustainable behaviors, aligning organisational objectives with broader societal and environmental needs.

However, realizing blockchain's full potential requires overcoming significant barriers, including regulatory ambiguities, technological interoperability issues, and resistance to change within organisations. Addressing these challenges necessitates a collaborative effort among businesses, policymakers and technology developers. Organisations must also cultivate a culture of sustainability, invest in education and training, and develop clear frameworks for blockchain implementation.

As businesses increasingly prioritize sustainability, blockchain offers a robust foundation for innovation and growth. By adopting this technology, organisations can not only meet current sustainability demands but also position themselves as leaders in a rapidly evolving global landscape. While challenges remain, the opportunities provided by blockchain are compelling and underscore its critical role in shaping a sustainable and resilient future.

References

Ahmad, M. S., & Abu Talib, N. B. (2015). Empowering local communities: Decentralization, empowerment and community driven development. *Quality & Quantity, 49*, 827–838. https://doi.org/10.1007/s11135-014-0025-8

Akella, G., Wibowo, S., Grandhi, S., & Mubarak, S. (2023). A systematic review of blockchain technology adoption barriers and enablers for smart and sustainable agriculture. *Big Data and Cognitive Computing, 7*(2), 86. https://doi.org/10.3390/bdcc7020086

Aljumah, A. (2023). Investigating the impact of blockchain technology on social sustainability and the mediating role of ethics and csr. *Sustainability, 15*(21), 15510. https://doi.org/10.3390/su152115510

Alsagheer, D. (2023). Decentralized machine learning governance: Overview, opportunities, and challenges. *IEEE Access, 11*, 96718–96732. https://doi.org/10.1109/access.2023.3311713

Arruñada, B., & Garicano, L. (2018). Blockchain: The birth of decentralized governance. *SSRN Electronic Journal*. https://doi.org/10.2139/ssrn.3160070

Atisa, G., Zemrani, A., & Weiss, M. (2020). Decentralized governments: Local empowerment and sustainable development challenges in africa. *Environment Development and Sustainability, 23*(3), 3349–3367. https://doi.org/10.1007/s10668-020-00722-0

Baklaga, L. (2024). Synergizing ai and blockchain: Innovations in decentralized carbon markets for emission reduction through intelligent carbon credit trading. *Journal of Computer Science and Technology Studies, 6*(2), 111–120. https://doi.org/10.32996/jcsts.2024.6.2.13

Barclay, I., Cooper, M., Hackel, J., & Perrin, P. (2022). Tokenizing behavior change: A pathway for the sustainable development goals. *Frontiers in Blockchain, 4*. https://doi.org/10.3389/fbloc.2021.730101

Bartoletti, M., & Pompianu, L. (2017). An empirical analysis of smart contracts: Platforms, applications, and design patterns., 494–509. https://doi.org/10.1007/978-3-319-70278-0_31.

Beck, R., Müller-Bloch, C., & King, J. (2018). Governance in the blockchain economy: A framework and research agenda. *Journal of the Association for Information Systems*, 1020–1034. https://doi.org/10.17705/1jais.00518

Bendarag, A., Boutkhoum, O., Abada, D., & Hanine, M. (2022). Blockchain adoption barriers in moroccan sustainable supply chain: A proposed approach. *Indonesian Journal of Electrical Engineering and Computer Science, 27*(2), 892. https://doi.org/10.11591/ijeecs.v27.i2.pp892-899

Bhushan, M., & Singh, R. (2024). Talent management facilitates net-zero transition through employee green behavior. In R. Singh, S. Khan, A. Kumar, S. Luthra, & H. Chokshi (Eds.), *Net zero economy, corporate social responsibility and sustainable value creation. CSR, sustainability, ethics & governance*. Springer. https://doi.org/10.1007/978-3-031-55779-8_7

Böhmecke-Schwafert, M., Wehinger, M., & Teigland, R. (2022). Blockchain for the circular economy: Theorizing blockchain's role in the transition to a circular economy through an empirical investigation. *Business Strategy and the Environment, 31*(8), 3786–3801. https://doi.org/10.1002/bse.3032

Bressanelli, G., Adrodegari, F., Pigosso, D., & Parida, V. (2022). Towards the smart circular economy paradigm: A definition, conceptualization, and research agenda. *Sustainability, 14*(9), 4960. https://doi.org/10.3390/su14094960

Brun, A., Karaosman, H., & Barresi, T. (2020). Supply chain collaboration for transparency. *Sustainability, 12*(11), 4429. https://doi.org/10.3390/su12114429

Bułkowska, K. (2023). Implementation of blockchain technology in waste management. *Energies, 16*(23), 7742. https://doi.org/10.3390/en16237742

Cao, B., Wang, Z., Zhang, L., Feng, D., Peng, M., & Zhang, L. (2023). Blockchain systems, technologies, and applications: A methodology perspective. *IEEE Communications Surveys & Tutorials, 25*(1), 353–385. https://doi.org/10.1109/comst.2022.3204702

Centobelli, P., Cerchione, R., Del Vecchio, P., Oropallo, E., & Secundo, G. (2022). Blockchain technology for bridging trust, traceability and transparency in circular supply chain. *Information & Management, 59*(7), 103508. https://doi.org/10.1016/j.im.2021.103508

Chen, Y., Pereira, I., & Patel, P. (2020). Decentralized governance of digital platforms. *Journal of Management, 47*(5), 1305–1337. https://doi.org/10.1177/0149206320916755

Cho, I. (2018). Fiscal decentralization in Korea. *Asian Education and Development Studies, 7*(3), 279–290. https://doi.org/10.1108/aeds-11-2017-0113

Dong, S., Abbas, K., Li, M., & Kamruzzaman, J. (2023). Blockchain technology and application: An overview. *PeerJ Computer Science, 9*, e1705. https://doi.org/10.7717/peerj-cs.1705

Elliott, F. (2023). The advantages of decentralized government from the viewpoint of public service innovation. *Journal of Management and Administration Provision, 2*(3), 65–72. https://doi.org/10.55885/jmap.v2i3.172

Fobbe, L. (2020). Analysing organisational collaboration practices for sustainability. *Sustainability, 12*(6), 2466. https://doi.org/10.3390/su12062466

Francisco, K., & Swanson, D. (2018). The supply chain has no clothes: Technology adoption of blockchain for supply chain transparency. *Logistics, 2*(1), 2. https://doi.org/10.3390/logistics2010002

Ghobakhloo, M. (2023). Blockchain technology as an enabler for sustainable business ecosystems: A comprehensive roadmap for socioenvironmental and economic sustainability. *Business Strategy & Development, 7*, 1. https://doi.org/10.1002/bsd2.319

Govindan, K. (2022). Tunneling the barriers of blockchain technology in remanufacturing for achieving sustainable development goals: A circular manufacturing perspective. *Business Strategy and the Environment, 31*(8), 3769–3785. https://doi.org/10.1002/bse.3031

Hamal, S. (2018). Understanding education decentralization and 'sustained' elite capture—Explicating cases from two public schools. *Journal of Education and Research, 8*(1), 29–45. https://doi.org/10.3126/jer.v8i1.25478

Harish, A. R., Wu, W., Li, M., & Huang, G. Q. (2023). Blockchain-enabled digital asset tokenization for crowdsensing in environmental, social, and governance disclosure. *Computers & Industrial Engineering, 185*, 109664. https://doi.org/10.1016/j.cie.2023.109664

Hellani, H., Sliman, L., Samhat, A., & Expósito, E. (2021). On blockchain integration with supply chain: Overview on data transparency. *Logistics, 5*(3), 46. https://doi.org/10.3390/logistics5030046

Holloway, S. (2024). Narratives of transparency: How supply chain communication shapes brand image in marketing. https://doi.org/10.20944/preprints202406.1124.v1.

Huzaifa, I. (2024). Impact of blockchain technology on supply chain management efficiency and transparency in Pakistan. *International Journal of Supply Chain Management, 9*(1), 16–26. https://doi.org/10.47604/ijscm.2320

Jin, Y., Lee, H., & Goerlitz, K. (2015). Transparency of global apparel supply chains: Quantitative analysis of corporate disclosures. *Corporate Social Responsibility and Environmental Management, 23*(5), 308–318. https://doi.org/10.1002/csr.1378

Joo, J., & Han, Y. (2021). An evidence of distributed trust in blockchain-based sustainable food supply chain. *Sustainability, 13*(19), 10980. https://doi.org/10.3390/su131910980

Juma, H., Shaalan, K., & Kamel, I. (2019). A survey on using blockchain in trade supply chain solutions. *IEEE Access, 7*, 184115–184132. https://doi.org/10.1109/access.2019.2960542

Karim, S., Naeem, M., Tiwari, A., & Ashraf, S. (2023). Examining the avenues of sustainability in resources and digital blockchains backed currencies: Evidence from energy metals and cryptocurrencies. *Ann. Oper. Res.* https://doi.org/10.1007/s10479-023-05365-8

Karunarathna, K. S. S. N., Dewasiri, N. J., Singh, R., & Rathnasiri, M. S. H. (2024). In R. Singh, S. Khan, A. Kumar, & V. Kumar (Eds.), *Artificial intelligence enabled management: An emerging economy perspective*. Walter de Gruyter GmbH & Co KG.

Kashmanian, R. (2017). Building greater transparency in supply chains to advance sustainability. *Environmental Quality Management, 26*(3), 73–104. https://doi.org/10.1002/tqem.21495

Kayıkçı, Y., Kazançoğlu, Y., Lafçı, Ç., Gozacan-Chase, N., & Mangla, S. (2021). Smart circular supply chains to achieving sdgs for post-pandemic preparedness. *Journal of Enterprise Information Management, 35*(1), 237–265. https://doi.org/10.1108/jeim-06-2021-0271

Khan, S., Singh, R., Sá, J. C., Santos, G., & Ferreira, L. P. (2022a). Modelling of determinants of logistics 4.0 adoption: Insights from developing countries. *Machines, 10*(12), 1242. https://doi.org/10.3390/machines10121242

Khan, S., Ali, S. S., & Singh, R. (2022b). Determinants of remanufacturing adoption for circular economy: A causal relationship evaluation framework. *Applied System Innovation, 5*(4), 62. https://doi.org/10.3390/asi5040062

Khan, S., Khan, M. I., & Singh, R. (2023). Modeling the barriers of Blockchain technology implementation in supply chain. *Journal of Industrial Integration and Management, 8*(03), 367–385. https://doi.org/10.1142/S2424862223500070

Khan, S., Singh, R., Alnahas, J., Abbate, S., & Centobelli, P. (2024). Navigating the smart circular economy: A framework for manufacturing firms. *Journal of Cleaner Production, 480*, 144007. https://doi.org/10.1016/j.jclepro.2024.144007

Khlaifat, A., Taleb, H., Makhamreh, Z., & Qtiashat, D. (2018). Analysis of the urban management practices and sustainable development in the local municipalities in Jordan: Case study - salt city. *Modern Applied Science, 12*(4), 109. https://doi.org/10.5539/mas.v12n4p109

Kılıç, C. (2024). Sustainable human development and human resource management strategies in building sustainable cities. *Kent Akademisi, 17*(Sürdürülebilir İnsani Kalkınma ve Kent), 153–173. https://doi.org/10.35674/kent.1501835

Kim, M., & Chung, J. (2018). Sustainable growth and token economy design: The case of steemit. *Sustainability, 11*(1), 167. https://doi.org/10.3390/su11010167

Kim, H., & Kim, D. (2024). Methodological advancements in standardizing blockchain assessment. *IEEE Access*. https://doi.org/10.1109/ACCESS.2024.3372578

Klapper, R., Berg, L., & Upham, P. (2020). Probing alignment of personal and organisational values for sustainability: An assessment of Barrett's organisational consciousness model. *Sustainability, 12*(18), 7584. https://doi.org/10.3390/su12187584

Ko, T., Lee, J., Park, D., & Ryu, D. (2022). Supply chain transparency as a signal of ethical production. *Managerial and Decision Economics, 44*(3), 1565–1573. https://doi.org/10.1002/mde.3765

Kolb, J., Becker, L., Fischer, M., & Winkelmann, A. (2019). The role of blockchain in enterprise procurement. https://doi.org/10.24251/hicss.2019.554.

Kondova, G., & Barba, R. (2019). Governance of decentralized autonomous organizations. *Journal of Modern Accounting and Auditing, 15*(8). https://doi.org/10.17265/1548-6583/2019.08.003

Kontopanou, M. (2023). A review of sustainability concerns in the use of blockchain technology: Evidence from the Agri-food and the pharmaceutical sectors. *E3s Web of Conferences, 436*, 07006. https://doi.org/10.1051/e3sconf/202343607006

Kristoffersen, E., Blomsma, F., Mikalef, P., & Li, J. (2020). The smart circular economy: A digital-enabled circular strategies framework for manufacturing companies. *Journal of Business Research, 120*, 241–261. https://doi.org/10.1016/j.jbusres.2020.07.044

Kumar, K., Singh, R., Sharma, A., & Lehri, B. P. S. (2024). Climate change adaptation strategies for achieving net-zero economy. In R. Singh & D. Crowther (Eds.), *Transition towards a sustainable future. Approaches to global sustainability, markets, and governance*. Springer. https://doi.org/10.1007/978-981-97-5756-5_8

Lam, W. (2023). Conceptual framework for blockchain business success factors with fuzzy analytic hierarchy process model. *Itm Web of Conferences, 57*, 01006. https://doi.org/10.1051/itmconf/20235701006

Leema, R. G., Rajmohan, R., Usharani, S., Kiruba, K., & Manjubala, P. (2021). Fundamentals of Blockchain and distributed ledger technology (DLT). In *Recent trends in Blockchain for information systems security and privacy* (pp. 3–38). CRC Press. https://doi.org/10.1201/9781003139737-2

Leible, S., Schlager, S., Schubotz, M., & Gipp, B. (2019). A review on blockchain technology and blockchain projects fostering open science. *Frontiers in Blockchain, 2*, 486595. https://doi.org/10.3389/fbloc.2019.00016

Liu, H. (2023). The impact of social capital on socially responsible supply chain performance: The moderating role of supply chain transparency. *Food, 12*(19), 3624. https://doi.org/10.3390/foods12193624

Lobo, A., Trevisan, A., Liu, Q., Yang, M., & Mascarenhas, J. (2021). Barriers to transitioning towards smart circular economy: A systematic literature review., 245-256. https://doi.org/10.1007/978-981-16-6128-0_24.

Lozano, R. (2018). Proposing a definition and a framework of organisational sustainability: A review of efforts and a survey of approaches to change. *Sustainability, 10*(4), 1157. https://doi.org/10.3390/su10041157

Luzzani, G., Grandis, E., Frey, M., & Capri, E. (2021). Blockchain technology in wine chain for collecting and addressing sustainable performance: An exploratory study. *Sustainability, 13*(22), 12898. https://doi.org/10.3390/su132212898

Malawi, N. (2024). Analysis of decentralization policy implementation in improving the quality of public services in rural areas. *International Journal of Economics and Management Research, 3*(2), 166–180. https://doi.org/10.55606/ijemr.v3i2.215

Maria, E., Marchi, V., & Galeazzo, A. (2022). Industry 4.0 technologies and circular economy: The mediating role of supply chain integration. *Business Strategy and the Environment, 31*(2), 619–632. https://doi.org/10.1002/bse.2940

Massaro, M., Mas, F., Jabbour, C., & Bagnoli, C. (2020). Crypto-economy and new sustainable business models: Reflections and projections using a case study analysis. *Corporate Social Responsibility and Environmental Management, 27*(5), 2150–2160. https://doi.org/10.1002/csr.1954

Meneguzzo, S., Selabi, E. G. K., Favenza, A., Gatteschi, V., & Schifanella, C. (2024). Enabling citizen sustainable behaviors in urban mobility through Blockchain and tokenization. In *Blockchain technology in the automotive industry* (pp. 245–279). CRC Press. https://doi.org/10.1201/9781003450306-17

Mercuri, F., Corte, G., & Ricci, F. (2021). Blockchain technology and sustainable business models: A case study of devoleum. *Sustainability, 13*(10), 5619. https://doi.org/10.3390/su13105619

Mik, E. (2017). Smart contracts: Terminology, technical limitations and real world complexity. *Law, Innovation and Technology, 9*(2), 269–300. https://doi.org/10.1080/17579961.2017.1378468

Mohite, J., & Kale, V. (2023). Blockchain technology based supply chain management system., 641–653. https://doi.org/10.2991/978-94-6463-136-4_55.

Moiana, D. (2023). Emerging technologies: A catalyst for sustainable business model innovation. *European Conference on Innovation and Entrepreneurship, 18*(1), 636–643. https://doi.org/10.34190/ecie.18.1.1699

Munir, M., Habib, M., Hussain, A., Shahbaz, M., Qamar, A., Masood, T., et al. (2022). Blockchain adoption for sustainable supply chain management: Economic, environmental, and social perspectives. *Frontiers in Energy Research, 10*. https://doi.org/10.3389/fenrg.2022.899632

Narayan, R., & Tidström, A. (2020). Tokenizing coopetition in a blockchain for a transition to circular economy. *Journal of Cleaner Production, 263*, 121437. https://doi.org/10.1016/j.jclepro.2020.121437

Nurgazina, J., Pakdeetrakulwong, U., Moser, T., & Reiner, G. (2021). Distributed ledger technology applications in food supply chains: A review of challenges and future research directions. *Sustainability, 13*(8), 4206. https://doi.org/10.3390/su13084206

Oriekhoe, O. (2024). Blockchain in supply chain management: A review of efficiency, transparency, and innovation. *International Journal of Science and Research Archive, 11*(1), 173–181. https://doi.org/10.30574/ijsra.2024.11.1.0028

Örnek, İ., & Kabak, S. (2023). Yenilenebilir enerji tüketimi ve ekonomik büyüme arasındaki ilişki: Türkiye üzerine bir inceleme. *Artuklu Kaime Uluslararası İktisadi ve İdari Araştırmalar Dergisi, 6*(2), 87–108. https://doi.org/10.55119/artuklu.1389880

Paik, H., Xu, X., Bandara, H., Lee, S., & Lo, S. (2019). Analysis of data management in blockchain-based systems: From architecture to governance. *IEEE Access, 7*, 186091–186107. https://doi.org/10.1109/access.2019.2961404

Park, A., & Li, H. (2021). The effect of blockchain technology on supply chain sustainability performances. *Sustainability, 13*(4), 1726. https://doi.org/10.3390/su13041726

Parmentola, A., Petrillo, A., Tutore, I., & Felice, F. (2021). Is blockchain able to enhance environmental sustainability? A systematic review and research agenda from the perspective of sustainable development goals (sdgs). *Business Strategy and the Environment, 31*(1), 194–217. https://doi.org/10.1002/bse.2882

Pirrone, L. (2023). How digital technologies can support sustainability of the waterborne passenger mobility ecosystem: A case study analysis of smart circular practices in northern europe. *Sustainability, 16*(1), 353. https://doi.org/10.3390/su16010353

Rejeb, A., Keogh, J., & Treiblmaier, H. (2019). Leveraging the internet of things and blockchain technology in supply chain management. *Future Internet, 11*(7), 161. https://doi.org/10.3390/fi11070161

Reshi, I. (2023). The blockchain conundrum: An in-depth examination of challenges, contributing technologies, and alternatives. *Concurrency and Computation Practice and Experience, 36*(8). https://doi.org/10.1002/cpe.7987

Richardson, L., Durose, C., & Dean, R. J. (2019). 'Why decentralize decision making? English local actors' viewpoints. *Governance, 32*(1), 159–176. https://doi.org/10.1111/gove.12365

Ronaghi, M., & Mosakhani, M. (2021). The effects of blockchain technology adoption on business ethics and social sustainability: Evidence from the middle east. *Environment Development and Sustainability, 24*(5), 6834–6859. https://doi.org/10.1007/s10668-021-01729-x

Saberi, S., Kouhizadeh, M., Sarkis, J., & Shen, L. (2019). Blockchain technology and its relationships to sustainable supply chain management. *International Journal of Production Research, 57*(7), 2117–2135. https://doi.org/10.1080/00207543.2018.1533261
Schädler, L. (2023). Analyzing decision-making in blockchain governance. *Frontiers in Blockchain, 6*. https://doi.org/10.3389/fbloc.2023.1256651
Schmidt, J. (2023). Blockchain and the transition to the circular economy: A literature review. *Corporate Social Responsibility and Environmental Management, 31*(3), 2010–2032. https://doi.org/10.1002/csr.2674
Singh, R., Khan, S., & Centobelli, P. (2022). Investigating the interplay between social performance and organisational factors supporting circular economy practices. *Sustainability, 14*(24), 16781. https://doi.org/10.3390/su142416781
Singh, R., Khan, S., & Dsilva, J. (2023). A framework for assessment of critical factor for circular economy practice implementation. *Journal of Modelling in Management, 18*(5), 1476–1497. https://doi.org/10.1108/JM2-06-2021-0145
Singh, R., Crowther, D., & Ravesangar, K. (2024a). Harnessing human Capital for a Sustainable Future: HR'S pivotal role toward net zero policies. In R. Singh & D. Crowther (Eds.), *Transition towards a sustainable future. Approaches to global sustainability, markets, and governance*. Springer. https://doi.org/10.1007/978-981-97-5756-5_2
Singh, R., Gupta, A., Bajpai, A., & Kandpal, V. (2024b). Exploring challenges of circular economy initiatives for smart cities. In *Smart cities and circular economy: The future of sustainable urban development* (pp. 59–69). Emerald Publishing Limited. https://doi.org/10.1108/978-1-83797-957-820241005
Singh, R., Kumar, K., & Khan, S. (2024c). A comprehensive view of artificial intelligence (AI)-based technologies for sustainable development goals (SDGs). In R. Singh, S. Khan, A. Kumar, & V. Kumar (Eds.), *Artificial intelligence enabled management: An emerging economy perspective*. Walter de Gruyter GmbH & Co KG. https://doi.org/10.1515/9783111172408-012
Smoke, P. (2018). Fiscal decentralisation frameworks for agenda 2030: Understanding key issues and crafting strategic reforms. *Commonwealth Journal of Local Governance*, 3, 3–23. https://doi.org/10.5130/cjlg.v0i20.6024
Sodhi, M., & Tang, C. (2019). Research opportunities in supply chain transparency. *Production and Operations Management, 28*(12), 2946–2959. https://doi.org/10.1111/poms.13115
Sunyaev, A., Kannengießer, N., Beck, R., Treiblmaier, H., Lacity, M., Kranz, J., et al. (2021). Token economy. *Business & Information Systems Engineering, 63*(4), 457–478. https://doi.org/10.1007/s12599-021-00684-1
Tišma, S., & Škrtić, M. (2023). Blockchain technology in the environmental economics: A service for a holistic and integrated life cycle sustainability assessment. *Journal of Risk and Financial Management, 16*(3), 209. https://doi.org/10.3390/jrfm16030209
Tomlinson, B., Boberg, J., Cranefield, J., Johnstone, D., Luczak-Roesch, M., Patterson, D., et al. (2020). Analyzing the sustainability of 28 'blockchain for good' projects via affordances and constraints. *Information Technology for Development, 27*(3), 439–469. https://doi.org/10.1080/02681102.2020.1828792
Touloupou, M., Themistocleous, M., Iosif, E., & Christodoulou, K. (2022). A systematic literature review toward a blockchain benchmarking framework. *IEEE Access, 10*, 70630–70644. https://doi.org/10.1109/access.2022.3188123
Udeh, E. O., Amajuoyi, P., Adeusi, K. B., & Scott, A. O. (2024). The role of IoT in boosting supply chain transparency and efficiency. *Magna Scientia Advanced Research and Reviews, 12*, 178–197. https://doi.org/10.30574/msarr.2024.11.1.0081
Udokwu, C. (2024). Formalizing and simulating the token aspects of blockchain-based research collaboration platform using game theory. *Mathematics, 12*(20), 3252. https://doi.org/10.3390/math12203252
Umar, M. (2024). The nexus between green intellectual capital, blockchain technology, green manufacturing, and sustainable performance. *Environmental Science and Pollution Research, 31*(10), 15026–15038. https://doi.org/10.1007/s11356-024-31952-8

Vendette, S. (2023). Decentralization in blockchain: Reconsidering change management theories. *American Journal of Management, 23*(3). https://doi.org/10.33423/ajm.v23i3.6376

Wang, Q., Li, R., & Zhan, L. (2021). Blockchain technology in the energy sector: From basic research to real world applications. *Computer Science Review, 39*, 100362. https://doi.org/10.1016/j.cosrev.2021.100362

Wang, S., Tian, G., Islam, T., Khaskhelly, F., & Shaikh, M. (2023). Effect of blockchain technology for sustainable performance in supply chain management. *Problemy Ekorozwoju, 18*(2), 159–166. https://doi.org/10.35784/preko.3954

Wang, Z., & Zhong, X. (2022). Stimulative coordination models for cooperative and competitive enterprise alliances based on token economy. *IEEE Access, 10*, 43454–43472. https://doi.org/10.1109/ACCESS.2022.3169598

Wijethilake, C., Upadhaya, B., & Lama, T. (2021). The role of organisational culture in organisational change towards sustainability: Evidence from the garment manufacturing industry. *Production Planning & Control, 34*(3), 275–294. https://doi.org/10.1080/09537287.2021.1913524

Xu, P., Lee, J., Barth, J., & Richey, R. (2021). Blockchain as supply chain technology: Considering transparency and security. *International Journal of Physical Distribution & Logistics Management, 51*(3), 305–324. https://doi.org/10.1108/ijpdlm-08-2019-0234

Yu, W. (2024). The effects of blockchain technology adoption on business ethics and social sustainability: Evidence from chinese fintech companies. *IJSASR, 4*(2), 299–308. https://doi.org/10.60027/ijsasr.2024.3932

Zhang, L., Xu, H., Onireti, O., Imran, M., & Cao, B. (2021). How much communication resource is needed to run a wireless blockchain network?. https://doi.org/10.48550/arxiv.2101.10852.

Zhang, Y., Gong, B., & Zhou, P. (2024). Centralized use of decentralized technology: Tokenization of currencies and assets. *Structural Change and Economic Dynamics, 71*, 15–25. https://doi.org/10.1016/j.strueco.2024.06.006

Chapter 4
Navigating the Future and Integration of Blockchain and Smart Circular Economy

Wasswa Shafik, Ali Tufail, Chandratilak De Silva Liyanage, Rosyzie Anna Awg Haji Mohd Apong, and Rubee Singh

Abstract Given the increasing use and importance of blockchain platforms and the circular economy concept in practice, this chapter will first address the trends and their dependencies. Secondly, it will identify the essentials of a new economic model that can be considered in these significant times, its status, and the tools that are and could be used. Thirdly, the structure of the essay will be explained. Arguably, information is valuable when there is an expectation that it is useful for better decision-making. Despite technology being theorized to degrade or diminish in importance over time, the information gathered with the help of methods or techniques has maintained consistent interest. Today, two of the trending applications and understandings are blockchain and the circular economy. While one is digitally derived, the other is concerned with the better use of resources. The arrival of digital systems has caused a drastic shift in our daily routines. The capability of society to adapt rapidly is a point of interest today. One of these shifts focuses on new waste management. The main streams form the circular economy model: development, design, production, and consumption; recovery and recycling provide the basis for the revolution in the waste cycle. The substitution of linear usage of products has turned into an opportunity for value by analyzing the residues. Furthermore, critics argue about the benefits of the model today or in the near future for these approaches and their compatibility with current economic practices. Digital systems can also support a better management approach. A new system has been suggested, which enables us to advocate all approaches as mentioned earlier. Blockchain, a technol-

W. Shafik (✉)
Dig Connectivity Research Laboratory (DCRLab), Kampala, Uganda

School of Digital Science, Universiti Brunei Darussalam, Gadong, Brunei

A. Tufail · C. D. S. Liyanage · R. A. A. H. M. Apong
School of Digital Science, Universiti Brunei Darussalam, Gadong, Brunei

R. Singh
Institute of Business Management, GLA University, Mathura, India

R. Singh et al. (eds.), *Blockchain Technologies for Smart Circular Economy and Organisational Sustainability*, Environmental Footprints and Eco-design of Products and Processes, https://doi.org/10.1007/978-3-031-89183-0_4

ogy derived from Bitcoin, has been widely discussed in terms of its success story and future in the area where it used to run digital footprints, despite the failures anticipated in the upcoming years.

Keywords Blockchain technology · Digital product passports · Product as a service · Smart circular economy · Supply chain transparency · Sustainable resource management · Tokenization and incentivization · Waste traceability

4.1 Introduction

Blockchain technology has recently shown significant promise as an innovative and transformative technology. Growing globally, it has shifted from being predominantly recognized through cryptocurrencies to becoming a platform for numerous business areas. Among these areas are supply chain management, healthcare, and intellectual property rights. Traditional economic models are based on the "use, profit, and dispose of culture," which has resulted in a number of inefficiencies and environmental problems (Santhuja & Anbarasu, 2023). One new proposed solution is the implementation of blockchain technology as the basis of a smart system to support the operability of the future. Most publications believe that the application of blockchain in a circular economy can contribute to either resource or waste management or enhance transparency and tracking. The innovation of blockchain is expected to achieve a variety of objectives, such as improving energy efficiency, material management, environmental protection, full lifecycle management, remanufacturing, and product afterlife use (Shishehgarkhaneh et al., 2023). Innovation in the field is deemed urgent given the need to conserve the supply of resources in a "closed-loop" pattern and reduce the volume and quality of waste. Given the significance of blockchain technology, policymakers should support these cross-cutting technologies, given the potential benefits they can offer to society through connecting business, nature, and human society and enhancing the value of the raw materials market (Castiglione et al., 2023).

A number of organizations have initiated industrial trials and projects in parallel with new regulations to experiment with various business models and technologies inspired by the principles of sustainability management. A closer look at the trends above indicates an increased awareness and concern for sustainability issues and environmental footprints. This is reflected in the use of terminologies that indicate the need for "green" or "sustainable" solutions (Kouhizadeh et al., 2023). Enterprises are aiming to address the growing pressure of fundamental global uncertainties by reinventing themselves as "environmentally sustainable." They embrace triple-bottom-line strategies to optimize their approach to "people, planet, and profit" by researching more sustainable materials and processes and embedding an outward focus on future product end-of-life as they design new products and services with hopes of reducing harm and creating less waste. These long-term strategies include lifecycle assessments with the intention of promoting better decisions, business practices, and processes that are "good" for the environment (Shafik, 2024a).

As the scope of this chapter has been partly established, it is time to explicate further our aims and objectives. The stated aims and purposes of this essay are indeed, for one, to provide a theoretical analysis of the trends and way ahead for the technological advancement and integration of blockchain and distributed ledger technologies within the normative assets of a Smart Circular Economy model. Secondly, the investigations delve further by discussing a number of possible use cases and applications specifically designed to embellish the outcomes of an empirical model of an ideal Smart Circular Economy (Hong & Xiao, 2024). A third and not less important aim is to discuss, at least to a certain extent, the current and ongoing integrated and cross-purposed applications, those being in particular platforms, not directly dealing with a circular economy but, once again, offering a broader, wider view eventually supporting the constructive synergy between cutting-edge distributed technologies and the socio-economic value chains embedding loosely knitting circular economy uses and features (Shafik et al., 2024).

It can be assumed that the primary readership will have a firm grasp of the fundamentals of both the current state and trends from blockchain on one side and smart cities and circular economies on another. It is also safe to assume that they possess some knowledge of recent trends and state-of-the-art in the data-driven and digital services-oriented city and urban infrastructure and a general knowledge of the state-of-the-art methods and application expertise developed inside circular economy projects (Gökbunar & Soylu, 2024). The entire main content of the essay will aim at corroborating the above discussion and complementing it through a series of technological applications that show the complementarity of blockchain and circular economy. The section then goes a bit further: at this stage, it is possible to aid some interest and pave the way further by stating that these applications will be characterizing both environmental, economic, and societal levels in a hypothetically distributed socio-economic model (Grati et al., 2024). A detailed discussion of these applications will then be described on the following page.

4.1.1 Understanding Blockchain Technology

Blockchain technology is the backbone and one of the key technologies that led to the establishment of one of the most significant innovations. For the uninitiated, the blockchain is a decentralized system, which is essentially a ledger for recording and managing transactions, but with improvements. For one, there is no central element that holds or manages the record; instead, each participating computer, or node, has a copy of the entire ledger. When one node updates the transactions, it updates the other nodes in the network at the same time. The system is based on consensus: if a transaction is deemed valid, it will be included in the ledger nearly instantaneously and will be nearly impossible to reverse. The underlying principles of a blockchain are decentralization, distribution, transparency, and mutability (Shafik, 2024b). Each of these features adds value to various fields of industry and serves as fundamental points for integrating blockchain into systems. Also, applying blockchain technology to sectors of economics involves both its features and their impacts. The core behind investing in blockchain starts with understanding the basic features of blockchain. The four demandable features are immutable, secure, transparent, and decentralized (Castiglione et al., 2023). There are currently two ways of integrating blockchain with traditional technologies. The first option blockchain allows is add-on complementary technology. In this case, the conventional database system runs while blockchain is implemented for enhancement in time-sensitive document verifiability, thus maintaining legal integrity. The second approach is collaborative middleware technology. In this option, blockchain functions as a hub that integrates multiple databases with dynamic technologies and systems, including integrating different blockchains and distributed ledgers (Hong & Xiao, 2024). Identifying the features of blockchain is a basic necessity to integrate blockchain with smart systems. Consequently, the features need to be materialized for better insights. Thus, 'integration' is identified as the interconnection of context awareness as three main points.

4.1.2 Definition and Core Principles

In 2008, the mysterious figure conceptualized a new decentralized digital currency protocol named in the first whitepaper of Bitcoin: A Peer-to-Peer Electronic Cash System and developed the first functioning prototype that used the blockchain system based on a peer-to-peer network. Since then, blockchain has been flourishing for more than a decade. In simple terms, blockchain is a digital distributed ledger that fully records and replicates all the transactions among connected peers on a peer-to-peer network maintained by computers, referred to as nodes (Corsini et al., 2023). Each block of transaction data in the ledger is linked to the previous block. The fundamental element of blockchain lies in the network architecture of decentralization, the consensus mechanism design principle to verify and validate the majority agreement of cryptographic transactions, and the cryptographic security measure for the double-entry bookkeeping system with trustworthy entities. The decentralization core of blockchain refers to a large network of nodes over a peer-to-peer computer network on the open internet without central points of failure or control of data storage in a single computer facility or data centers like traditional cloud computing (Gökbunar & Soylu, 2024). The consensus mechanism is the process of validating the unique majority agreed identical database version among the connected peer nodes for the authenticity of cryptocurrency transactions to eliminate the double-spending issue. Security with cryptography is the encryption and decryption protocol implemented in the blockchain system to secure the metadata structure and the identity of the record ownership by using hash functions and public-private key cryptographic elements. The technical working of the blockchain system requires an in-depth understanding of the decentralization nature of the blockchain network infrastructure and the process of transaction verification and validation through the distributed consensus mechanism executed by the blockchain software protocol embedded in the blockchain platform system (Castiglione et al., 2023). The effective adoption of blockchain technology in various application developments relies on a deep understanding and a comprehensive review of its fundamental technology elements and functionalities. The implication of decentralization in facilitating the blockchain system aligns with the conceptual rationale of the consensus mechanism and the decentralized internet of value that ensures the transaction assets are recorded permanently and immutably to create an immutable digital ledger of transactions across multiple networks of nodes without the participation of third-party payment system administrators. The decentralization of the blockchain essentially ensures information integrity that enhances the trustworthiness of the users and information proprietors over the Internet of data (Truant et al., 2024). Decentralized blockchain architecture provides the exact framework for the development of crypto storage, payment, rewards, and credits in the development of the smart circular economy in the future.

4.1.3 Key Features and Benefits

The key features that make blockchain a likely candidate to be used in recent and future years are based on a number of attributes including transparency, security, traceability, immutability, and so on. However, transparency in blockchain, which indicates that there is no central authority, implies that it is free from single-point failure. This high level of trust gained from the transparency in the distribution structure leads organizations to trust other network participants. Consequently, blockchain technology eliminates the role of central authorities for trust generation, which in turn gives power to each user over the veracity of transactions (Corsini et al., 2023). As a result, the main contributing factor that allows blockchain to be quite successful in practice is traceability. Using blockchain technology, it is feasible to trace the entire history of any specific item. Once material and data are contributed to the blockchain, its location is stored in the form of a chain, which can further be used for traceability. Being immutable ensures that a transaction that has been added can no longer be removed. Blockchain establishes transactions that can never be modified or deleted, which ensures that transactions remain authentic over time, leading to tamper resistance (Kouhizadeh et al., 2023). Implementing blockchain technology in different areas leads to a number of advantages. Owing to its characteristics like security, safety, elimination of intermediaries, tamper resistance, traceability, and transparency, it eliminates redundant activities and data storage, affecting process efficiency. Due to the disparate nature of transactions, media, contracts, and records associated with diverse industries that are based on digital data, the cost of intermediaries to validate and vouch for the authenticity of transactions is significantly reduced (Gökbunar & Soylu, 2024). The potential of blockchain is inevitable in different critical domains such as healthcare, financial sectors, and governance, addressing challenges related to supra-national jurisdiction, anti-money laundering, digital transactions, currency exchange, double spending, and so on. Nowadays, most centralized solutions are initiated in the marketplace with the hope of fostering waste minimization and a culture of reuse, and hence, these applications are able to enhance the management effectiveness of obsolete electronic appliances toward a circular economy (Hong & Xiao, 2024).

4.2 Factoring Circular Economy Concepts

One of the recent trends in the field of technological, economic, and environmental changes is the concern and utilization of the development and integration of waste management systems, where it is very important to create a systemic flow. The circular economy is very popular nowadays, and discussions concerning this notion appear in global-scale newspapers, scientific magazines, and research reports (Shafik, 2024c). Very often, the term circular economy is identified with sustainable development and the green economy. The circular economy is not just about recycling products; it is a cyclic process that should end as a production process with transferred materials to a natural feedstock source or full energy recovery. This approach is completely different from a traditional production and consumption model (linear economy), based on the "take-make-consume-dispose" pattern. In this case, a traditional value is a new product. Presently, production could occur based on products with stored information within blockchain technology (Shafik, 2024d). Blockchain technology presents a public ledger system in the network, working on the basis of trust techniques and containing reliable information. The circular economy is a concept of waste management and rational consumption of resources. Its basic principles are: "Waste minimization including the use of legal methods," "Recovery of the maximum amount of valuable materials from waste," and "Optimal saving and sustainable resource consumption." In conclusion, the transition to the new model of the economy is connected with the concept of the circular economy, which is based on new principles of production and waste management (Santhuja & Anbarasu, 2023; Shafik & Azrour, 2024). The development of closed-loop systems and closing industrial waste into the economic system will significantly contribute to the reduction of the demand for natural resources without the global economic slowdown. Apart from the increasing awareness about the shortage of resources, extended and fair market advantages of the segregation and treatment of EEEW have influenced the constantly increasing trends for EEEW recovery and treatment. The implementation of the considerations can result in better protection of the environment and tackle sustainability challenges with special emphasis on policy guidelines for the EEEW sector (Grati et al., 2024).

4.2.1 Definition and Principles

A circular economy can be defined as an economic and industrial system that is designed to be restorative, regenerative, and sustainable. This concept focuses on the revision of production and consumption patterns by reconsidering two main factors. The first one is the design for productive longevity; goods should be designed to follow renewable cycles, emphasizing clear and easy reuse, upgrade, refurbishing, and remanufacturing. The second focuses on drastically diminishing waste, releasing value from ecosystems, and restoring the exploited balance of renewable resource usage (Truant et al., 2024). Moving from linear development models to economic operations based on circular principles like longevity, refurbishing, recycling, and renewable resource input not only benefits the environment but also helps restore the long-term future of growth and social equity. The idea behind a smart circular economy lies in defining the principles that guide such a change and the technological innovation that will make circularity as profitable and convenient as a linear economy, if not more so. This defines three main principles of circularity and explains why and how innovation in product lifecycle management can help transform an extractive system into a regenerative one. Cutting off the company's solid waste is a short-term solution; instead, the organization must work to redesign the way waste is viewed and generate value from waste material produced (Amjad & Diaz-Elsayed, 2024). This change will initiate innovative considerations toward waste production, impacting many areas of the organization, most notably the way products are brought in. The thinking in terms of waste and resource reduction is not just a sustainable practice; it is also a vital element of a successful lean system. The broad outline of a transitioning economy goes beyond reducing waste. Larger social, economic, and environmental issues are driving interest in creating a system that emulates a living system, one that reuses inputs, reduces waste and aims for lasting value. A circular economy is a part of a regenerating system; it imitates natural processes by creating closed-loop systems to store or reintroduce extracted resources back into processes, reducing the environmental load (Koshariya et al., 2024). Less resource usage in a lifespan will also save production costs.

4.2.2 Importance and Benefits

The projection of the future leads to a discussion about the integration between blockchain and the digital economy. A new economic model that quickly gains acceptance in the world is also expected to drive the future. A circular economy is an approach that seeks to reduce waste and lock up the value of products, materials, and resources rather than consume them. It contrasts with a linear economy, which considers the depletion of natural resources and the disposal of valuable materials, a model that is very damaging to the environment due to significant waste (Truant et al., 2024). Although recycling has been around for a long time, the integration of a number of global factors, such as climate and energy transition, urbanization, rising prices of raw materials, increasing pollution, and the growing threat of resource scarcity, has raised the issue of transforming conventional linear economic principles into priorities. Restricting climate change, ensuring the complete use of valuable materials, and ultimately creating a better society are the goals to be achieved. The transition to new economic approaches, such as the circular economy, has both economic benefits and environmental and social advantages (Santhuja & Anbarasu, 2023). At the center of the philosophy of the circular economy is the goal of improving the efficient use of resources. Using resources more efficiently and managing waste flows will improve companies' margins and reduce the environmental impact. However, the goals are broader, as the principles of the circular economy allow companies to combine limited resources, reduce waste, and innovate in sustainable development, social issues, and policies that benefit companies and share wealth with regions. The circular economy affects all major functions of a company, from procurement, production, maintenance, and new product design to end-of-life processing and marketing of recovered products (Hong & Xiao, 2024). A circular economy can be innovative.

4.3 Intersection of Blockchain and Circular Economy

The intersection of blockchain technology and the principles of a circular economy is an interesting and promising topic. The ideas behind both concepts are relatable. While a circular economy tries to keep the products, components, and materials at their highest level of value creation by reducing the consumption or waste of these items through the reuse and remanufacturing of components, blockchain technology allows data to be stored in a decentralized way. The decentralization and storage characteristics of blockchain make it an interesting technology for traceability in circular economy concepts, especially in chaotic fields like second-hand products or materials management (Truant et al., 2024). By providing better traceability and transparency in the product and material flow, blockchain can contribute to the integration of circular practices. We already see some practical applications in this sense. For instance, blockchain can be used for the resale traces of fashion products after the original buyer sells them. We also see several proofs of concept and use cases. Most of the applications are related to food and product traceability. Products from such use cases are recorded on the blockchain at several steps in the value process. A vote of the majority from the consortium is required to delete or update the component details. The opportunities for further integration of the two concepts are shown in several discussions (Grati et al., 2024). However, there are still some challenges for blockchain and the integration of the circular economy to achieve the desired level of integration. The perimeter of data detected and stored by blockchain is of great importance not only for companies trying to integrate blockchain with the circular economy but also for blockchain service providers. Several obstacles and inhibitors can be listed for real saturation. In order to provide a sustainable and resilient blockchain ecosystem, the integration of blockchain with circular economy strategies should be taken into account by companies and relevant policymakers (Castiglione et al., 2023).

4.3.1 Current Applications and Use Cases

Applying blockchain for circular economy solutions is an emerging research track. In the following section, the latest results on this topic are presented and introduced based on the focus areas: waste management, circular and resource efficiency-oriented services, etc. Definitions of circular economy can include reducing resource consumption, adding value to resources, or considering product life cycles. In some cases, the inclusion of the earlier mentioned Industry 4.0 paradigm into the circular economy is also indicated. In the examined use cases, the focus was placed on applications currently working at the POC, MVP, or even commercial (fully or partially) level provided by different actors, mainly companies but also universities and research centers (Bułkowska et al., 2023; Kouhizadeh et al., 2023). Apart from the focus areas, we consider the categorization based on the current progress of the presented solutions, from the ideation stage through a few different stages of the prototype, like POC or MVP, to fully operational systems. In the extant literature on blockchain for a sustainable and circular economy, the technological aspects, standards, and regulations are discussed. There are also publications mentioning only the possible leverage of this technology, interconnected with IoT, data exchange, and combinations with multicellularity or shifting business paradigms. In all presented use cases, blockchain technology was chosen to provide more transparency, in particular along the entire supply chain (Alexandrova, 2024; Hong & Xiao, 2024). Blockchain is also presented as smart contracts or DLT for secure communication and data storage. In all use cases, process monitoring and asset management during the lifespan, from production to the end of the use phase and further, are topics that are mentioned in the use cases, interconnected with the more general management of the product life cycle. Blockchain and other supporting technologies include identification through IoT devices, NFC tags with QR codes, sensors, drones, etc. The reverse supply chain is also a part of or at least interconnected with the whole supply chain (Hong & Xiao, 2024). The use of resources or waste handling is interconnected with the optimization of logistic processes and circular solutions like recycling or sorting.

4.3.2 Potential Synergies and Opportunities

Blockchain, with its decentralized and trust-based properties, allows the secure coordination of diverse actors and assets, as well as the reliable documentation of all value transactions within a circular economy. This facilitates quality control, collaboration, and transparency—all vital for the integrity of a circular economy, in which resources will re-enter value loops, thus avoiding waste. At the same time, blockchain can be leveraged to incentivize sustainability through traceable transactions financially and agreed terms, including smart contracts (Santhuja & Anbarasu, 2023). Blockchain complements and enhances the promotion of circular economy business models, pools of resources, and incentives to close cycles in goods and materials. This can add new business opportunities and markets to the framework of resources that can help the developing economy establish a strong foothold in a variety of industries. Blockchain's role in supporting circular economy activity is ripe for further innovation and collaboration. To date, a proportion of blockchain activity in the circular economy has arisen in an exploratory sense, with new business models and ideas seeking to prove feasibility; in delivery and operationalization terms, the existing activity focuses on the business-to-business marketplace (Castiglione et al., 2023). The sheer diversity of innovative blockchain solutions working with circular economy actors to improve dynamics across a variety of waste streams and quality separators suggests that, as well as an ongoing activity, new technical and digital assets are also being improved continuously. Integrating blockchain into circular economy operations could form an even greater part of blockchain's core services in tomorrow's circular economy, meaning that there is potential for further system enhancement in an already transforming market for circular businesses and digital enablers (Elghaish et al., 2023; Khan et al., 2024). Government departments and regulators must be engaged to discuss how such further system enhancements can be governed and communicated to stakeholders.

4.4 Challenges and Limitations

In this regard, we can divide the limitations and challenges into two aspects. The first aspect is blockchain incorporation problems. From a technological perspective, the current state of the Ethereum blockchain has latency problems, as the mining process takes at least 1 min and 15 s. In addition, most of the blockchain projects are in the proof-of-concept stage; hence, they may take time to reach real-world scenarios. Furthermore, the smart contract scalability of the two-layer approach with consensus, which depends on the number of smart contracts, is a main signal of slowdowns in the cryptographic currency scene. Regulatory limitations and lack of standards are also big problems, as the insurance industry does not have a set of standards for the industry (Khan et al., 2022, 2023). In addition to the victims, technology.

4.4.1 Technical and Scalability Issues

The most important technical challenges of blockchain include transaction speed and scalability. As a consequence of the growing number of blockchain users, the available block size and block interval are inadequate, which then becomes a bottleneck for the excessive expansion of the blockchain. Bitcoin transactions generally take 10 min, the Ethereum network runs at a high capacity with a block time of 30 s, and a series of large-scale distributed ledgers based on the classical mechanism can process transactions with far lower speeds compared to traditional financial systems (Corsini et al., 2023; Singh et al., 2022). Given that the current mechanism has become too sluggish to suit the next round of large-scale economic activity, a new mechanism has been presented to boost on-chain transaction speeds, and the limits on block capacity are expected to be reached with Bitcoin's Central Side Chain. Massive amounts of ongoing data processing and verification by thousands of cryptocurrencies also result in enormous energy consumption. At present, the Bitcoin and Ethereum blockchains together consume almost 80 million kWh per day (Hong & Xiao, 2024).

A number of technical issues that need to be addressed for blockchain to be developed have been identified. First, the single-node response reduces the linear expansion of storage levels; the response rate slows down when the tower structure turns into a hatted tower structure. Improper storage processes can degrade performance. Second, the number of consensus nodes is excessive, and the response time delay also grows rapidly, directly leading to a stall in commercial applications. As a result of the extensive application of blockchain, a huge storage and transmission accounting system is created (Truant et al., 2024; Singh et al., 2023). Consequently, big data's storage and transmission capabilities are needed to analyze various sectors that the blockchain requires. Furthermore, it requires the rapid upgrading of accelerators and prices in the market to match the price of the storage part of the blockchain. Since users require an upper limit of confidentiality on this fact, it is more difficult to increase or decrease the storage, thereby preventing the demand for transactions and applications on the blockchain. Given the above reasons, blockchain is confronted with the core difficulties of sticking the "winged feet" and "iron ears" of both Achilles' heels (Grati et al., 2024; Singh et al., 2024). Based on the literature, some lessons learned were extracted and presented in bullet points when it comes to the integration of blockchain technology as a facilitator in the establishment of a smart circular economy system:

- Engaging stakeholders early in the process.
- Defining the requirements of both IoT and blockchain implementation from the start.
- Examining the systems based on functional and technological fitness.
- Ensuring an upper layer of security in data exchange.
- You should not utilize blockchain if it is no better than centralized or conventional best practices.

- Evaluating the blockchain at regular time intervals or upon detecting atypical behavior.
- Having a framework in place that enables you to streamline all hypothesis work and manage a large-scale experiment with a large number of participants.
- Do not be afraid to abandon blockchain technology integration as your circumstances demand it simply.

4.5 Conclusion

This chapter provides knowledge and understanding of what blockchain technology and smart circular economy are and how to integrate blockchain technology into the circle. Based on this, we provide an overview of the theoretical framework of the integration of blockchain technology and the circular economy. In addition, we explained the future of the circular economy and how blockchain technology influences the sustainability of the natural environment and the allocation of resources. Similarly, the application of a smart circular economy in various industries has begun to exert its significant influence, and the profound impact of the circular economy on smart community construction is demonstrated. The door to the full implementation of this mode of energy trading has now been opened widely by the rise of blockchain technology, a driving force behind its transactions. An in-depth understanding of the development of new business models based on blockchain technology and their application in social practice has also been systematically demonstrated. In the future, businesses need to reduce costs and raw materials, design products as services, repair, recycle, and share, thus gradually setting up circular business models. In the age of the "Internet of Everything," commodities such as cars, scooters, bicycles, washing machines, televisions, and computers will not be owned and used exclusively by individuals but will evolve into shared consumer models. In the digital-first model, such consumer-sharing behavior is likely to break through the constraints of traditional leasing behaviors by being partially unused and wasted. In this model, product producers and service providers have a strong motivation to reduce the failure rate of sharing products. Thus, reducing usage and maintenance costs by increasing the durability of the product will be an area where circular business models create a natural competitive advantage. In summary, blockchain technology can bring significant changes to the social share economy based on the virtual integration of the real world. Researchers can continue to delve deeply into the internal mechanisms of the circular economy to study how to more accurately use innovative technologies to determine the effective operation of the algorithm and to explore more practical applications in the industry.

References

Alexandrova, E. (2024). Digital Technology in Luxury Industry Development. *Lecture Notes in Networks and Systems, 733*. https://doi.org/10.1007/978-3-031-37978-9_38

Amjad, M. S., & Diaz-Elsayed, N. (2024). Smart and sustainable urban manufacturing for a circular economy. *Environment, Development and Sustainability*. https://doi.org/10.1007/s10668-024-04671-w

Bułkowska, K., Zielińska, M., & Bułkowski, M. (2023). Implementation of Blockchain Technology in Waste Management. *Energies, 16*, 23. https://doi.org/10.3390/en16237742

Castiglione, A., Cimmino, L., Di Nardo, M., & Murino, T. (2023). A framework for achieving a circular economy using the blockchain technology in a sustainable waste management system. *Computers and Industrial Engineering, 180*. https://doi.org/10.1016/j.cie.2023.109263

Corsini, F., Gusmerotti, N. M., & Frey, M. (2023). Fostering the circular economy with Blockchain technology: Insights from a bibliometric approach. *Circular Economy and Sustainability, 3*, 4. https://doi.org/10.1007/s43615-023-00250-9

Elghaish, F., Hosseini, M. R., Kocaturk, T., Arashpour, M., & Bararzadeh Ledari, M. (2023). Digitalised circular construction supply chain: An integrated BIM-Blockchain solution. *Automation in Construction, 148*. https://doi.org/10.1016/j.autcon.2023.104746

Gökbunar, H. B., & Soylu, B. (2024). Blockchain enabled lateral transshipment system for the redistribution of unsold textile products in a circular economy. *Lecture Notes in Mechanical Engineering*. https://doi.org/10.1007/978-981-99-6062-0_58

Grati, R., Loukil, F., Boukadi, K., & Abed, M. (2024). A blockchain-based framework for circular end-of-life vehicle processing. *Cluster Computing, 27*(1). https://doi.org/10.1007/s10586-023-03981-4

Hong, Z., & Xiao, K. (2024). Digital economy structuring for sustainable development: The role of blockchain and artificial intelligence in improving supply chain and reducing negative environmental impacts. *Scientific Reports, 14*(1). https://doi.org/10.1038/s41598-024-53760-3

Khan, S., Ali, S. S., & Singh, R. (2022). Determinants of remanufacturing adoption for circular economy: A causal relationship evaluation framework. *Applied System Innovation, 5*(4), 62. https://doi.org/10.3390/asi5040062

Khan, S., Khan, M. I., & Singh, R. (2023). Modeling the barriers of Blockchain technology implementation in supply chain. *Journal of Industrial Integration and Management, 8*(03), 367–385. https://doi.org/10.1142/S2424862223500070

Khan, S., Singh, R., Alnahas, J., Abbate, S., & Centobelli, P. (2024). Navigating the smart circular economy: A framework for manufacturing firms. *Journal of Cleaner Production, 480*, 144007. https://doi.org/10.1016/j.jclepro.2024.144007

Koshariya, A. K., Kumar, V., Ahmad, V., Babu, B. H., Umarani, B., & Ramesh, S. (2024). Blockchain technology for agricultural data sharing and sustainable development of the ecosystem. *Artificial Intelligence, Blockchain, Computing and Security—Proceedings of the International Conference on Artificial Intelligence, Blockchain, Computing and Security, ICABCS 2023, 1*. https://doi.org/10.1201/9781003393580-40.

Kouhizadeh, M., Zhu, Q., & Sarkis, J. (2023). Circular economy performance measurements and blockchain technology: An examination of relationships. *International Journal of Logistics Management, 34*(3). https://doi.org/10.1108/IJLM-04-2022-0145

Santhuja, P., & Anbarasu, V. (2023). Blockchain-enabled IoT solution for e-waste management and environmental sustainability through tracking and tracing. *International Journal of Engineering Trends and Technology, 71*(12). https://doi.org/10.14445/22315381/IJETT-V71I12P216

Shafik, W. (2024a). Artificial intelligence and Blockchain technology enabling cybersecurity in telehealth systems. In *Artificial Intelligence and Blockchain Technology in Modern Telehealth Systems*. https://doi.org/10.1049/pbhe061e_ch11.

Shafik, W. (2024b). Blockchain-based internet of things (B-IoT): Challenges, solutions, opportunities, open research questions, and future trends. In *Blockchain-based internet of things* (pp. 35–58). Chapman and Hall/CRC. https://doi.org/10.1201/9781003407096-3

Shafik, W. (2024c). Exploring the potential of internet of things (IoT) and challenges in enabling circular economy practices in smart cities. In *Smart cities and circular economy* (pp. 125–138). Emerald Publishing Limited. https://doi.org/10.1108/978-1-83797-957-820241012

Shafik, W. (2024d). Industry 4.0 technologies' opportunities and challenges for Realising net-zero economy. In *Net zero economy, corporate social responsibility and sustainable value creation: Exploring strategies, drivers, and challenges* (pp. 19–41). Springer. https://doi.org/10.1007/978-3-031-55779-8_2

Shafik, W., & Azrour, M. (2024). Building a greener world: Harnessing the power of IoT and smart devices for sustainable environment. In *Technical and technological solutions towards a sustainable society and circular economy* (pp. 35–58). Springer. https://doi.org/10.1007/978-3-031-56292-1_3

Shafik, W., Kalinaki, K., & Zakari, R. Y. (2024). Blockchain's motivation for IoT-enabled Smart City. In *Secure and intelligent IoT-enabled smart cities* (pp. 195–221). IGI Global. https://doi.org/10.4018/979-8-3693-2373-1.ch010

Shishehgarkhaneh, M. B., Moehler, R. C., & Moradinia, S. F. (2023). Blockchain in the construction industry between 2016 and 2022: A review, bibliometric, and network analysis. *Smart Cities, 6*, 2. https://doi.org/10.3390/smartcities6020040

Singh, R., Khan, S., & Centobelli, P. (2022). Investigating the interplay between social performance and Organisational factors supporting circular economy practices. *Sustainability, 14*(24), 16781. https://www.mdpi.com/2071-1050/14/24/16781

Singh, R., Khan, S., & Dsilva, J. (2023). A framework for assessment of critical factor for circular economy practice implementation. *Journal of Modelling in Management, 18*(5), 1476–1497. https://doi.org/10.1108/JM2-06-2021-0145

Singh, R., Gupta, A., Bajpai, A., & Kandpal, V. (2024). Exploring challenges of circular economy initiatives for smart cities. In *Smart cities and circular economy: The future of sustainable urban development* (pp. 59–69). Emerald Publishing Limited. https://doi.org/10.1108/978-1-83797-957-820241005

Truant, E., Giordino, D., Borlatto, E., & Bhatia, M. (2024). Drivers and barriers of smart technologies for circular economy: Leveraging smart circular economy implementation to nurture companies' performance. *Technological Forecasting and Social Change, 198*. https://doi.org/10.1016/j.techfore.2023.122954

Part II
Blockchain Applications in Resource Management

Chapter 5
Leveraging Blockchain Applications for Corporate Social Responsibility (CSR) Towards Circular Economy: A Review

M. C. R. Jayalath, P. G. S. Amila Jayarathne, Amit Joshi, and Damith Gangodawilage

Abstract This chapter discusses how much Blockchain has evolved and encourages contemporary industries toward better transparency, accountability, and sustainability. Hence, this chapter deals with a review of the concept of CSR, after which a discussion is given on how Blockchain works in harmony with CSR values in promoting responsible and sustainable business behaviour. It describes Blockchain as a potent tool to boost stakeholder trust by allowing verifiable, tamper-proof data records that assure transparency in reporting and compliance. These are holistically categorised into public, private, and hybrid Blockchains. The paper provides an in-depth view of the advantages, limitations, and applications of each kind of Blockchain to enable further comprehension of how it works in various industries. Particular attention is paid to supply chain management, where Blockchain allows for real-time tracking, authenticity, and reduction of inefficiencies, and as such will be a significant enabler for digital transformation. Further, the work looks at the integration of Blockchain in digital technologies to show its capability to enable streamlining operations, cost reduction, and enhancement of decision-making. In the higher education environment, Blockchain is a game-changing invention that allows for the safe verification of credentials and collaboration and ensures academic integrity. However, despite these advantages, the chapter addresses several

M. C. R. Jayalath (✉)
Department of Commerce and Financial Management, University of Kelaniya, Kelaniya, Sri Lanka

P. G. S. Amila Jayarathne
De Montfort University, Leicester, UK
e-mail: amila.jayarathne@dmu.ac.uk

A. Joshi
Kumaun University, Nainital, Uttrakhand, India

D. Gangodawilage
Villa College, Malé, Maldives
e-mail: damith.gangodawilage@villacollege.edu.mv

R. Singh et al. (eds.), *Blockchain Technologies for Smart Circular Economy and Organisational Sustainability*, Environmental Footprints and Eco-design of Products and Processes, https://doi.org/10.1007/978-3-031-89183-0_5

challenges: scalability issues, high energy consumption, regulatory gaps, and resistance to adoption. It also outlines several viable solutions that transcend these barriers: energy-efficient Blockchain models, establishing global regulatory standards, and stakeholder awareness and collaboration. The discussion illuminates how Blockchain could harmoniously develop into an industrial transformation with transparency, sustainability, and innovation.

Keywords Blockchain · Corporate social responsibility · Sustainability · Transparency and accountability

5.1 Introduction to Blockchain Technology

There are many data structures, and Blockchain is the most critical technology. It is an open, distributed, and decentralised data structure. Blockchain has become a worldwide known technology since 2009, when the Bitcoin network, the first of many modern cryptocurrencies in the financial industry, launched (Bidry et al., 2023). This is the first moment that Blockchain technology has been introduced to the world. Blockchain has also been shown as a new descriptive and foundational technology. This kind of decentralised structure removes the intermedium elements, saves and reduces transaction costs, enhances the efficiency and effectiveness of activities, and establishes the data's trust and traceability (Jimoh et al., 2019). Blockchain plays a crucial role in the industry that promotes environmental, social, and economic sustainability.

Blockchain technology will have more strengths. Transparency is the most crucial strength, as transactions are visible and predictable to all participants engaging on the network; it always enhances accountability, especially in supply chain management, finance, health care, and public services. This makes tracking products from origin to delivery possible in industries related to logistics, ensuring authenticity and fraud reduction (Woodside et al., 2017). Blockchain Technology is used without human intervention in pure contract agreements like compliance and payment. Blockchain thus allows faster and borderless transactions, providing real help for underbanked communities in developing nations (Taherdoost, 2022).

Blockchain thus establishes trust in processes prone to human error or manipulation. More and more research points to its increased importance in IoT, where devices in decentralised ecosystems securely share data, enabling smart cities and industry automation (Jain et al., 2024). Despite the advantages, Blockchain also draws criticism related to regulatory uncertainty, data privacy laws, and interoperability issues (Joshi et al., 2024). Using Blockchain technology requires massive investment in infrastructure, which may be difficult for small businesses to handle. Besides, public awareness and technical complexity are the challenges to mainstream acceptance (Janssen et al., 2020).

Also, Blockchain Technology supports the creation of corporate social responsibility (CSR) that is efficient, accountable, and transparent. It grants ethical practices, tracks the supply chain, and validates sustainability goals, ensuring the trust of stakeholders. By ensuring an accurate climate on emission reduction and supporting sustainability efforts, Blockchain provides a track for Carbon Footprint. It records carbon credits and energy use in the world. Also, agriculture and garment manufacturing industries are using Blockchain in labour rights. Companies can also use Blockchain Technology to provide an accurate CSR performance report by recording sustainability metrics like waste management or energy use (Agi & Jha, 2022).

5.2 Methodology

This section explains the method used in the study to explore the Blockchain applications for CSR. The systematic literature review method is employed in conducting this review as employed by Jayawardena and Dewasiri (2023), Dewasiri et al. (2024), Dewasiri (2024), Karunarathna et al. (2024), Mishra et al. (2024), Rathnasiri and Gunathilaka (2024), Rathnasiri et al. (2024), Vijerathne et al. (2024), Chathuranga et al. (2024), and Bandara et al. (2024). The primary data analysis technique employed in this present study is content analysis, which involves a more practical analysis and review of the available literature. This approach allows the involvement of existing literature and practices in evaluation to find patterns, trends, and insights in the existing knowledge domain and paw future investigations. The study's research design was qualitative, and a content analysis was deemed appropriate for systematically evaluating the findings obtained from the different data sources. To this end, the study focuses on exploring the application of Blockchain technology in advancing CSR initiatives. The key steps involved in the research design are as follows:

5.2.1 Selection of Sources

An extensive search of the published literature was conducted to find the relevant document. This comprised scholarly journal articles, policy documents, government documents, and case studies. Keywords, namely, 'corporate social responsibility', 'categories of Blockchain', 'transparency and accountability of Blockchain', 'environmental sustainability', 'higher education and Blockchain', 'EduCTX platform', 'supply chain and Blockchain', 'technological innovations', 'digital technology and supply chain, and 'Blockchain for inventory management' in the internet search helped in sourcing the literature. As inclusion criteria, peer-reviewed articles containing information about the applications of Blockchain with CSR were considered. Both theoretical and empirical works were included. Scholars who did not relate their work to the research questions or had insufficient information on applications of Blockchain for CSR were excluded. Non-English sources were also excluded.

5.2.2 Data Collection

Such published material was actively searched for and classified. The data collection included academic databases, governmental archives, and information on organisation websites. One hundred eighty-four sources were found initially and then refined based on the defined inclusion/exclusion criteria to focus on 38 sources.

5.3 Discussion of the Findings

The major themes derived from the content analysis and validated through inter-researcher and inter-coder reliability are discussed as follows.

5.3.1 Overview of Corporate Social Responsibility

Corporate social responsibility is an organisation's way of operating to add value to the economy, quality of life, and the environment while pursuing business operations. It depicts the sense that business entities bear responsibilities beyond profit maximisation. The significant dimensions of CSR include economic, legal, ethical, and philanthropic responsibilities. These domains guide companies on how to add economic value, keep within the law, act appropriately, and return to society through philanthropic activities. Companies could be involved in supporting educational programs, contributing to disaster management, and investing in community health programs (Wang et al., 2016).

Focusing on stakeholders is essential in CSR. The businesses must address the needs and concerns of various groups, including customers, employees, investors, and the community. Balancing these interests requires transparency, fairness, and accountability. For instance, companies commit to stakeholders through fair labour practices, ethical sourcing, and reducing environmental harm. By adopting such practices, businesses can build trust and strengthen their reputation (Godfrey, 2009). CSR also contributes to long-term success. According to the *Journal of Business Ethics* (2004), CSR helps companies build better relations with stakeholders and attain loyalty. This approach improves the brand image and enhances customer and employee satisfaction. When customers find businesses responding to social and environmental challenges, they are bound to support such companies. For instance, every company that goes green in production through renewable energy and reducing plastic waste garners favour from eco-friendly consumers (Bhushan & Singh, 2024; Nwokolo et al., 2023a, b, c).

5.3.2 Categorization of Blockchain

It is possible to broadly categorise Blockchain technology into its structure and usage for different purposes, not to mention the unique advantages of each. Public Blockchains are open to all; any participant can read, write, and validate transactions (Casino et al., 2019). This kind of Blockchain allows excellent insight and is very secure; hence, it is ideal for cryptocurrencies such as Bitcoin and Ethereum, which involve a decentralised network of nodes to maintain the ledger's integrity. The private Blockchain is allowed for the responsible and authorised person. Such Blockchains are often intra-organization and help automate processes toward better data security, aside from ensuring that internal policies are adhered to. An important example includes Hyper Ledger Fabric and R3 Corda, which are extensively used in enterprise applications for secure data sharing and internal audits (Gadekallu et al., 2022).

Consortium Blockchains, or federated Blockchains, lie somewhere between public and private. This kind of Blockchain is mainly applied to applications specific to a particular industry that require collaboration among several organisations. Excellent examples of consortium Blockchain include Quorum, developed by JPMorgan, and Energy Web Foundation. The other important category includes sidechains, where Blockchains run parallel to the main Blockchain. They are the mechanisms for transferring assets between the main chain and the sidechain, thus allowing scalability, which permits experimentation without affecting the leading network. The best example is the Liquid Network: a Bitcoin sidechain with faster transactions and more liquidity (Huynh-The et al., 2023). Each of these Blockchains has different features, from public networks for complete transparency and security to private and consortium environments offering controlled access for collaboration. These are the categories on which Blockchain will soon base digital transactions, data security, and decentralised applications and foster innovation across industries (Singh et al., 2024).

5.3.3 Blockchain for Transparency and Accountability

However, Blockchain technology quickly transforms transparency and accountability from various sectors to others, creating a secure, decentralised, and tamper-proof method for recording and verifying transactions. The Blockchain model offers trust through cryptographic algorithms and consensus mechanisms instead of conventional models, where trust is usually bestowed on some central authority (Rizal Batubara et al., 2019). Each transaction becomes a block in the chain of events and, as soon as it becomes a part of it, cannot be amended without the network's consensus, making it such a suitable model for increasing transparency and lowering fraud. According to studies published by ACM Digital Library, Blockchain's distributed ledger structure eliminates information asymmetry, allowing stakeholders access to accurate, real-time data and trust in the processes.

Blockchain's most significant contribution to transparency is financial systems (Kumar et al., 2024). Blockchain-based platforms trace financial transactions for accountability and transparency in the private and public sectors. This enhances transparency by allowing all transactions to be visible and traceable. Other companies also use Blockchain to improve transparency in their supply chains (Tyma et al., 2022). The Blockchain would enable tracking goods, from production to delivery, ensuring ethical sourcing and fair-trade practices.

Moreover, Blockchain increases accountability in corporate governance. Frontiers in Blockchain indicates that "smart contracts," self-executing contracts coded onto the Blockchain, hold parties accountable by implementing contractual terms automatically without human intervention. This tends to reduce conflicts or disagreements.

5.3.4 Blockchain for Supply Chain Management

The supply chain ecosystem involves all the processes from suppliers through consumers in designing, engineering, manufacturing, and distributing products or services. These processes are usually governed by regulations that safeguard consumer rights and entail the movement of goods, information, and money. Eight of these are fundamental consumer rights adopted by the United Nations, including the right to safety, the right to be informed, the right to object, and the right to a wholesome environment. Products sold to the USA market are regulated, supervised, and enforced without exception by the FDA to ensure that they meet the guidelines set forth for public health safety.

Nonetheless, many occurrences over time have raised doubts about the efficiency of the supply chain and the integrity of product information. Meanwhile, March 2018 in the United States became the peak month for 210 counts of E. coli infections. Within months after that incident, health officials discovered the contamination source, which turned out to be romaine lettuce, tainted due to being irrigated by dirty water in the Yuma growing region.

Transparency and traceability have to be improved along manufacturing supply chains. The supply chain risk is mainly along the product journey. More information about the product, origin, processing, and shipping journey must be provided. Consumers cannot check the integrity of the acquired product. They have to trust the certification logo printed on products. Verifying this certification integrity requires strenuous auditing. Transparency must be enabled to regain the consumer's trust, help the producer get a better perspective of supply chain breaches, and understand how management product decisions and environmental circumstances can affect a product. Achieving transparency requires accurate data collection and secure data storage, a problematic task currently entrusted to third parties through centralised information depositories.

As disclosed by Satoshi Nakamoto (Nakamoto, 2008), Blockchain technology appeared in 2008 to become the shared ledger of the cryptocurrency Bitcoin. Unlike traditional banking, Bitcoin does not need intermediaries, and it is a great way to record information about transactions effectively (Gupta, 2018). Blockchain has brought security, immutability, transparency, and the ability to excise the middleman to financial services. Blockchain, used to record transactions and trace asset movements, revolutionised the conventional business network. Many sectors, especially healthcare, insurance, government, supply chain management, and the Internet of Things, are most likely to be transformed by the Blockchain.

5.3.5 *Blockchain and Environmental Sustainability*

Blockchain technology is in a headlong rush of growth; new systems are trying to adopt new methods, such as PoS, that consume much less energy. This change is crucial for supply chain operations with three important objectives: efficiency, transparency, and sustainability. More traditional Blockchain networks, including Bitcoin and early Ethereum, implemented the Proof of Work consensus method, which requires significant amounts of energy to confirm transactions. The shift-over to PoS-as is happening in Ethereum 2.0, and green chains like Avalanche's AVAX represent the giant leap toward eco-friendly environmental impacts.

Supply chains range from extracting raw materials to delivering finished products to consumers. This naturally creates inefficiencies, waste, and massive carbon emissions. With Blockchain, tracking, recording, and verifying transactions securely and transparently can address these challenges. Moving on, adopting proof-of-stake-based Blockchain systems no doubt adds another layer, drastically reducing the footprint of digital transactions and data management for supply chains (Chohan, 2019).

For instance, companies can use Blockchain in such a way that it identifies ethically sourced raw materials in real-time, improves transportation routes, reduces optimised inventory levels, minimises waste, and eventually has superfluous emission savings. Further, Blockchain transparency does contribute to businesses proving their claims about the environment, be it in the form of carbon offset projects or certifications of environment-friendly products. Moreover, this is about the proofs of sustainability consumers have asked about and to which Blockchain lets one confidently deliver.

There are many environmental benefits of proof-of-stake Blockchains. Without that wasted electricity, the supply chain traditionally took heaps of paper to exchange much other resource-consuming junk. For example, smart contracts- so-called 'self-executed' contracts coded on Blockchains—perform the transactions automatically to enforce such standards on the path to sustainability without human intervention. For example, Avalanche is a Blockchain network that runs entirely on the PoS system, further showing how such technology can help in ecological ventures (Miraz et al., 2021). It is a perfect fit for supply chain applications since it can allow for fast and low-power consumption of transactions. Organisations integrating this into their operations can enjoy other benefits from using Blockchain, such as security and transparency. Correspondingly, Ethereum 2.0's movement into PoS has given wider industry signals toward sustainability. These developments prove that Blockchain is not just about increased efficiencies but is a catalyst for positive environmental change (Gola & Sedlmeir, 2022).

Moreover, Blockchain can enable supply chains to adopt models such as the circular economy. A circular economy adopts a product-material approach: reusing, recycling, and repurposing for minimal waste (Khan et al., 2024). Using Blockchain, businesses can track product and material life cycles, ensuring that products or materials are disposed of responsibly or recycled. This is particularly important for industries like electronics, whose products, if not disposed of properly, can become environmental hazards. Blockchain details each product component, which helps build a more sustainable and responsible supply chain.

As Blockchain technology evolves, energy consumption will decrease further. That is especially true for the old systems running on PoW, which are mostly upgraded to PoS or replaced by newer and more efficient ones. This would, therefore, mean that Blockchain's ecological impact will probably decrease over time, with consistency in the goals pursued by humanity. Supply chains show that Blockchain use can be pragmatically and morally reasonable (Sapra et al., 2023).

Integrating Blockchain into the supply chain has modernised how companies do business in transparency, efficiency, and environmental care. Shifting to energy-efficient consensus methods, such as proof of stake, also cements how Blockchain positively affects environmental conditions. With businesses and governments' increasing focus on sustainability, Blockchain technology provides a solid tool to help meet such demands while tackling challenges regarding climate change (Wendl et al., 2023). Supply chains become more effective with the adoption of Blockchain while contributing to a greener future for one and all.

5.4 Blockchain and Higher Education

This, therefore, gives a clear overview of the potential of Blockchain in the management, sharing, and verification of academic records, especially in higher education (Raimundo & Rosário, 2021). Speaking to the possibility of such Blockchain adoption for higher education, a 2018 study published by IEEE Access, "EduCTX: A Blockchain-Based Higher Education Credit Platform," identified Blockchain as one helpful, decentralised and secure approach to managing student data. Traditional academic record management systems have been either paper-based or centralised digital systems, which are often late and erroneous and may sometimes face fraud. More importantly, the Blockchain solves these problems by making a tamper-proof digital ledger on which academic merits such as course credits, degrees, and certifications can be stored and accessed safely (Ayub Khan et al., 2021). One example of Blockchain use in higher education includes the EduCTX platform: EduCTX makes exchanging academic credits easier and more standardised worldwide. EduCTX is based on the European Credit Transfer and Accumulation System, thus enabling universities to record their students' achievements on the Blockchain. These recorded achievements become a digital portfolio that students can instantly share with other universities, employers, or institutions. By design, Blockchain is decentralised, with no single institution owning the data. This enhances fairness and security. Blockchain integration into higher education encourages further collaboration across borders, raising efficiency and the overall student experience (Turkanović et al., 2018).

This part underlines that platforms like EduCTX assign a role to Blockchain in making higher education innovative and effective, overcoming some inefficiencies towards global standardisation. While Blockchain itself is still evolving, possible applications to education could further interlink and smooth the system for the benefit of students, institutions, and employers—a step that higher education would need on the road to modernisation for the digital era.

5.5 Challenges and Limitations

While Blockchain technology has immense potential to enhance corporate social responsibility initiatives, it also has limitations and disadvantages that must be appropriately weighed (Upadhyay et al., 2021). Among the significant issues concerning Blockchains, scalability is a concern: most of them cannot handle high volumes of data or transactions, limiting their application in large-scale initiatives that cover a range of activities touching on CSR issues. Further, there is no clear regulatory framework for Blockchain applications. This means that there are challenges on the regulatory front. They lead to legal uncertainty that creates difficulties in compliance for organisations interested in deploying the technology (Venkatesh et al., 2020) . Another pressing issue relates to energy use since traditional Blockchain using Proof of Work mechanisms has been proven to utilise high energy levels, raising environmental concerns that are antithetical to the aims of most CSR initiatives.

Public awareness, and hence understanding, of Blockchain technology is relatively low, which hampers the wide diffusion necessary for practical use in many CSR applications. Several misunderstandings may hamper stakeholders' easy buy-in, which is necessary for leveraging its full benefits. Integrating Blockchain technology into existing systems can be tricky and resource-intensive, entailing significant integration challenges for organisations (Aljumah, 2023). Most of the applications on the Blockchain are expensive in their development and maintenance, especially for smaller organisations or NGOs with limited resources.

Another challenge is that trust in the Blockchain must be built up. Wide acceptance of Blockchain technology will be built upon stakeholder trust and confidence in the benefit of Blockchain solutions, which might be hard to come by as the technology is perceived by many as relatively new and complex. Besides, the technical intricacy of Blockchain requires expert knowledge of its creation and handling, hence shrinking the pool of skilled talent and boosting the challenges of implementation. It is in the overcoming of such challenges that the true potential of Blockchain technology can be realised in bringing transparency, accountability, and efficiency into CSR initiatives (Ltifi & Mesfar, 2022). The organisation will be able to exploit Blockchain to effectively drive social change and sustainable practices only after the limitations have been overcome.

5.6 Suggestions for the Challenges

According to Aljumah (2023), Blockchain technology has great potential to enhance Corporate Social Responsibility; several challenges and limitations are noted: the scalability issues most face include an inability to process large volumes of transactions or data. Should an extended CSR project seek such usage, that would weaken its effectiveness. Secondly, a lack of explicit regulatory frameworks concerning Blockchain applications may cause high regulatory challenges, legal uncertainty, and possible compliance problems when an organisation wants to use this technology. Another critical issue is power consumption, which in more traditional Blockchain systems applies the Proof of Work consensus mechanisms that, it is known, consume gigantic amounts of energy (Dierksmeier & Seele, 2020). Something that raises environmental concerns opposite to what most CSR initiatives are trying to achieve.

The public is still unaware of Blockchain technology, which is a barrier to its wide-scale adoption and effective use for CSR purposes. Such a lack of understanding may impede organisations' ability to gain stakeholder buy-in and leverage the full potential of Blockchain. Besides, integrating Blockchain technology with existing systems is very complex and resource-intensive, posing significant integration challenges to organisations. These costs prohibit developing and maintaining Blockchain applications, especially in smaller organisations or NGOs with minimal resources.

Another challenge is gaining trust in Blockchain technologies. The key stakeholders critical for larger-scale adoption need to gain greater confidence in the added value of Blockchain solutions (Sansone et al., 2023). This is not easy to develop because of the perceived complexity and novelty of Blockchain. Moreover, Blockchain is a technically complex technology to develop and manage; it requires specialised knowledge that shrinks the available talent further and amplifies the implementation challenges. Overcoming such challenges is crucial if the full potential of Blockchain technology is to be tapped for enhanced transparency, accountability, and efficiency in the field of CSR. Overcoming such limitations allows an organisation to use Blockchain to bring social change and promote sustainable practices effectively.

5.7 Conclusions and Implications

This chapter shows the applications of Blockchain for CSR. Integrating Blockchain and CSR values has proven responsible and sustainable business behaviour. It builds trust in stakeholders with verifiable data records that are impossible to tamper with. Classifying Blockchains as public, private, and hybrid will give complete insight into their various applications and their advantages and disadvantages. This will, therefore, be major in driving digital transformation with Blockchain's underlying benefits in supply chain management, primarily on real-time tracking, authenticity, and reduction of inefficiencies.

The chapter has also explored Blockchain's role in how digital technologies work seamlessly: it saves costs and builds better decision-making. Blockchain became an innovative disruption regarding higher education, ensuring secure verification of earned credentials and promoting academic integrity. Despite the many positives, the chapter has sought to address some heavy challenges such as scalability, high energy use, regulatory gaps, and unwillingness to adopt. He proposed possible solutions to surmount these barriers: energy-efficient Blockchain models, establishing globally uniform regulatory standards, awareness, and collaboration among stakeholders.

In general, it is expected that Blockchain technology will drive industry transformation with transparency, sustainability, and innovation. This chapter presented a holistic assessment of Blockchain applications and their limitations to further advances in the technology and its greater diffusion.

References

Agi, M. A., & Jha, A. K. (2022). Blockchain technology in the supply chain: An integrated theoretical perspective of organisational adoption. *International Journal of Production Economics, 247*, 108458. https://doi.org/10.1016/j.ijpe.2021.108458

Aljumah, A. I. (2023). Investigating the impact of blockchain technology on social sustainability and the mediating role of ethics and CSR. *Sustainability, 15*(21), 15510. https://doi.org/10.3390/su152115510

Ayub Khan, A., Laghari, A. A., Shaikh, A. A., Bourouis, S., Mamlouk, A. M., & Alshazly, H. (2021). Educational blockchain: A secure degree attestation and verification traceability architecture for higher education commission. *Applied Sciences, 11*(22), 10917. https://doi.org/10.3390/app112210917

Bandara, C., Rathnasiri, M. S. H., Dewasiri, N. J., Chathuranga, B., & Singh, N. (2024). Greening the economy for a sustainable future: Opportunities and perceptions. *Greening Our Economy for a Sustainable Future*, 197–213. https://doi.org/10.1016/B978-0-443-23603-7.00016-9

Bhushan, M., & Singh, R. (2024). Talent management facilitates net-zero transition through employee green behavior. In R. Singh, S. Khan, A. Kumar, S. Luthra, & H. Chokshi (Eds.), *Net zero economy, corporate social responsibility and sustainable value creation. CSR, sustainability, ethics & governance*. Springer. https://doi.org/10.1007/978-3-031-55779-8_7

Bidry, M., Ouaguid, A., & Hanine, M. (2023). Enhancing e-learning with blockchain: Characteristics, projects, and emerging trends. *Future Internet, 15*(9), 293. https://doi.org/10.3390/fi15090293

Casino, F., Dasaklis, T. K., & Patsakis, C. (2019). A systematic literature review of Blockchain-based applications: Current status, classification and open issues. *Telematics and Informatics, 36*, 55–81. https://doi.org/10.1016/j.tele.2018.11.006

Chathuranga, B., Rathnasiri, M. S. H., Dewasiri, N. J., Jayarathne, P., Bandara, C., & Bakari, M. A. (2024). Direct and indirect effect on industries and services (A). Greening our economy for a sustainable future, 215–226. https://doi.org/10.1016/B978-0-443-23603-7.00017-0

Chohan, U. W. (2019). Blockchain and environmental sustainability: Case of IBM's blockchain water management. Notes on the 21st Century (CBRI).

Dewasiri, N. J. (2024). Sustainable strategies for net zero: An Asia Pacific perspective. In R. Singh & D. Crowther (Eds.), *Transition towards a sustainable future. Approaches to global sustainability, markets, and governance*. Springer. https://doi.org/10.1007/978-981-97-5756-5_1

Dewasiri, N. J., Gnanapala, W. K. A. C., Rathnasiri, M. S. H., Yadav, M., Tewari, L. M., & Mishra, U. (2024). The way forward for net zero: Aligning nationally determined contributions (NDCs) and holistic sustainable development in South Asia. In R. Singh & D. Crowther (Eds.), *Transition towards a sustainable future. Approaches to global sustainability, markets, and governance*. Springer. https://doi.org/10.1007/978-981-97-5756-5_5

Dierksmeier, C., & Seele, P. (2020). Blockchain and business ethics. *Business Ethics: A European Review, 29*(2), 348–359. https://doi.org/10.1111/beer.12258

Gadekallu, T. R., Huynh-The, T., Wang, W., Yenduri, G., Ranaweera, P., Pham, Q. V., … Liyanage, M. (2022). Blockchain for the metaverse: A review. arXiv preprint arXiv:2203.09738.

Godfrey, P. C. (2009). Corporate social responsibility in sport: An overview and key issues. *Journal of Sport Management, 23*(6), 698–716. https://doi.org/10.1123/jsm.23.6.698

Gola, C., & Sedlmeir, J. (2022). Addressing the sustainability of distributed ledger technology. Bank of Italy Occasional Paper, (670). https://doi.org/10.32057/0.DP.670.2022.

Gupta, S. (2018). Blockchain transaction processing. In *Blockchain technology and applications* (pp. 1–11). Springer International Publishing.

Huynh-The, T., Gadekallu, T. R., Wang, W., Yenduri, G., Ranaweera, P., Pham, Q. V., et al. (2023). Blockchain for the metaverse: A review. *Future Generation Computer Systems, 143*, 401–419. https://doi.org/10.1016/j.future.2023.02.010

Jain, B., Ghosal, I., Chotia, A., & Jayarathne, P. A. (2024). Blockchain and IoT devices: A futuristic digital and smart supply chain approach. In *Applications of blockchain technology and artificial intelligence: Lead-ins in banking, finance, and capital market* (pp. 73–88). Springer International Publishing. https://doi.org/10.1007/978-3-031-36666-1_5

Janssen, M., Weerakkody, V., Ismagilova, E., Sivarajah, U., & Irani, Z. (2020). A framework for analysing blockchain technology adoption: Integrating institutional, market and technical factors. *International Journal of Information Management, 50*, 302–309. https://doi.org/10.1016/j.ijinfomgt.2019.08.012

Jayawardena, N. S., & Dewasiri, N. J. (2023). Food acquisition and consumption issues of south Asian countries: A systematic literature review and future research agenda. *FIIB Business Review*. https://doi.org/10.1177/23197145231194113

Jimoh, F. O., Abdullahi, U. G., & Ibrahim, I. A. (2019). An overview of Blockchain technology adoption. *Journal of Computer Science and Information Technology, 7*(2), 26–36. https://doi.org/10.15640/jcsit.v7n2a4

Joshi, A., Singh, R., Kumar, V., & Rani, S. (2024). Net zero strategy: Exploring opportunities, confronting challenges for developing countries. In R. Singh & D. Crowther (Eds.), *Transition towards a sustainable future. Approaches to global sustainability, markets, and governance*. Springer. https://doi.org/10.1007/978-981-97-5756-5_11

Karunarathna, K. S. S. N., Dewasiri, N. J., Singh, R., & Rathnasiri, M. S. H. (2024). What does artificial intelligence–powered ChatGPT bring to academia? A review. In R. Singh, S. Khan, A. Kumar, & V. Kumar (Eds.), *Artificial intelligence enabled management: An emerging economy perspective* (pp. 3–13). De Gruyter. https://doi.org/10.1515/9783111172408

Khan, S., Singh, R., Alnahas, J., Abbate, S., & Centobelli, P. (2024). Navigating the smart circular economy: A framework for manufacturing firms. *Journal of Cleaner Production, 480*, 144007. https://doi.org/10.1016/j.jclepro.2024.144007

Kumar, K., Singh, R., Sharma, A., & Lehri, B. P. S. (2024). Climate change adaptation strategies for achieving net-zero economy. In R. Singh & D. Crowther (Eds.), *Transition towards a sustainable future. Approaches to global sustainability, markets, and governance*. Springer. https://doi.org/10.1007/978-981-97-5756-5_8

Ltifi, M., & Mesfar, S. (2022). Does the corporate social responsibility of the service based on Blockchain technology affect the real behaviour of the consumer? *Journal of Air Transport Management, 104*, 102256. https://doi.org/10.1016/j.jairtraman.2022.102256

Miraz, M. H., Excell, P. S., & Rafiq, M. K. S. B. (2021). Evaluation of green alternatives for blockchain proof-of-work (PoW) approach. *Annals of Emerging Technologies in Computing (AETiC)*, 54–59. https://doi.org/10.33166/AETiC.2021.04.005

Mishra, U., Rathnasiri, M. S. H., & Dewasiri, N. J. (2024). Net-zero economy and energy sustainability: A Nepalese perspective. In R. Singh, S. Khan, A. Kumar, S. Luthra, & H. Chokshi (Eds.), *Net zero economy, corporate social responsibility and sustainable value creation. CSR, sustainability, Ethics & Governance* (pp. 145–161). Springer. https://doi.org/10.1007/978-3-031-55779-8_9

Nakamoto, S. (2008). *Bitcoin: A peer-to-peer electronic cash system.* Retrieved from https://bitcoin.org/bitcoin.pdf. https://nakamotoinstitute.org/library/bitcoin/

Nwokolo, S. C., Singh, R., Khan, S., Kumar, A., & Luthra, S. (2023a). Remedies to the challenges of renewable energy deployment in Africa. In *Africa's path to net-zero. CSR, sustainability, ethics & governance.* Springer. https://doi.org/10.1007/978-3-031-44514-9_3

Nwokolo, S. C., Singh, R., Khan, S., Kumar, A., & Luthra, S. (2023b). Africa's awakening to climate action. In *Africa's path to net-zero. CSR, sustainability, ethics & governance*. Springer. https://doi.org/10.1007/978-3-031-44514-9_9

Nwokolo, S. C., Singh, R., Khan, S., Kumar, A., & Luthra, S. (2023c). Impacts of climate change in Africa. In *Africa's path to net-zero. CSR, sustainability, ethics & governance*. Springer. https://doi.org/10.1007/978-3-031-44514-9_7

Raimundo, R., & Rosário, A. (2021). Blockchain system in higher education. *European Journal of Investigation in Health, Psychology and Education, 11*(1), 276–293. https://doi.org/10.3390/ejihpe11010021

Rathnasiri, M. S. H., & Gunathilaka, D. H. P. M. (2024). Net zero policies for sustainable future: Issues and challenges in the Asia Pacific region. In R. Singh & D. Crowther (Eds.), *Transition towards a sustainable future. Approaches to global sustainability, markets, and governance*. Springer. https://doi.org/10.1007/978-981-97-5756-5_12

Rathnasiri, M. S. H., Dewasiri, N. J., & Kumar, A. A. (2024). Policy framework and implementation strategies for Sri Lanka's transition to a net-zero economy. In R. Singh, S. Khan, A. Kumar, S. Luthra, & H. Chokshi (Eds.), *Net zero economy, corporate social responsibility and sustainable value creation. CSR, sustainability, ethics & governance* (pp. 43–60). Springer. https://doi.org/10.1007/978-3-031-55779-8_3

Rizal Batubara, F., Ubacht, J., & Janssen, M. (2019). Unraveling transparency and accountability in blockchain. In *Proceedings of the 20th annual international conference on digital government research (pp. 204–213)*. Association for Computing Machinery. https://doi.org/10.1145/3325112.3325247

Sansone, G., Santalucia, F., Viglialoro, D., & Landoni, P. (2023). Blockchain for social good and stakeholder engagement: Evidence from a case study. *Corporate Social Responsibility and Environmental Management, 30*(5), 2182–2193. https://doi.org/10.1002/csr.2421

Sapra, N., Shaikh, I., & Dash, A. (2023). Impact of proof of work (PoW)-based blockchain applications on the environment: A systematic review and research agenda. *Journal of Risk and Financial Management, 16*(4), 218. https://doi.org/10.3390/jrfm16040218

Singh, R., Gupta, A., & Kandpal, V. (2024). ESG excellence in the digital world. In D. R. Ernesto, S. Gonzalez, V. Kandpal, P. K. Ozili, & P. Chatterjee (Eds.), *Fintech for ESG and the circular* (pp. 93–204). Economy, Scrivener Publishing LLC. https://doi.org/10.1002/9781394234035.ch10

Taherdoost, H. (2022). A critical review of Blockchain acceptance models—Blockchain technology adoption frameworks and applications. *Computers, 11*(2), 24. https://doi.org/10.3390/computers11020024

Turkanović, M., Hölbl, M., Košič, K., Heričko, M., & Kamišalić, A. (2018). EduCTX: A blockchain-based higher education credit platform. *IEEE Access, 6*, 5112–5127. https://doi.org/10.1109/ACCESS.2018.2789929

Tyma, B., Dhillon, R., Sivabalan, P., & Wieder, B. (2022). Understanding accountability in blockchain systems. *Accounting, Auditing & Accountability Journal, 35*(7), 1625–1655. https://doi.org/10.1108/AAAJ-09-2021-5301

Upadhyay, A., Mukhuty, S., Kumar, V., & Kazancoglu, Y. (2021). Blockchain technology and the circular economy: Implications for sustainability and social responsibility. *Journal of Cleaner Production, 293*, 126130. https://doi.org/10.1016/j.jclepro.2021.126130

Venkatesh, V. G., Kang, K., Wang, B., Zhong, R. Y., & Zhang, A. (2020). System architecture for blockchain-based transparency of supply chain social sustainability. *Robotics and Computer-Integrated Manufacturing, 63*, 101896. https://doi.org/10.1016/j.rcim.2019.101896

Vijerathne, D. T., Wahala, W., Rathnasiri, M. S. H., De Silva, V., & Dewasiri, N. J. (2024). Literature review on sustainability. *Greening Our Economy for a Sustainable Future, 33-45*. https://doi.org/10.1016/B978-0-443-23603-7.00004-2

Wang, H., Tong, L., Takeuchi, R., & George, G. (2016). Corporate social responsibility: An overview and new research directions: Thematic issue on corporate social responsibility. *Academy of Management Journal, 59*(2), 534–544. https://doi.org/10.5465/amj.2016.5002

Wendl, M., Doan, M. H., & Sassen, R. (2023). The environmental impact of cryptocurrencies using proof of work and proof of stake consensus algorithms: A systematic review. *Journal of Environmental Management, 326*, 116530. https://doi.org/10.1016/j.jenvman.2023.116530

Woodside, J. M., Augustine, F. K., Jr., & Giberson, W. (2017). Blockchain technology adoption status and strategies. *Journal of International Technology and Information Management, 26*(2), 65–93. https://doi.org/10.58729/1941-6679.1300

Chapter 6
Harnessing Blockchain Technology in Renewable Energy Systems: A Path to Organizational Sustainability

Mananage Shanika Hansini Rathnasiri, Ali Najeeb, Seema Rani, and Bhaskara Srinivas

Abstract This chapter discusses how blockchain technology can be applied to renewable energy systems for organisational sustainability. The chapter mainly uses a Systematic Literature Review (SLR) that follows the Preferred Reporting Items for Systematic Reviews and Meta-Analyses (PRISMA) framework. The various applications, issues, and opportunities concerning blockchain-enabled renewable energy systems were ascertained using literature, reports, and case study analysis. The research findings show that blockchain increases renewable electric involvement by promoting decentralised electric networks, increasing peer-to-peer energy trading more reliant on renewables, and addressing transparency issues through renewable energy tracking and credentialing. However, challenges are also discernible in the technological scalability, regulatory equivocations, and organisational tensions in sufficiently realising the integration's enormous benefits. Nonetheless, the case experiences of blockchain-based microgrids and the management of carbon footprints suggest that these solutions are timely and potentially benefit organisations. These results have important implications for researchers, practitioners, and policymakers. Management needs to understand that blockchain can help organisations make efficiency improvements to energy usage, increase transparency, and

M. S. H. Rathnasiri (✉)
Department of Marketing Management, Faculty of Management Studies, Sabaragamuwa University of Sri Lanka, Belihuloya, Sri Lanka
e-mail: shanika@mgt.sab.ac.lk

A. Najeeb
Qasim Ibrahim School of Business, Villa College, Malé, Maldives
e-mail: najeeb@villacollege.edu.mv

S. Rani
Chandigarh School of Business, Chandigarh Group of Colleges, Jhanjeri, Punjab, India

B. Srinivas
Department of Management, University School of Business, Chandigarh University, Mohali, India

R. Singh et al. (eds.), *Blockchain Technologies for Smart Circular Economy and Organisational Sustainability*, Environmental Footprints and Eco-design of Products and Processes, https://doi.org/10.1007/978-3-031-89183-0_6

meet sustainability objectives. To remove economic and regulatory enablers, policymakers should implement favourable conditions; to remove the constraints, researchers work to fill the gaps in the literature, such as the scalability and energy consumption of blockchain.

Keywords Blockchain technology · Renewable energy systems · Organizational sustainability · Decentralized energy trading · Smart grids

6.1 Introduction

Greening has emerged as a requirement for global organisations, governments, and communities due to increased climate change, resources, and environmental degradation (Bhushan & Singh, 2024). There is a rising call on organisations to embrace practices that are on par with the United Nations' Sustainable Development Goals (SDGs). These demands are no longer simply environmental but are linked with economic and social factors as they develop sustainable solutions to help organisations be both sustainable and competitive (Saxena et al., 2021). During this change, technology has become vital as one of the tools that help bring more sustainable solutions to the difficulty of challenges (Khan et al., 2024).

Blockchain was formerly designed to support cryptocurrencies such as Bitcoin, but it has now grown to become a solution in various industries such as energy. Combined decentralisation, output transparency, and data immutability present fresh avenues for combatting inefficiencies and building credibility in energy systems. Nevertheless, renewable energy systems, such as solar, wind, and hydroelectric power systems, are strong support structures required to combat climate change (Kumar et al., 2024). They are supposed to offer path-breaking clean energy solutions and provide scalability, efficiency, and distribution equity issues. Blockchain solutions for decentralised data and transactions can fill these gaps and bring harmony between technology and sustainability.

Blockchain technology can enhance renewable energy systems for organisational sustainability (Gawusu et al., 2022). They may facilitate accurate energy trade, efficient stewardship of energy as a resource, and adherence to environment-friendly practices. It is evident that through blockchain, many organisations can minimise existing operational struggles, decrease carbon emissions, and establish trust with stakeholders. While the world is transitioning from fossil-based systems, separating blockchain from renewable energy systems is insufficient for technological advancement and sustainability. This chapter examines these intersections to reveal how blockchain technology for renewable energy systems forms the basis of the sustainability of organisational change.

6.1.1 Blockchain Technology

Blockchain technology is a distributed and decentralised registry for storing and managing simultaneously decentralised transactional and data records. Some of the most important features include decentralisation, where instead of centralised control, the control is spread over several nodes; transparency, which means that all members involved can check the transactions in real time; and security, which results from the using cryptography techniques which make it almost impossible to alter the data (Dewasiri et al., 2023). These features make blockchain systems most appropriate for solving challenges in complex systems where trust, traceability, and efficiency are needed.

The following are some exciting opportunities for blockchain in terms of energy and sustainability. This allows peer-to-peer (P2P) energy trading, which means producers and consumers of renewable energy can buy and sell energy to fellow consumers without involving a third party (Wongthongtham et al., 2021). Energy management is better served by smart contracts, which originate from blockchain technology and semiautomated energy transactions. Another area in which blockchain is used to avoid green energy fraud is tracking and certifying the generated renewable energy (Liu et al., 2024). Statistics show how blockchain promotes reforms of sustainable systems and organisations by solving issues associated with liberalising demand and supply.

6.1.2 Renewable Energy Systems

Renewable energy systems tap other natural resources, including the sun, wind, and water, to produce energy (Mishra et al., 2024a, b). Solar energy involves tapping sunlight through solar panels; wind energy uses wind turbines; and hydroelectric energy harnesses water flow to produce power. These resources are abundant and naturally occurring, making them suitable for use in the shift from fossil fuels.

However, many challenges affect renewable energy systems, including their drawbacks, such as their intermittent nature, high initial costs, and structural constraints. In addition, when implementing renewable electricity, problems of supply-demand control and equal access to grids are susceptible, especially if the grids have been initially designed as non-renewable sources of electricity. These challenges call for energy generation, distribution, and demand efficiency through blockchain technologies.

6.1.3 Sustainability in Organizations

Organisation sustainability encompasses interrelated dimensions, such as organisational, societal, economic, ecological, and social (Hahn et al., 2015). The former concept is referred to as economic sustainability, which means maintaining the financial stability of an organisation; the latter is environmental sustainability, which aims at decreasing the effects of the organisation on the environment and using resources wisely (Singh et al., 2024). The last is social sustainability, which implies improving the individual's quality of life and raising the stakes of communities involved in the organisation's business processes (Davidson, 2010). Attaining these objectives calls for due consideration and, more so, strategies.

This study implies that technological platforms are important instruments for overcoming sustainability issues. In particular, implementing blockchain technology in renewable energy systems can become an effective response to organisations' efforts to achieve sustainability. This integration enables organisations to decrease greenhouse gas emissions, balance resource utilisation, and comply with legal requirements due to increased clarity and responsibility. Thus, sustainability, blockchain, and renewable energy systems create the perfect foundation to help organisations make a positive global difference.

6.2 Methodology

This chapter follows a systematic literature review (SLR) methodology based on Preferred Reporting Items for Systematic Reviews and Meta-Analyses (PRISMA) guidelines to analyse blockchain integration with renewable energy systems for organisational sustainability. The SLR methodology was selected because it focuses on methodological stringency in the literature review, the assessment of the current state of research, and the emergence of avenues for further studies and initiatives, as highlighted by Dewasiri (2024) and Dewasiri et al. (2024), Iddagoda et al. (2024); Chathuranga et al. (2024), Bandara et al. (2024), Karunarathna et al. (2024), Vijerathne et al. (2024), and Jayawardena and Dewasiri (2023). The PRISMA framework is a transparent and structured literature selection, evaluation, and synthesis process described below.

6.2.1 Search Strategy and Data Collection

Academic databases such as Google Scholar and Scopus were exhaustively consulted using search terms such as 'blockchain technology, renewable energy systems, energy management, sustainability', and organisational sustainability. These keywords were entered into Boolean operators to help search for relevant publications that examined the use of blockchain in renewable energy contexts, such as trading, smart grids, and carbon credit. The search was restricted to publications in the last 15 years (2010–2024) to include the most recent developments in blockchain technology and renewable energy systems.

6.2.2 Inclusion and Exclusion Criteria

A list of inclusion and exclusion criteria was used to select only the most relevant and high-quality research articles. Academic papers, research papers, articles published in peer-reviewed scholastic journals, business and industry reports, case studies, or investigative reports on using blockchain in the renewable energy sector or organisational sustainability were included. The inclusion criteria for the papers selected were that all the papers were required to present empirical findings, case studies, or theoretical models. The study excluded sources unrelated to blockchain technology, renewable energy, or organisational sustainability and articles not written in English.

6.2.3 Study Selection and Data Extraction

For detailed analysis, titles and abstracts were used to select articles that passed the initial screening for further full-text review. To maintain transparency, the PRISMA flowchart was used to record the rationale for the systematic review inclusion process. The information search was guided by topics such as what has been developed in the context of using blockchains in energy systems, peer-to-peer trading and decentralised grids, the issues and limitations of blockchain integration, and its influence on the sustainability of organisations. Data on case studies, pilot projects, and actual applications were also collected to present the strengths and weaknesses of applying blockchain technology to renewable energy. The final sample of this study consists of forty-three (43) seminal studies.

6.3 Findings

6.3.1 Blockchain in Renewable Energy Systems

Blockchain is rapidly changing the renewable energy industry because it provides new tools that address the significant challenges of modernisation. Using energy certification in connection with other energy systems causes energy problems, such as inefficient distribution, lack of trust in certificates, and low energy market involvement. The following are the areas of blockchain application in renewable energy systems and examples of places where the technology has been applied to eliminate these challenges.

6.3.1.1 Peer-to-Peer (P2P) Energy Trading

In the case of blockchain, the platform is decentralised, allowing almost anyone, whether an individual or organisation, to engage directly in trading renewable energy. For instance, producers such as households with solar panels can sell excess energy to other people or consumers without the services of an intermediary, thereby enhancing communal approaches to energy. Organisations such as Power Ledger and LO3 Energy have developed blockchain-based P2P energy trading systems that enhance decentralised markets and empower prosumers (Mohammed & Yakubu, 2024).

6.3.1.2 Smart Contracts for Energy Management

Energy trading powered by smart contracts or self-executing legal agreements coded in a blockchain allows for activating such an alternative involving predefined triggers. Such contracts can improve energy usage efficiency by providing flexible real-time payments for energy loads and efficient energy distribution. For example, smart contracts can support automated billing in microgrids—the organisation of energy use and payments (Vieira & Zhang, 2021). This cuts bureaucracies and improves the measures of dependability on energy systems.

6.3.1.3 Decentralised Energy Grids

Blockchain enables the formulation of distributed power networks, where power generation and use are undertaken at regional levels, thereby minimising power station dependencies. They increase reliability mainly in rural and calamity-susceptible regions through these microgrids. When integrated with blockchain technology, autonomous consumers and producers of electricity in a decentralised grid can securely exchange data, map out supply and demand, and ensure fairness in energy sharing. This is why current projects, such as Brooklyn Microgrid in New York, can illustrate that decentralised grids based on blockchain technology can bring energy sovereignty and sustainability (Mishra et al., 2024a, b).

6.3.1.4 Tracking and Certifying Renewable Energy Production

Through decentralisation and record sharing, blockchain creates efficiency and a trustworthy market in renewables. It enables consumers and organisations to exercise their derivative demand by ensuring that the energy they purchase stems from renewable sources. For instance, WePower and Energy Web platforms, based on a blockchain, can track and certify renewable energy certificates, thereby decreasing fraud and increasing market credibility (Mohammed & Yakubu, 2024). This traceability also assists organisations in attaining sustainable reporting proficiency and green standards.

6.3.1.5 Case Studies and Examples

Several successful blockchain applications highlight their transformative potential in renewable energy systems.

Power Ledger (Australia): This application enables P2P energy trade for communities and tracks renewable energy certificates using Power Ledger's blockchain technology (Jamil et al., 2021).

LO3 Energy (USA): In its Brooklyn Microgrid venture, LO3 Energy has effectively utilised blockchain to disconnect energy grids and strengthen local energy tact (Huang et al., 2020).

Energy Web Foundation (Global): The Energy Web incorporates blockchain to solve real-world problems and create a more open, trustless market for energy products, which can accelerate global transitions to cleaner energy sources.

WePower (Lithuania): WePower provides renewable energy producers with a platform to tokenise their energy production, allowing other businesses to purchase energy directly from the producer while keeping track (Choobineh et al., 2023).

Blockchain applications in renewable energy systems demonstrate how blockchain technology can be used to build effective, productive, and transparent energy systems. Beyond its application as a technical solution in renewable energy systems, the proposed blockchain model addresses various organisational sustainability challenges.

6.3.2 Integration for Organisational Sustainability

Adding blockchain technology allows for modifying renewable energy systems for organisational sustainability. This integration provides new solutions that can help organisations improve their effectiveness, openness, and responsibility by solving important issues in energy management and enhancing sustainability performance. Here are milestones and details of how blockchain renewable energy integration can affect organisational sustainability, with examples.

6.3.2.1 Reducing Operational Inefficiencies

Energy management in many organisations is problematic because of procurement, distribution, and consumption management systems. Blockchain technology solves all these problems within an extra-centralised system of tracking real-time energy consumption and automating transactions. For example, smart contracts facilitate energy billing and supply chain activities, where much time is spent and many mistakes occur due to manual work (Li & Kassem, 2021). This leads to rational energy use that fits an organisation's operational agenda, thus enhancing an energy management system.

6.3.2.2 Promoting Transparency in Energy Usage

Another advantage of blockchain technology is that it maintains full transparency within energy transactions and consumption (Wu & Tran, 2018). An important application of blockchain technology is that organisations can monitor their energy utilisation patterns and be held responsible for fulfilling sustainable goals. It develops credibility with the concerned customers, investors, and regulators and rises with sustainable commitments. Unlike traditional websites, blockchain-based platforms such as Energy Web contain tools that can help organisations improve the transparency of energy usage so that they can make informed decisions about performance improvement.

6.3.2.3 Enhancing Carbon Footprint Management

Corporate social responsibility (CSR) is one of the most significant factors in organisational development, and carbon footprint determination is one of the most critical CSR elements (Gao et al., 2014). Blockchain allows organisations to track their energy sources and measure their carbon footprint. Renewable energy certificates or carbon offset certificates can be applied to blockchains, rendering sustainability claims authentic. Blockchain enables organisations to tackle the climate crisis and prove their compliance by creating an open and easily traceable information record of energy and emissions.

6.3.2.4 Driving Cost-Effectiveness

Due to the lack of central control, some of which are characteristic of blockchain, there are reduced intermediary costs in energy markets. P2P energy trading using blockchain technology can allow organisations to buy renewable energy directly from producers, reducing the markup costs of conventional energy markets (Padmanaban & Dhanabalan, 2022). Smart energy management systems utilising blockchain capability also have the potential to cut energy wastage and expenditure. These points make brainstorming for sustainability strategies more effective and economically feasible for organisations.

6.3.2.5 Real-World Examples of Adoption

Several organisations are leveraging blockchain-based renewable energy systems to advance their sustainability agendas:

Iberdrola (Spain): This major energy company also employs technology to authenticate renewable power generation while guaranteeing transparency in its commitment to green energy (Parmentola & Tutore, 2023).

Siemens and LO3 Energy (USA): Siemens decided to collaborate with LO3 Energy to introduce blockchain to manage microgrids, facilitating P2P electricity trading in specific neighbourhoods. This integration improves energy security and lowers expenses (Woodhall, 2017).

South Pole and Poseidon (UK): These organisations use blockchain to track purchases made with carbon offsets and validate emission cuts to ensure that global businesses follow the agreed climate targets.

Acciona (Global): This renewable energy company uses blockchain to provide energy provenance and minimise the problem of fake green energy worldwide (Delardas & Giannos, 2022).

These examples prove blockchain and renewable energy integration are possible and highly practical. Organisations implementing these technologies are now shaping the model of sustainability, operation efficiency, stakeholder confidence, and environmental consciousness. This integration allows organisations to lead innovation as the world calls for strict compliance with sustainable resource use and managing climate change impacts (Nwokolo et al., 2023a, b, c).

6.3.3 Challenges and Barriers

Despite the benefits that can be unlocked when blockchain technology is employed in renewable energy systems for organisational sustainability, this chapter seeks to note that it comes with challenges. Such challenges range from technological, economic, regulatory, and organisational to ethical, all showing that this integration requires strategic planning and innovation to harness its full potential.

6.3.3.1 Technological Challenges

Scalability is the leading technological issue in blockchain applications (Khan et al., 2021). The effectiveness of blockchain networks may significantly decrease when transaction degrees rise owing to low operating speed, information throughput, and transaction fees. These limitations may limit energy systems in which real-time transactions and data exchanges are important (Joshi et al., 2024).

Another critical challenge is the high energy demand for blockchains and associated networks (Wang et al., 2022), mainly if they are built using proof-of-work (PoW) consensus algorithms. Bitcoin and other such networks have been scrutinised for their egregious environmental impacts. Although newer, better methods are emerging, such as proof-of-stake (PoS), these are not yet fully implemented. This leads to the state that such a technology aimed at supporting sustainability harms the environment when poorly controlled.

6.3.3.2 Economic and Regulatory Barriers

The cost-benefit of blockchain in renewable energy systems can challenge organisations, institutes, and small and medium enterprises (Biswas & Gupta, 2019). The implementation costs are high at first due to necessities such as establishing proper infrastructure and a skilled workforce. Furthermore, some ambiguity about earnings realisation in blockchain-supported energy projects may cause indecision.

Legal systems, specifically rules, also remain docile to innovative development (Prewett et al., 2020). Blockchain and renewable energy markets currently function in various jurisdictions with ambiguous or outdated legislation and regulatory and compliance issues. Some matters concerning blockchain applications that must be addressed include cross-border energy trading, data privacy, and the acceptance of renewable energy credits backed up by the blockchain.

6.3.3.3 Organisational Resistance to Adoption

Organisations face challenges in transforming to disruptive technologies, such as blockchain (Frizzo-Barker et al., 2020). This resistance may be due to ignorance of the technology, fear of change, or scepticism about the technology's advantages. Consumers may not want to involve themselves in a system believed to be complicated or dangerous; this is especially the case in those sectors of the economy recognised as backward innovatively.

Furthermore, incorporating blockchain into renewable energy systems calls for the close engagement of information technology (IT), energy management, and sustainability departments. Discrepancies between these organisational groups hinder implementation or may lead to significant delays. However, this resistance can be addressed by providing sufficient training for the personnel involved, adequate communication, and purchase by the leaders.

6.3.3.4 Ethical and Environmental Considerations

As with any new technology, blockchain has ethical and environmental issues (Tang et al., 2020). Some blockchain networks are power-heavy, which contrasts with the use of renewable energy and raises questions about whether the technology yields a positive environmental impact or whether renewable energy sources are used to power blockchain systems important for supporting the sustainability objective.

The decentralised nature of blockchain is good and bad—ethically—because it makes it difficult to assign responsibility and govern. A decentralised system may result in disagreement, controversy, or frog and misuse where authorised large-scale energy trading systems are involved. This is why there needs to be more transparency in governance structures and ethical rules within organisations.

These issues involve stakeholders, technological advancement, appropriate laws, organisational culture, and ethical standards. Blockchain technology will require the cooperation of governments, industries, and academic institutions to invent effective large-scale blockchain technologies, establish relatively definite legal requirements, and provide conditions to support blockchain applications. Of course, the barriers are high, but they are not insurmountable, and their removal will be key to activating blockchain-based renewable energy systems for organisational sustainability.

6.4 Opportunities and Future Directions

Blockchain's integration with renewable energy systems means organisations have a unique chance to align with global sustainability goals and examine energy issues. The integration of IT systems can be propelled through technological advancements, enabling policies, and incorporating multi-organizational work that will enable both organisations and society to have a healthier state.

6.4.1 Innovations in Blockchain and Renewable Energy Technologies

Technology has evolved to meet the potential of blockchain and renewable energy systems. New technologies, such as the second generation of blockchain, have alternative advantages such as high scalability, low power consumption, and higher numbers of transactions made per second (Nasir et al., 2022). They partly solved some of the problems generally linked to blockchain to open it for more extensive use in energy systems.

In renewable energy, improvements in storage technology, such as batteries and hydrogen fuel cells, make renewables favourable because they deal with the problem of unreliability. These blockchain-enabled P2P energy trading and smart grids enable confident, efficient localised energy management. For instance, quantum computing and artificial intelligence can refine renewable energy and blockchain platform innovation, creating a prognosis of energy demand and balance of the grid system.

6.4.2 Policy Frameworks to Support Integration in Organizations

Adopting renewable energy solutions based on blockchain technology relies on government and international policies (Juszczyk & Shahzad, 2022). Government incentives to increase renewable energy demand, including subsidies, tax credits, and feed-in tariffs, should accompany regulatory incentives encouraging blockchain deployment in energy markets. For instance, energy transparency and carbon reporting requirements push organisations to use blockchain to track energy and provide certificates.

Furthermore, procedures and standards for blockchain integration and security in the energy field need to be defined. Cooperative arrangements among governments, industries, and technology suppliers can lead to the formulation of policies that remove barriers while promoting the deployment of sustainable energy technologies.

6.4.3 Collaborative Models for Blockchain-Based Energy Solutions

Effective blockchain implementation in renewable energy systems requires cooperation among key market players, producers, technology providers, governments, and consumers. Energy cooperatives and public-private partnerships with engaged key players and stakeholders can help build and expand blockchain solutions for energy systems.

For example, using blockchain technology, several energy cooperatives have obtained optimal means for implementing decentralised power generation and sharing renewable energy resources across local communities. Likewise, cooperation between governments and private businesses to finance and develop blockchain-based microgrids empowers the sustainability of electricity in otherwise unreached areas. Blockchain platforms, such as Energy Web, are also open-sourced to encourage cooperation by offering relevant tools to involved players in sustainable energy markets.

6.4.4 Implications for Global Sustainability Targets

Blockchain and renewable energy integration align with sustainable development goals, starting with the numerous goals proposed by the United Nations and its recent Paris Agreement on Climate Change. Opening up energy systems in this way can catalyse efforts to meet specific targets faster: affordable and clean energy (SDG 7), climate change (SDG 13), and sustainable cities and communities (SDG 11).

The increased amplitude and decentralised nature of blockchain-based renewable energy infrastructure create the opportunity to decarbonise global industries and end energy poverty for communities worldwide. By adopting these solutions, organisations achieve global sustainability, improve their image and brand, and build stakeholder trust and competitive advantage, especially given the dynamic nature of the business environment today.

Consequently, using blockchain technology in renewable energy systems can be a disruptive opportunity for organisations to redefine themselves as leaders in sustainable practice. They found that organisational change through technology, policy, and people is key to creating sustainable and equitable power systems in the future.

6.5 Conclusions

The combination of blockchain technology with renewable energy systems is a revolution in organisational sustainability improvement. This chapter has examined how these innovations can be integrated to remove bottlenecks, promote more disclosure, and bring organisational practices into compliance with international sustainable development objectives. When considering the opportunities created by combining the specific characteristics of the blockchain and the further development of renewable energy solutions, companies and organisations can approach significant problems that arise to generate long-term value.

The evaluation focuses on how applications such as P2P energy trading, decentralised energy grids, and RECs hinge on the blockchain for enhancing energy systems. Moreover, concerning renewable energy systems, it highlights that their literature also focuses on their elemental role in environmental and economic sustainability and discusses issues such as intermittency and scalability of renewable energy systems. The combination of blockchain and renewable energy offers an excellent solution for organisational sustainability regarding process optimisation, costs, and accountability. The understanding and necessity of this integration in some organisational settings are additionally explained and supported by examples and case studies.

6.5.1 Strategic Recommendations for Organizations

To harness the potential of blockchain and renewable energy systems, organisations should adopt a strategic approach that includes the following:

Investing in innovation: Invest in research and pilot projects on blockchain applications in energy-constrained organisation-specific applications.

Building partnerships: Engage with other technology industries and energy-relevant authorities to ensure further incorporation and increase these innovations.

Focusing on capacity building: Empower teams to understand and apply blockchain-enabled renewable energy systems and manage them properly.

Aligning with policies: Therein are newly emerging principal regulations that embrace renewable energy and blockchain incentives.

Monitoring and reporting: Several tools adopted in blockchain technology deal with energy usage and sustainability reportage to foster stakeholders' trust and/or compliance.

6.5.2 Call to Action for Researchers, Practitioners, and Policymakers

However, this suggests that all shareholders' efforts should be coordinated for blockchain renewable energy systems to succeed. The authors should consider technological issues such as scalability and power consumption when foraging for new applications. Members are urged to try these methods and report their experiences to improve their practices.

In contrast, the latter requires policymakers to develop favourable policies, provide funding for pilot project implementation, and develop benchmarks for blockchain applications in energy trading. Such ventures can only be scaled up with global cooperation's support, so these solutions become valuable in climate change efforts and sustainability goals.

Thus, applying blockchain to renewable energy systems provides a new direction for working toward organisational sustainability. It is possible to demonstrate how using decisions at various levels and broad collaboration between organisations and stakeholders can create better solutions to ensure a sustainable future for all. The process towards realising this possibility starts here, with every advance leading us to the future of interlinked technologies and sustainability.

References

Bandara, C., Rathnasiri, M. S. H., Dewasiri, N. J., Chathuranga, B., & Singh, N. (2024). Greening the economy for a sustainable future: Opportunities and perceptions. *Greening Our Economy for a Sustainable Future, 197–213*. https://doi.org/10.1016/B978-0-443-23603-7.00016-9

Bhushan, M., & Singh, R. (2024). Talent management facilitates net-zero transition through employee Green behavior. In R. Singh, S. Khan, A. Kumar, S. Luthra, & H. Chokshi (Eds.), *Net zero economy, corporate social responsibility and sustainable value creation. CSR, sustainability, ethics & governance*. Springer. https://doi.org/10.1007/978-3-031-55779-8_7

Biswas, B., & Gupta, R. (2019). Analysis of barriers to implement blockchain in industry and service sectors. *Computers & Industrial Engineering, 136*, 225–241.

Chathuranga, B., Rathnasiri, M. S. H., Dewasiri, N. J., Jayarathne, P., Bandara, C., & Bakari, M. A. (2024). Direct and indirect effect on industries and services (A). *Greening Our Economy for a Sustainable Future, 215-226*. https://doi.org/10.1016/B978-0-443-23603-7.00017-0

Choobineh, M., Arabnya, A., Sohrabi, B., Khodaei, A., & Paaso, A. (2023). Blockchain technology in energy systems: A state-of-the-art review. *IET Blockchain, 3*(1), 35–59.

Davidson, M. (2010). Social sustainability and the city. *Geography Compass, 4*(7), 872–880.

Delardas, O., & Giannos, P. (2022). The ripple effects of the energy crisis on academia: Europe's energy crisis deepens concerns about the future of academia and research laboratories are in dire need of sustainable solutions. *EMBO Reports, 23*(12), e56287.

Dewasiri, N. J. (2024). Sustainable strategies for net zero: An Asia Pacific perspective. In R. Singh & D. Crowther (Eds.), *Transition towards a sustainable future. Approaches to global sustainability, markets, and governance*. Springer. https://doi.org/10.1007/978-981-97-5756-5_1

Dewasiri, N. J., Karunarathne, K. S. S. N., Menon, S., Jayarathne, P. G. S. A., & Rathnasiri, M. S. H. (2023). Fusion of artificial intelligence and Blockchain in the banking industry: Current application, adoption, and future challenges. In A. Saini & V. Garg (Eds.), *Transformation for sustainable business and management practices: Exploring the Spectrum of Industry 5.0* (pp. 293–307). Emerald Publishing Limited. https://doi.org/10.1108/978-1-80262-277-520231021

Dewasiri, N. J., Dharmarathna, D. G., & Choudhary, M. (2024). Leveraging artificial intelligence for enhanced risk management in banking: A systematic literature review. In R. Singh, S. Khan, A. Kumar, & V. Kumar (Eds.), *Artificial intelligence enabled management: An emerging economy perspective* (pp. 197–213). De Gruyter. https://doi.org/10.1515/9783111172408

Frizzo-Barker, J., Chow-White, P. A., Adams, P. R., Mentanko, J., Ha, D., & Green, S. (2020). Blockchain as a disruptive technology for business: A systematic review. *International Journal of Information Management, 51*, 102029.

Gao, T., Liu, Q., & Wang, J. (2014). A comparative study of carbon footprint and assessment standards. *International Journal of Low-Carbon Technologies, 9*(3), 237–243.

Gawusu, S., Zhang, X., Ahmed, A., Jamatutu, S. A., Miensah, E. D., Amadu, A. A., & Osei, F. A. J. (2022). Renewable energy sources from the perspective of blockchain integration: From theory to application. *Sustainable Energy Technologies and Assessments, 52*, 102108.

Hahn, T., Pinkse, J., Preuss, L., & Figge, F. (2015). Tensions in corporate sustainability: Towards an integrative framework. *Journal of Business Ethics, 127*, 297–316.

Huang, L. Y., Cai, J. F., Lee, T. C., & Weng, M. H. (2020). A study on the development trends of the energy system with blockchain technology using patent analysis. *Sustainability, 12*(5), 2005.

Iddagoda, A., Abeysinghe, R., Dissanayake, H., Dewasiri, N. J., & Wickramaratne, W. P. R. (2024). Cradle of research gaps: Reflections on Green orientation. In S. Gupta, S. Taneja, E. Özen, K. Sood, & S. Grima (Eds.), *Green management: A new paradigm in the world of business (emerald studies in finance, insurance, and risk management)* (pp. 339–357). Emerald Publishing Limited. https://doi.org/10.1108/978-1-83797-442-920241018

Jamil, F., Iqbal, N., Ahmad, S., & Kim, D. (2021). Peer-to-peer energy trading mechanism based on blockchain and machine learning for sustainable electrical power supply in smart grid. *IEEE Access, 9*, 39193–39217.

Jayawardena, N. S., & Dewasiri, N. J. (2023). Food acquisition and consumption issues of south Asian countries: A systematic literature review and future research agenda. *FIIB Business Review*. https://doi.org/10.1177/23197145231194113

Joshi, A., Singh, R., Kumar, V., & Rani, S. (2024). Net zero strategy: Exploring opportunities, confronting challenges for developing countries. In R. Singh & D. Crowther (Eds.), *Transition towards a sustainable future. Approaches to global sustainability, markets, and governance*. Springer. https://doi.org/10.1007/978-981-97-5756-5_11

Juszczyk, O., & Shahzad, K. (2022). Blockchain technology for renewable energy: Principles, applications and prospects. *Energies, 15*(13), 4603.

Karunarathna, K. S. S. N., Dewasiri, N. J., Singh, R., & Rathnasiri, M. S. H. (2024). What does artificial intelligence–powered ChatGPT bring to academia? A review. In R. Singh, S. Khan, A. Kumar, & V. Kumar (Eds.), *Artificial intelligence enabled management: An emerging economy perspective* (pp. 3–13). De Gruyter. https://doi.org/10.1515/9783111172408

Khan, D., Jung, L. T., & Hashmani, M. A. (2021). Systematic literature review of challenges in blockchain scalability. *Applied Sciences, 11*(20), 9372.

Khan, S., Singh, R., Alnahas, J., Abbate, S., & Centobelli, P. (2024). Navigating the smart circular economy: A framework for manufacturing firms. *Journal of Cleaner Production, 480*, 144007. https://doi.org/10.1016/j.jclepro.2024.144007

Kumar, K., Singh, R., Sharma, A., & Lehri, B. P. S. (2024). Climate change adaptation strategies for achieving net-zero economy. In R. Singh & D. Crowther (Eds.), *Transition towards a sustainable future. Approaches to global sustainability, markets, and governance*. Springer. https://doi.org/10.1007/978-981-97-5756-5_8

Li, J., & Kassem, M. (2021). Applications of distributed ledger technology (DLT) and Blockchain-enabled smart contracts in construction. *Automation in Construction, 132*, 103955.

Liu, W. J., Chiu, W. Y., & Hua, W. (2024). Blockchain-enabled renewable energy certificate trading: A secure and privacy-preserving approach. *Energy, 290*, 130110.

Mishra, U., Rathnasiri, M. S. H., & Dewasiri, N. J. (2024a). Net-zero economy and energy sustainability: A Nepalese perspective. In R. Singh, S. Khan, A. Kumar, S. Luthra, & H. Chokshi (Eds.), *Net zero economy, corporate social responsibility and sustainable value creation. CSR, sustainability, ethics & governance* (pp. 145–161). Springer. https://doi.org/10.1007/978-3-031-55779-8_9

Mishra, S., Khatami, R., & Chen, Y. C. (2024b). Decentralisation of energy systems with Blockchain: Bridging top-down and bottom-up management of the electricity grid. In M. H. Amini (Ed.), *Distributed machine learning and computing. Big and integrated artificial intelligence* (Vol. 2). Springer. https://doi.org/10.1007/978-3-031-57567-9_6

Mohammed, A., & Yakubu, I. N. (2024). Revolutionising energy systems: The impact of blockchain on renewable energy integration. *The Intersection of Blockchain and Energy Trading, 143–170*. https://doi.org/10.1016/B978-0-443-23627-3.00007-7

Nasir, M. H., Arshad, J., Khan, M. M., Fatima, M., Salah, K., & Jayaraman, R. (2022). Scalable blockchains—A systematic review. *Future Generation Computer Systems, 126*, 136–162.

Nwokolo, S. C., Singh, R., Khan, S., Kumar, A., & Luthra, S. (2023a). Remedies to the challenges of renewable energy deployment in Africa. In *Africa's path to net-zero. CSR, sustainability, ethics & governance*. Springer. https://doi.org/10.1007/978-3-031-44514-9_3

Nwokolo, S. C., Singh, R., Khan, S., Kumar, A., & Luthra, S. (2023b). Africa's awakening to climate action. In *Africa's path to net-zero. CSR, sustainability, ethics & governance*. Springer. https://doi.org/10.1007/978-3-031-44514-9_9

Nwokolo, S. C., Singh, R., Khan, S., Kumar, A., & Luthra, S. (2023c). Impacts of climate change in Africa. In *Africa's path to net-zero. CSR, sustainability, ethics & governance*. Springer. https://doi.org/10.1007/978-3-031-44514-9_7

Padmanaban, S., & Dhanabalan, S. S. (2022). Peer-to-peer energy trading using blockchain technology. *Energy Reports, 8*, 2348–2350.

Parmentola, A., & Tutore, I. (2023). Unveiling the positive and negative effects of Blockchain technologies on environmental sustainability in practice. In *Industry 4.0 Technologies for Environmental Sustainability: Intended and unintended consequences* (pp. 59–78). Springer International Publishing.

Prewett, K. W., Prescott, G. L., & Phillips, K. (2020). Blockchain adoption is inevitable—Barriers and risks remain. *Journal of Corporate accounting & finance, 31*(2), 21–28.

Saxena, A., Ramaswamy, M., Beale, J., Marciniuk, D., & Smith, P. (2021). Striving for the United Nations (UN) sustainable development goals (SDGs): What will it take? *Discover Sustainability, 2*, 20. https://doi.org/10.1007/s43621-021-00029-8

Singh, R., Gupta, A., & Kandpal, V. (2024). ESG excellence in the digital world. In D. R. Ernesto, S. Gonzalez, V. Kandpal, P. K. Ozili, & P. Chatterjee (Eds.), *Fintech for ESG and the circular economy* (pp. 93–204). Scrivener Publishing LLC. https://doi.org/10.1002/9781394234035.ch10

Tang, Y., Xiong, J., Becerril-Arreola, R., & Iyer, L. (2020). Ethics of blockchain: A framework of technology, applications, impacts, and research directions. *Information Technology & People, 33*(2), 602–632.

Vieira, G., & Zhang, J. (2021). Peer-to-peer energy trading in a microgrid leveraged by smart contracts. *Renewable and Sustainable Energy Reviews, 143*, 110900.

Vijerathne, D. T., Wahala, W., Rathnasiri, M. S. H., De Silva, V., & Dewasiri, N. J. (2024). Literature review on sustainability. *Greening Our Economy for a Sustainable Future, 33-45*. https://doi.org/10.1016/B978-0-443-23603-7.00004-2

Wang, T., Hua, H., Wei, Z., & Cao, J. (2022). Challenges of blockchain in new generation energy systems and future outlooks. *International Journal of Electrical Power & Energy Systems, 135*, 107499.

Wongthongtham, P., Marrable, D., Abu-Salih, B., Liu, X., & Morrison, G. (2021). Blockchain-enabled peer-to-peer energy trading. *Computers & Electrical Engineering, 94*, 107299.

Woodhall, A. (2017). How Blockchain can democratize global energy supply. *Transforming Climate Finance and Green Investment With Blockchains, 65–82*. https://doi.org/10.1016/B978-0-12-814447-3.00005-7

Wu, J., & Tran, N. K. (2018). Application of blockchain technology in sustainable energy systems: An overview. *Sustainability, 10*(9), 3067.

Chapter 7
Blockchain for Sustainable Waste Management: Transforming Recycling and Circular Economy Practices

Narayanage Jayantha Dewasiri, Rubee Singh, and R. Gobinath

Abstract This chapter examines how blockchain technology revolutionises the waste management and recycling sectors. Expanding from this perspective, this chapter utilises a Systematic Literature Review (SLR) approach using Preferred Reporting Items for Systematic Reviews and Meta-Analyses (PRISMA) to systematically synthesise the current literature and comprehensively categorise applications based on blockchain technology. According to the research, the benefits identified in this area are as follows: increased traceability in the collection and recycling of wastes, fostering decentralised networks for recycling and using smart contracts, and tokenisation for incentive application. Combining it with other burgeoning technologies, such as the Internet of Things (IoT) and Artificial Intelligence (AI), strengthens the benefits of blockchain technology in waste management by allowing live monitoring and updating and deterministic and responsive optimisation of the management process. However, it has drawbacks in that numerous technological hurdles, lack of legislation, and steep start-up costs must be overcome to apply it on a large scale. Thus, the findings of this study offer numerous practical implications for further research and application by researchers, practitioners, and policymakers to undertake their work, respectively. Blockchain holds the potential for stakeholders to introduce positive change and collaboration and develop environmentally and socially sustainable systems. This chapter emphasises coordinated actions for the elimination of current deficiencies. This underlines the crucial role of blockchain in further developing arrangements for waste management and recycling.

N. J. Dewasiri (✉)
Sabaragamuwa University of Sri Lanka, Belihuloya, Sri Lanka

Qasim Ibrahim School of Business, Villa College, Malé, Maldives
e-mail: jayantha@mgt.sab.ac.lk

R. Singh
Institute of Business Management, GLA University, Mathura, India

Department of Management Studies, Kumaun University, Nainital, Uttrakhand, India

R. Gobinath
University Center for Research and Development, Chandigarh University, Mohali, India

R. Singh et al. (eds.), *Blockchain Technologies for Smart Circular Economy and Organisational Sustainability*, Environmental Footprints and Eco-design of Products and Processes, https://doi.org/10.1007/978-3-031-89183-0_7

Keywords Blockchain technology · Waste management · Recycling · Circular economy · Sustainability

7.1 Introduction

Waste disposal management and recycling are core basics of sustainable development goals, but we face numerous challenges that hamper their process (Zorpas, 2020). Even if each community manages its waste in the best way possible now, unsustainable population growth, climate change, resource depletion, and pollution mean there is an urgent need for innovation in waste management (Nwokolo et al., 2023a, b, c). This is because conventional waste management mechanisms entail poor record keeping, low accountability, and even the absence of motivation compensation for stewards. This leads to deterioration of the natural environment and more wastage of valuable natural resources.

Blockchain has a wholly decentralised and transparent structure that presents a solution for these perennial complications (Dewasiri et al., 2023). Initially known as a supporting technology for digital currencies such as Bitcoin, blockchain has expanded to include industries such as finance, healthcare, and even logistics. These attributes include immutability, transparency, and operation through smart contracts, making them more appropriate for solving current waste management and recycling challenges.

This chapter seeks to determine how blockchain technology can change and improve the waste management sector by making it more responsive, equitable, and efficient. Applying blockchain technology to recycling and the circular economy will lead to the development of systems that, in turn, will easily manage waste and foster sustainable consumption and production (Joshi et al., 2024).

The chosen topic is introduced by presenting the reader with general information on the use of blockchain technology and its base principles, as well as an analysis of the problems observed in the function of traditional waste management systems. It then explores how blockchain addresses these challenges and provides an overview of the disruptive potential of circular economy solutions. Through case studies, benefits analysis, and the discussion of limitations, this chapter provides a clear perspective on how blockchain must be leveraged to create a sustainable world through efficient waste management and recycling management.

Accordingly, this chapter encourages the widespread use of blockchain technology by researchers, practitioners, and policymakers in the waste management sector. As the world moves towards a circular economy, incorporating blockchain can create a cleaner, greener, and more sustainable world (Khan et al., 2024).

7.2 Methodology

Even though there is a paucity of academic research on how blockchain technology could revolutionise waste management and recycling, this chapter follows an SLR approach to analyse this possibility systematically. Through the SLR approach, this study pays close attention to the structure and rigour crucial to identifying the research themes, issues, and gaps in the current academic and industrial literature related to this relatively new discipline. PRISMA framework was used to guide the conduct of this review and increase the methodological rigour in the study's reporting, as highlighted by Vijerathne et al. (2024), Bandara et al. (2024); Chathuranga et al. (2024); Rathnasiri et al. (2024), Rathnasiri and Gunathilaka (2024), Iddagoda et al. (2024), Karunarathna et al. (2024), and Mishra et al. (2024).

The study developed testable research questions to discover blockchain's relevance in improving the organisation's waste management systems. We then performed an academic database and industry report search using the following terms: blockchain technology, waste management, recycling, and circular economy. In the case of data extraction, ensuring the relevance of insights gathered and information searched for when extracting was related to the features of blockchain, its applications, and other technologies integrating with blockchain and blockchain in the aspect of waste management: its merits, demerits, and difficulties experienced when implementing it. Finally, the analysed data were integrated to discover trends, present examples, and make relevant conclusions.

This study utilises the SLR methodology to offer a strong framework for analysing the nature of change presented by blockchain technology for global waste management through the PRISMA framework. This ensures that the points presented in this chapter are derived from a systematic analysis of the existing knowledge base. The final sample consists of thirty-five (35) seminal studies.

7.3 Findings

7.3.1 Challenges in Traditional Waste Management Systems

Although accurate for preserving both environmental and public health, existing obstacles hinder the applicability and efficiency of traditional waste management systems (Kumar et al., 2024). Such challenges can lead to poor operation, efficiency, and productivity, causing environmental harm. Therefore, overcoming these deficits is essential to elaborating a stable waste management regime, and knowing these peculiarities is a starting point for investigating how blockchain can support mitigating the challenges.

7.3.1.1 Lack of Accountability and Transparency

Among these weaknesses, the major one in conventional waste management systems is a lack of clarity. Another important characteristic of ineffective waste management is unawareness or lack of information (Breukelman et al., 2019). From when waste is produced to when it is collected or recycled, it usually moves through many stakeholders' hands. Lacking a single and transparent framework to monitor this process, there is significant potential for inefficient, unlawful, or untruthful handling of waste processing. For instance, recyclable substances can end up in landfills owing to a lack of appropriate supervision and, in the process, defeat the entire intention of recycling, seeking to protect the environment.

7.3.1.2 Inefficient Tracking of Waste Generation and Recycling

One of the main challenges is the lack of proper tracking of generated and recycled waste (Hannan et al., 2015). The collection of waste data is traditionally done manually or using several isolated devices, which is why it is inaccurate and time-consuming. This weakness negatively affects decision-making, resource management, and fulfilment of regulatory requirements. Furthermore, most waste tracking systems are not designed as real-time systems; hence, delays are experienced in the real-time detection of necessary improvements or adjustments. Thus, reusable materials are wasted, and the possibility of a circular economy remains unfulfilled.

7.3.1.3 Corruption and Fraud in Waste Recycling Value Chains

Bribery, corruption, and fraud are fundamental issues in traditional waste management systems. The absence of corporate social responsibility and accountable and autonomous decision-making led to some organisational misconduct, including manipulation of records in waste processing, interception and selling of recyclable materials, and misuse of funds allocated for waste management projects. Such practices undermine public confidence and disincentivise investment in appropriate waste management technologies. For instance, scams in the certification process of recycled content are bound to distort markets and erode the believability of green stories (Bhushan & Singh, 2024).

7.3.1.4 High Costs and Limited Scalability of Existing Solutions

Conventional waste management processes entail high operational costs; hence, their implementation is unfeasible within economies of scale, particularly in developing countries (Khetriwal et al., 2009). Some costs, including acquiring development infrastructure, waste collection, and recycling, may reach heights, especially for less-endowed municipalities. Furthermore, centralised systems need much administrative effort and cannot address the future growth of waste production and its heterogeneity in urban environments. This non-scalability worsens the situation, resulting in several communities being served inappropriately and not contributing to solving the global waste issue.

Innovative waste management solutions face challenges regarding efficiency, transparency, and scalability. This section summarises these pain points and shows how blockchain technology can restore waste management networks as sustainable, transparent, and accountable entities. Thus, by overcoming these challenges, stakeholders can achieve the goals of a circular economy in which resource reuse and effective waste reduction are dominant concepts (Singh et al., 2024).

7.3.2 *Role of Blockchain in Waste Management and Recycling*

Blockchain technology can significantly revolutionise current ineffective waste management systems. Considering blockchain's decentralised, transparent, and secure infrastructure, technology can disrupt and transform inefficient and ineffective waste collection, tracking, and recycling systems. It is not limited to technology transforming waste management's social, economic, and environmental face.

7.3.2.1 Enhancing Transparency and Traceability in Waste Collection and Recycling

Among the various opportunities that originate from using blockchain for waste management, one is perhaps its most valuable feature: improving transparency (França et al., 2020). Every transaction within a blockchain system is captured in an immutable register, making it possible for each stakeholder to follow the waste from when it was created to when it was disposed of or recycled. This transparency promotes the responsibility of generators, collectors, and recyclers not to dump waste illegally or indulge in other wrong deeds. For instance, firms can employ blockchain to ensure consumers and regulatory bodies that their waste has undergone the required processes of environmental standards.

7.3.2.2 Facilitating Decentralized Recycling Networks

Standard systems for waste removal are centralised and, therefore, not as effective, and their accessibility can be further limited. Through blockchain, decentralised recycling can be formed to involve the public, companies, and local governments. These networks enable communities to deal with waste and coordinate stakeholder responsibilities independently. Decentralisation also minimises the use of intercessors, helping lower the costs of provided waste management solutions and expand access to them, especially in regions with weak demand (Jiang et al., 2023).

7.3.2.3 Leveraging Smart Contracts for Waste Exchange and Incentives

Smart contracts are instrumental in managing waste exchange actions and encouraging sustainable behaviour (Gupta et al., 2022). These smart contracts work per pre-specified criteria. For instance, payment/disbursement/reward is automatic when a particular task is performed. For example, a simple application could release payment to a waste disposal services provider after a recycling plant confirms receiving recyclable items. Similarly, smart contracts can organise electronic platforms to sell recyclable waste, where businesses and consumers can exchange such materials with ease and credibility. This automation of processes makes the work easier and free from challenges such as human errors or fraud.

7.3.2.4 Tokenisation of Recycling Efforts: Rewarding Sustainable Practices

Blockchain also helps tokenise recycling, thus creating economic and social incentives for recycling (Gong et al., 2022). Coins, digital assets, can be issued to people or firms involved in recycling to encourage them. These tokens may be exchanged for goods, services, or even cash. Thus, the public is motivated to recycle their waste correctly. For instance, buyers who return plastic waste for recycling can gain tokens redeemed for free consumer goods or donations to the nobles. It also fosters bottom-up engagement and how economic and innovation processes can support ecological purposes.

Applying blockchain technology in waste management and recycling processes presents fascinating opportunities for solving deep-seated issues, improving strengths, and encouraging new advancements. The resulting systems were refined, open, and productive. Blockchain's value includes making stakeholders accountable, fostering sustainable practices, and promoting the community's rights to unleash their potential for a sustainable future.

7.3.3 Blockchain-Enabled Circular Economy Practices

Blockchain technology enables circular economy practices, including resource optimisation, improvement of waste management systems, and encouragement of multi-stakeholder engagement (Singh & Kumar, 2024). When applied alongside other emerging technologies, blockchain opens up a range of applications for retrofitting noncircular consumption paradigms. Such activities embrace the principles of waste avoidance and improve the management of a material's lifecycle according to international sustainability standards.

7.3.3.1 Promoting Resource Efficiency Through Better Material Tracking

Resource efficiency is the core of the circular economy, and blockchain technology significantly increases the possibility of material tracking (Khan et al., 2022). Blockchain provides firms with resource data from their sourcing, use, and recycling through a constant and unalterable account for each material. Hence, it is possible to determine shortcomings in other processes, minimise the loss of materials, and enable the reuse or recycling of valuable products rather than disposal in a landfill. For instance, in material sourcing, blockchain enables manufacturers to access materials with a quality guarantee of containing recycled material, as demanded by law and the market.

Blockchain also supports extended producer responsibility (EPR) programs in which producers are responsible for a product throughout its usage (Gui, 2020). By implementing blockchain in tracking the material, the company can more easily ensure that the EPR requirements are met and, in return, enhance the recovery rates to bring minimal effects to the environment.

7.3.3.2 Integrating Blockchain with IoT for Smart Waste Bins and Real-Time Data

Applying a technology such as blockchain for the Internet of Things (IoT) enhances the efficacy of waste management systems by compiling waste mitigation through real-time analysis. An IoT sensor-enabled smart waste bin can help track waste levels, types of waste, and disposal frequency. This means that these bins can send data to blockchain systems, such as records of waste generation or collection.

For instance, a smart bin in a residential area could notify authorities when it is complete, allowing the efficient organisation of waste pickup routes and, therefore, cost-cutting. In addition, blockchain technology can safely retain and transfer this data, including constant updates with other parties, to make efficient decisions and allocate resources.

7.3.3.3 Case Studies of Successful Blockchain Applications in Recycling

Several real-world examples demonstrate the impact of blockchain-enabled practices in promoting a circular economy.

Plastic Bank: A social business venture that leverages blockchain technology to motivate the recycling of plastics in emerging markets (Bhubalan et al., 2022).

Plastic materials are segregated by individuals and are exchanged for digital tokens in a blockchain system, which can be used later to purchase goods, services, or cash. A characteristic of this model is the flexibility of decreasing oceanic plastic waste and providing employment possibilities to susceptible groups.

VeChain: This is based on a blockchain platform that monitors and verifies the recycling of consumer products, such as electronics and apparel (Ying & Mok, 2023). Brand owners can use VeChain to guarantee consumers that their purchasing decisions support recycling initiatives related to products conforming to sustainable brand standards.

RecycleGO: A blockchain application to enhance recyclable supply chain information flow (Gong et al., 2022). RecycleGO allows businesses, recyclers, and consumers to get together efficiently to recover waste, thus enhancing their recycling process. The ability to make the platform transparent enables organisations to achieve sustainability goals and, at the same time, encourages stakeholders to recycle.

Blockchain technology is also critical to circular economic activities. It enhances resource productivity and real-time tracking while demonstrating good practice. This promotes transparency, accountability, and innovation, which enhance the shift to sustainable processes and optimum resource leveraging.

7.3.4 Benefits of Blockchain in Waste Management

Blockchain technology presents a revolutionary paradigm for correctly classifying waste-related challenges. It has many benefits, including enhanced productivity, responsibility, and shareholder management. These benefits overcome the drawbacks of conventional waste systems and mark a way to improve and implement effective waste management schemes.

7.3.4.1 Increased Transparency and Trust Among Stakeholders

Transparency is a very important factor in the implementation of waste management, and by deploying blockchain, every activity carried out in the waste lifecycle is made transparent owing to the use of blockchain's immutable ledger (Scott et al., 2023). This concrete integration of transparency assures other stakeholders, such as waste generators, collectors, recyclers, and regulators. For instance, a given blockchain platform enables the correct confirmation of dangerous waste disposal as environmental laws dictate. Blockchain promotes multiparty sustainability collaboration by offering proof and removing contentious issues, allowing coordination by giving those operating within achievable targets.

7.3.4.2 Reduction in Operational Inefficiencies and Costs

Organisational waste strategies typically fail because they are inefficient, process-based, lack timely information, and use intermediaries. This makes many operations technological and greatly depends on what smart contracts can achieve by minimising human input and bureaucracy roles. For instance, payments to waste collectors could be made as soon as the service has been delivered, saving time and reducing transaction costs. Furthermore, blockchain can improve waste collection routes through IoT with real-time data to minimise fuel consumption and operating costs.

7.3.4.3 Encouragement of Consumer Participation Through Incentivized Models

The importance of consumer participation in waste management to promote sustainability is apparent; blockchain enables incentivised models (Alarood et al., 2023). Blockchain, in a way, tokenises recycling work and incentivises people to practice proper disposal and recycling. These incentives, represented by tokens, may be converted to products, services, or cash, establishing a real and immediate motivation towards sustainable actions. For instance, people who return their plastic bottles at central pickup points can gain tokens that they can use to pay their bills or donate to some charities. This method includes the community and encourages shared responsibility towards the environment.

7.3.4.4 Support for Regulatory Compliance and Monitoring

Waste regulation is a key consideration in waste management, and with the blockchain's integrated transparency of waste activities, compounding invokes an efficient solution for regulatory compliance. Since they can obtain accurate accounts of waste produced, disposed of, and recycled in real time, they enhance efficiency in monitoring and often enforcement. Blockchain also helps ensure that waste processors meet certain environmental practices and have low compliance risks with related consequences. For example, corporations involved in electronic waste management can apply blockchain to meet extended producer responsibility (EPR) regulations to show how they deal with the relevant items disposed of or recycled.

7.3.5 Challenges and Limitations

Although blockchain technology has many benefits for waste management and recycling systems, its application has several challenges. These barriers must be overcome to fully realise blockchain as a tool for encouraging sustainable practices and developing an optimal waste disposal system.

7.3.5.1 Technological Barriers: Scalability, Energy Consumption, and Adoption

Scalability is one of the most essential challenges that holds the key to the future of blockchain technology (Kouhizadeh et al., 2021). This means the more transactions on the network per time, the more the system slows down and becomes inefficient. This issue can compound large-scale waste management, which requires real-time data processing. Thus, depending on the type of blockchain used, public blockchains, for example, can have congestion problems that make them unsuitable for highly active use.

A very important factor is energy consumption, including computational energy, during the functioning of blockchain networks, which apply proof-of-work (PoW) consensus algorithms. These energy-intensive processes make the sustainability objectives of blockchain, which seek to transform waste management systems, uncomfortable because of its higher environmental impact. However, other types of consensus mechanisms, such as proof of stake (PoS), are relatively energy efficient.

It also requires blockchain systems to be compatible with more established programs and procedures. This often involves substantial infrastructural development, which may present difficulties with a weak technological base.

7.3.5.2 Regulatory and Legal Issues in Blockchain Implementation

Since blockchain is a decentralised technology, implementing it faces some problems with compliance with the numerous rules and laws governing waste disposal. Most locations have no specific procedures for employing blockchain for waste management, recycling, or other commitments.

In addition, regional cross-border waste management may have extra legal complications because of differences in standards and compliance between countries. Blockchain is still in a legal grey area, especially in public and private investments; this uncertainty might slow down blockchain adoption and reduce stakeholders' willingness to invest in new technology (Karisma & Moslemzadeh Tehrani, 2023).

7.3.5.3 Resistance from Stakeholders and Lack of Technical Expertise

Another issue is the non-acceptance of change, which may be a primary problem when implementing blockchain technologies. Some value players in the waste management value chain, such as intermediaries, resist this because they consider blockchain to be a disruptor of their everyday operations.

Furthermore, blockchain technology is technical and might not readily apply to most practitioners and organisations engaging in waste management. Because of its complexity, there is bound to be slower adoption and practices in transition, and the causal result of implementing blockchain systems owing to this conceptual curve is less effective resource use during the transition period.

7.3.5.4 Initial Investment and Cost Considerations

The cost of integrating blockchain solutions may be relatively high because of purchasing equipment, software, and staff training (Habib et al., 2022). These costs might hinder implementing blockchain for waste management in countries, firms, SMEs, or local governments with low funding.

Therefore, it is important to note that blockchain carries the seed of reduced operational costs in the long run, although the benefits may not be directly traceable immediately. Investors may be reluctant to invest initially without expecting an inherent and tangible return on investment (ROI). Such economic instability can hinder blockchain integration, particularly in developing countries.

Future research efforts will help resolve technological challenges, develop correct legal requirements, facilitate stakeholder cooperation, and handle cost issues to harness blockchain's advantages. By addressing these challenges, blockchain elements can be transformed into the global foundation of sustainable waste management systems.

7.4 Conclusions

Blockchain technology offers revolutionary solutions to existing problems in waste management and recycling. By advocating transparency, accountability, and efficiency in what used to be complex systems, blockchain offers the potential to redesign the flow and stewardship of waste. It has emergently offered new solutions for implementing circular economies and minimising the harm of pollution and waste by decentralising operations and encouraging sustainability. The shifting application examples show that blockchain is a technology and an enabler for decoupling human societies from traditional thinking over sustainability.

As a result, research-practice deliberations and policy formulation and implementation require concerted orchestration to optimise this potential. Future studies must persistently examine blockchain for waste management by striving to overcome the current drawbacks and identify solutions best suited to various settings. Blockchain implementation requires the cooperation of practitioners who handle waste management and providers of technologies, such as AI and IoT, to improve it by incorporating it with blockchain systems. On the other hand, policymakers are responsible for implementing policies enabling companies to adopt blockchain technology without prejudicing irresponsible social practices.

Hence, it needs a vision and agreement to progress towards success. The three focus areas of collective innovation and investment are scalability, energy consumption, and stakeholder resistance. In addition, education and public involvement are important since they will help create awareness among the people needed to take measures and hence get involved in using the facilities to be developed across countries.

7.5 Future Directions and Opportunities

An exciting area for applying blockchain technology is its combination with waste management and recycling. By addressing the current problem areas and seizing present opportunities, blockchain can pave the way for the next generation of waste management and zero-waste societies.

7.5.1 Integration with Emerging Technologies: AI, IoT, and Big Data

Combining blockchain with other related technologies like AI, IoT, and big data potentially revamp the waste management sector. Smart bins and sensors are IoT-implemented devices that produce live-waste level, type, and bin location information. These data can then be stored on the blockchain, where its authenticity and correctness can be determined. When integrated with AI, this data can forecast the amount of waste generated, manageable collection routes, and areas under optimisation in the recycling process.

Big data analytics takes decision-making to another level by producing decision-support data on users' buying trends, resource consumption, and system effectiveness. For example, such technologies could allow flexible price structures for waste collection services on demand and in geographical areas, encouraging people to become more resource-efficient.

7.5.2 Potential for Creating Global Decentralized Waste Management Networks

This is because blockchain is a distributed database that can support waste management networks worldwide. Such networks could bring together stakeholders, such as waste generators, recyclers, manufacturers, and policymakers. Blockchain innovation can erode barriers to established waste management systems that have always worked in silos, isolating one another.

For instance, a global decentralised network can facilitate tracking recycled products and materials across countries and regions as part of the technology. A decentralised worldwide network may indicate compliance with global sustainable recycled products and materials standards. This approach might also benefit small-scale recyclers and informal workers in the waste management chain's productive and effective markets by providing remunerative compensation through blockchain-based payment systems.

7.5.3 *Role of Policymakers and Stakeholders in Driving Adoption*

Policymakers and stakeholders are the main actors in changing the status of blockchain for waste management. Governments can support the growth of this technology by carving out legal frameworks that benefit it while deterring misuse. Another way of catalysing adoption is by forming public-private partnerships through the synergy created on both sides.

For instance, governments can provide various organisations adapting to blockchain-based systems with incentives such as tax credits and/or grants to the organisations that have adapted to the blockchain-based waste management system. Simultaneously, all industry players, such as technology suppliers and recyclers, can reach a consensus on standardising relevant blockchain applications. Awareness generation and enhancement of technical knowledge can also be part of educational interventions so that future education will not meet the huddle of non-adoption and non-implementation observed in current times.

7.5.4 *Vision for Blockchain-Enabled Zero-Waste Societies*

Blockchain envisions a future where waste management societies are near zero waste and resources are fully cycled or closed-looped. Blockchain has the potential to offer the level of openness necessary for increasing circular economy patterns and for allowing the material to be traced and realised adequately in each stage of its usage.

In the future, customers can attach a barcode to the product to obtain all that information regarding its recyclability and the consequences of its production. It could make society more knowledgeable and able to make personal decisions, as well as be involved in the stewardship of resources, primarily natural resources. Further, such incentives on the blockchain could persuade communities to practice a culture of zero waste and leave a better world for future generations.

References

Alarood, A. A., Abubakar, A., Alzahrani, A., & Alsubaei, F. S. (2023). The electronic waste collection incentivisation scheme is based on the blockchain. *Sustainability, 15*(13), 10209.

Bandara, C., Rathnasiri, M. S. H., Dewasiri, N. J., Chathuranga, B., & Singh, N. (2024). Greening the economy for a sustainable future: Opportunities and perceptions. *Greening Our Economy for a Sustainable Future, 197-213*. https://doi.org/10.1016/B978-0-443-23603-7.00016-9

Bhubalan, K., Tamothran, A. M., Kee, S. H., Foong, S. Y., Lam, S. S., Ganeson, K., Vigneswari, S., Amirul, A. A., & Ramakrishna, S. (2022). Leveraging blockchain concepts as watermarks of plastics for sustainable waste management in progressing circular economy. *Environmental Research, 213*, e113631.

Bhushan, M., & Singh, R. (2024). Talent management facilitates net-zero transition through employee green behavior. In R. Singh, S. Khan, A. Kumar, S. Luthra, & H. Chokshi (Eds.), *Net zero economy, corporate social responsibility and sustainable value creation. CSR, sustainability, ethics & governance*. Springer. https://doi.org/10.1007/978-3-031-55779-8_7

Breukelman, H., Krikke, H., & Löhr, A. (2019). Failing services on urban waste management in developing countries: A review on symptoms, diagnoses, and interventions. *Sustainability, 11*(24), 6977.

Chathuranga, B., Rathnasiri, M. S. H., Dewasiri, N. J., Jayarathne, P., Bandara, C., & Bakari, M. A. (2024). Direct and indirect effect on industries and services (a). *Greening Our Economy for a Sustainable Future, 215-226*. https://doi.org/10.1016/B978-0-443-23603-7.00017-0

Dewasiri, N. J., Karunarathne, K. S. S. N., Menon, S., Jayarathne, P. G. S. A., & Rathnasiri, M. S. H. (2023). Fusion of artificial intelligence and Blockchain in the banking industry: Current application, adoption, and future challenges. In A. Saini & V. Garg (Eds.), *Transformation for sustainable business and management practices: Exploring the Spectrum of industry 5.0* (pp. 293–307). Emerald Publishing Limited. https://doi.org/10.1108/978-1-80262-277-520231021

França, A. S. L., Neto, J. A., Gonçalves, R. F., & Almeida, C. M. V. B. (2020). Proposing the use of blockchain to improve solid waste management in small municipalities. *Journal of Cleaner Production, 244*, 118529.

Gong, Y., Xie, S., Arunachalam, D., Duan, J., & Luo, J. (2022). Blockchain-based recycling and its impact on recycling performance: A network theory perspective. *Business Strategy and the Environment, 31*(8), 3717–3741.

Gui, L. (2020). Recycling infrastructure development under extended producer responsibility in developing economies. *Production and Operations Management, 29*(8), 1858–1877.

Gupta, Y. S., Mukherjee, S., Dutta, Y., & Bhattacharya, S. (2022). A blockchain-based approach using smart contracts to develop a smart waste management system. *International journal of Environmental Science and Technology, 19*, 7833–7856. https://doi.org/10.1007/s13762-021-03507-8

Habib, G., Sharma, S., Ibrahim, S., Ahmad, I., Qureshi, S., & Ishfaq, M. (2022). Blockchain technology: Benefits, challenges, applications, and integration of blockchain technology with cloud computing. *Future Internet, 14*(11), e341. https://doi.org/10.3390/fi14110341

Hannan, M. A., Al Mamun, M. A., Hussain, A., Basri, H., & Begum, R. A. (2015). A review on technologies and their usage in solid waste monitoring and management systems: Issues and challenges. *Waste Management, 43*, 509–523.

Iddagoda, A., Abeysinghe, R., Dissanayake, H., Dewasiri, N. J., & Wickramaratne, W. P. R. (2024). Cradle of research gaps: reflections on green orientation. In S. Gupta, S. Taneja, E. Özen, K. Sood, & S. Grima (Eds.), *Green management: A new paradigm in the world of business (emerald studies in finance, insurance, and risk management)* (pp. 339–357). Emerald Publishing Limited. https://doi.org/10.1108/978-1-83797-442-920241018

Jiang, P., Zhang, L., You, S., Van Fan, Y., Tan, R. R., Klemeš, J. J., & You, F. (2023). Blockchain technology applications in waste management: Overview, challenges and opportunities. *Journal of Cleaner Production, 421*, 138466.

Joshi, A., Singh, R., Kumar, V., & Rani, S. (2024). Net zero strategy: Exploring opportunities, confronting challenges for developing countries. In R. Singh & D. Crowther (Eds.), *Transition towards a sustainable future. Approaches to global sustainability, markets, and governance*. Springer. https://doi.org/10.1007/978-981-97-5756-5_11

Karisma, K., & Moslemzadeh Tehrani, P. (2023). Blockchain: Legal and regulatory issues. In *Sustainable oil and gas using Blockchain. Lecture notes in energy, vol 98*. Springer. https://doi.org/10.1007/978-3-031-30697-6_4

Karunarathna, K. S. S. N., Dewasiri, N. J., Singh, R., & Rathnasiri, M. S. H. (2024). What does artificial intelligence–powered ChatGPT bring to academia? A review. In R. Singh, S. Khan, A. Kumar, & V. Kumar (Eds.), *Artificial intelligence enabled management: An emerging economy perspective* (pp. 3–13). De Gruyter. https://doi.org/10.1515/9783111172408

Khan, S. A. R., Yu, Z., Sarwat, S., Godil, D. I., Amin, S., & Shujaat, S. (2022). The role of blockchain technology in circular economy practices to improve organisational performance. *International Journal of Logistics Research and Applications, 25*(4–5), 605–622.

Khan, S., Singh, R., Alnahas, J., Abbate, S., & Centobelli, P. (2024). Navigating the smart circular economy: A framework for manufacturing firms. *Journal of Cleaner Production, 480*, 144007. https://doi.org/10.1016/j.jclepro.2024.144007

Khetriwal, D. S., Kraeuchi, P., & Widmer, R. (2009). Producer responsibility for e-waste management: Key issues for consideration–learning from the Swiss experience. *Journal of Environmental Management, 90*(1), 153–165.

Kouhizadeh, M., Saberi, S., & Sarkis, J. (2021). Blockchain technology and the sustainable supply chain: Theoretically exploring adoption barriers. *International Journal of Production Economics, 231*, 107831.

Kumar, K., Singh, R., Sharma, A., & Lehri, B. P. S. (2024). Climate change adaptation strategies for achieving net-zero economy. In R. Singh & D. Crowther (Eds.), *Transition towards a sustainable future. Approaches to global sustainability, markets, and governance*. Springer. https://doi.org/10.1007/978-981-97-5756-5_8

Mishra, U., Rathnasiri, M. S. H., & Dewasiri, N. J. (2024). Net-zero economy and energy sustainability: A Nepalese perspective. In R. Singh, S. Khan, A. Kumar, S. Luthra, & H. Chokshi (Eds.), *Net zero economy, corporate social responsibility and sustainable value creation. CSR, sustainability, ethics & governance* (pp. 145–161). Springer. https://doi.org/10.1007/978-3-031-55779-8_9

Nwokolo, S. C., Singh, R., Khan, S., Kumar, A., & Luthra, S. (2023a). Remedies to the challenges of renewable energy deployment in Africa. In *Africa's path to net-zero. CSR, sustainability, ethics & governance*. Springer. https://doi.org/10.1007/978-3-031-44514-9_3

Nwokolo, S. C., Singh, R., Khan, S., Kumar, A., & Luthra, S. (2023b). Africa's awakening to climate action. In *Africa's path to net-zero. CSR, sustainability, ethics & governance*. Springer. https://doi.org/10.1007/978-3-031-44514-9_9

Nwokolo, S. C., Singh, R., Khan, S., Kumar, A., & Luthra, S. (2023c). Impacts of climate change in Africa. In *Africa's path to net-zero. CSR, sustainability, ethics & governance*. Springer. https://doi.org/10.1007/978-3-031-44514-9_7

Rathnasiri, M. S. H., & Gunathilaka, D. H. P. M. (2024). Net zero policies for sustainable future: Issues and challenges in the Asia Pacific Region. In R. Singh & D. Crowther (Eds.), *Transition towards a sustainable future. Approaches to global sustainability, markets, and governance*. Springer. https://doi.org/10.1007/978-981-97-5756-5_12

Rathnasiri, M. S. H., Dewasiri, N. J., & Kumar, A. A. (2024). Policy framework and implementation strategies for Sri Lanka's transition to a net-zero economy. In R. Singh, S. Khan, A. Kumar, S. Luthra, & H. Chokshi (Eds.), *Net zero economy, corporate social responsibility and sustainable value creation. CSR, sustainability, ethics & governance* (pp. 43–60). Springer. https://doi.org/10.1007/978-3-031-55779-8_3

Scott, I. J., de Castro Neto, M., & Pinheiro, F. L. (2023). Bringing trust and transparency to the opaque world of waste management with blockchain: A Polkadot parathread application. *Computers & Industrial Engineering, 182*, 109347.

Singh, A. K., & Kumar, V. P. (2024). Integrating blockchain technology success factors in the supply chain of circular economy-driven construction materials: An environmentally sustainable paradigm. *Journal of Cleaner Production, 460*, 142577.

Singh, R., Gupta, A., & Kandpal, V. (2024). ESG excellence in the digital world. In D. R. Ernesto, S. Gonzalez, V. Kandpal, P. K. Ozili, & P. Chatterjee (Eds.), *Fintech for ESG and the circular economy* (pp. 93–204). Scrivener Publishing LLC. https://doi.org/10.1002/9781394234035.ch10

Vijerathne, D. T., Wahala, W., Rathnasiri, M. S. H., De Silva, V., & Dewasiri, N. J. (2024). Literature review on sustainability. *Greening Our Economy for a Sustainable Future, 33-45*. https://doi.org/10.1016/B978-0-443-23603-7.00004-2

Ying, N. S., & Mok, P. Y. (2023). A review of Blockchain technology for sustainable fashion. *Digital Fashion Innovations*, 179–204.

Zorpas, A. A. (2020). Strategy development in the framework of waste management. *Science of the Total Environment, 716*, 137088.

Chapter 8
Blockchain Technology and the Circular Economy: Applications and Opportunities in Supply Chain Management

Md Ishtiak Uddin, Abhiraj Malia, Biswajit Das, and Bhubaneswari Bisoyi

Abstract This study provides an in-depth look at blockchain technology's applications, opportunities, and latent future developments in supply chain management (SCM). In the context of the dynamic global supply chain environment, the incorporation of blockchain technology stands out as a significant advancement, presenting profound possibilities for improving transparency and security. The key aim is to conduct an in-depth assessment of the current research on blockchain in supply chain management operations, with a particular emphasis on its applications in terms of transparency and future opportunities. The paper comprises four sections: an overview of blockchain, its application in supply chain operations, the transparency of information in the supply chain, the integration of blockchain in various sectors, and the future prospects of blockchain in supply chain management. The research uncovered herein elucidates the key aspects of blockchain technology as they pertain to supply chain networks. The study focuses on its prospects, potential repercussions, recent technological advances, and major trends. Additionally, it examines various industries such as real estate, healthcare, banking, education, government, and public organizations, which blockchain-based technology can effectively transform by enhancing visibility and simplifying business processes.

Keywords Blockchain · Supply Chain Management · Transparency · Applications · Opportunities

M. I. Uddin · B. Das
KIIT School of Management, KIIT University, Bhubaneswar, India
e-mail: biswajit@ksom.ac.in

A. Malia (✉)
School of Tribal Resource Management, KISS University, Bhubaneswar, India

B. Bisoyi
KIIT School of Law, KIIT University, Bhubaneswar, India

R. Singh et al. (eds.), *Blockchain Technologies for Smart Circular Economy and Organisational Sustainability*, Environmental Footprints and Eco-design of Products and Processes, https://doi.org/10.1007/978-3-031-89183-0_8

8.1 Introduction

The accelerated advancement of technology gives a potential chance for supply chains to take advantage of that possibility (Sunmola & Burgess, 2023). Adapting supply chains to new technology is crucial in today's world, driven by globalization and other external forces (Nurmilaakso, 2008). Industry 4.0, the fourth industrial revolution, incorporates numerous technological advances in the digital realm. When horizontal and vertical integration, automation, collaboration, and data exchange permeate an industry's technology and operations, we say that we've entered Industry 4.0 (Alcácer & Cruz-Machado, 2019). The fundamental idea behind Industry 4.0 is that businesses may drastically increase their efficiency by automating routine tasks, moving towards autonomous processes powered by AI and learning, and decreasing their need for human labor (Sunmola & Burgess, 2023). Most importantly, innovations in Industry 4.0 technology are changing supply networks around the world. Technologies such as sophisticated human-machine interfaces, virtual and augmented reality, machine learning (ML), the global web and blockchain are examples of these advancements (Rajput & Singh, 2019). In a big way, blockchain technology is influencing the fourth industrial revolution.

Blockchain is an innovative technology that protects all data's privacy, information and resources. This is a very transparent and decentralized ledger that facilitates a repository of all data and transactions underpinned by encryption within a distributed network of peers (Hughes et al., 2019). Blockchain technology provides a high level of security for business transactions because, with the approval of all or almost all parties involved, it is impossible to modify records without also changing the records that came before them (Risius & Spohrer, 2017; Dutta et al., 2020). It has endless potential applications in a variety of fields, including medical care record sharing and the development of smart contracts to identify financial fraud (Dutta et al., 2020). Numerous studies (Ghobakhloo et al., 2024) show that companies of all sizes are eager to implement blockchain technology in order to stay ahead of the digitization race and get an advantage in the highly saturated and volatile market. This technology is anticipated to provide positive aspects across various industries, considering all its features. Blockchain technology is predicted to have a big impact on logistics networks, with information exchange and storage being a crucial component (Song et al., 2019). Despite the plethora of successful transformations within supply chains, challenges persist regarding information transparency, cost, and privacy (Cole et al., 2019). Blockchain technology has made it possible to improve every aspect of supply chain management, from procuring the essential ingredients in order to provide products to final consumers (Agrawal et al., 2021; Pazaitis et al., 2017; Kumari et al., 2024). Additionally, it makes it possible to reengineer supply chain operations by building a framework for business process reengineering (BPR) that is based on blockchain technology (Tripathi & Gupta, 2020; Chang et al., 2019). This blockchain framework, designed in a manner that guarantees the security and information transparency of supply chains (Wang et al., 2019), facilitates a number of advantages that are projected to be provided by blockchain technology to supply chain networks. These advantages include the enhancement of efficiency and the reduction of costs, which are the major goals of supply chains. Supply chain systems built on blockchain technology are the subject of growing research (Nandi et al., 2020).

In regards to the supply chain that uses blockchain technology, the real-time tracking of the locations of items is enabled. Blockchain technology allows relevant parties to verify all transaction information instantly. As a result, a supply chain that utilizes blockchain technology enhances transparency significantly (Saberi et al., 2019; Pradhan et al., 2023). It is well known that blockchain technology reduces the risk of fake goods in countries (Park & Li, 2021). This arises from the nature of blockchain as a decentralized record-keeping system (Park & Li, 2021). Blockchain technology is capable of efficiently managing inventories, as well as accumulating and supervising data and information history. Additionally, blockchain technology streamlines the contract process inside supply chains by eliminating unnecessary and complicated paperwork via the use of smart contracts (Lin et al., 2022). Moreover, supply chains empowered by blockchain technology enhance inventory management efficiency and play a substantial role in reducing costs (Wong et al., 2020). Therefore, blockchain is especially pertinent to supply chain management as one of the most intriguing avenues for the use of disruptive digital technology; however, it is necessary to do further research to determine the specific ways in which blockchain technology affects supply chain transparency since there is a dearth of knowledge about the technology and few real-world instances.

8.1.1 Blockchain and the Circular Economy: A Connection

The renewable and healing idea of the Circular Economy is becoming a reality due to increased digitalization. Industry 4.0 technologies, such as artificial intelligence, cloud computing data analysis, the Internet of Things (IoT), and 3D printing, are specifically among them. The introduction of blockchain technology has made it possible to deploy the Internet of Things, 3D and additive production, food production, medical care, and other tamper-proof and environmentally friendly processes in a secure, distributed, and integrated manner (Upadhyay et al., 2021). Blockchain technology has the potential to create and manage innovative circular economy networks. The components of sharing, optimization, virtualization, and exchange, in particular, can be efficiently accommodated by blockchain platforms in connection to the Resolve architecture. Consequently, a wealth of regenerative digital opportunities for the circular economy's sustainability objective became available (Tripathi & Gupta, 2020).

Blockchain can be viewed as a form of social technology in the context of the circular economy. Connecting and coordinating several dispersed databases that can all be updated at once and made available to everyone is beneficial. In addition, the circular economy can embrace a decentralized concept of value production and circulation rather than creating value and value appropriation by fostering concurrent rivalry and cooperation within the loop (Nandi et al., 2020). With its decentralization, distributiveness, and tamper-resistance principles, as well as its smart contract and tokenization capabilities, this fundamental aspect of blockchain technology can help achieve the aforementioned circular economy goals (Lohmer et al., 2020).

Therefore, blockchain technology can make a substantial contribution by offering the foundational technology that connects intricate networks and databases with pre-built designs and configurations that allow all linked databases to be updated simultaneously, irrevocably, and to enable automation when necessary. For such innovation to be successful, several businesses within the chain or network must adopt the new technology and change their configurations.

8.1.2 Untapped Dimensions of Blockchain Technology in Supply Chain Management

The study of blockchain technology within the realm of supply chain management (SCM) has uncovered notable research gaps, especially in areas that remain unexamined and in potential applications yet to be explored. Research by Centobelli et al. (2022) has brought attention to the importance of studying how blockchain technology helps public and private companies in developing nations manage trust, traceability, and transparency. This study will highlight both the similarities and differences between developing countries and developed countries, as well as the research advancements that have been made in this area (Singh et al., 2023).

To better comprehend the concepts behind blockchain-based supply chain architecture, Sunmola and Burgess (2023) have made a valuable contribution to transparency in these systems, which will aid in future study and development. Their research is on studying the effects of using the design-by-transparency paradigm in an empirical setting. Furthermore, the ideas and dimensions of framework should be validated by supply chain stakeholders and experts. In another way, Dutta et al. (2020) stressed that a robust benchmark for the adoption framework and reengineering system should be established to make the most of the opportunities given by blockchain.

It is imperative to conduct empirical research that analyses the organizational and technological aspects that are indispensable for the successful application of blockchain technology in the supply chain network (Khan et al., 2023). Queiroz et al. (2020) gave several instances of how technology in blockchain may be integrated with supply chain management (SCM), highlighting the need to rethink business strategies in order to embrace blockchain technology.

Therefore, this research is essential in the development of the future of blockchain applications in supply chain processes, as it guarantees that the technology's greatest upside is realized in the improvement of transparency and incorporation within the supply chain ecosystem.

8.2 Scope of the Study

This research intends to include all of the publications that have been published, articles that are currently being published, review papers, and surveys that are related to the Applications and deployment of blockchain-based technologies in logistics and supply chains. It sheds light on the present trends of blockchain technology in SC operations and paves the way for further research.

8.3 An Overview of Blockchain Technology

Since its introduction, blockchain technology—which served as the basis for cryptocurrencies like Bitcoin—has developed beyond its original intent to become a game-changing force in a number of sectors. The decentralized and transparent characteristics of blockchain have garnered significant attention due to its inherent ability to address various challenges across multiple sectors (Kumari et al., 2024).

Blockchain represents a sophisticated distributed ledger technology characterized by its immutability and resistance to tampering. It operates in a synchronized and shared environment where all transactions are validated by users and kept traceable. It enables a system characterized by decentralization, allowing participants within the network to engage securely without dependence on a central authority for trust (Centobelli et al., 2022).

Blockchain is composed of a series of interconnected nodes that facilitate the unambiguous monitoring of transaction histories, thereby increasing the technology's dependability and transparency (Dutta et al., 2020). In order to safeguard the safety of transactions, each block is uniquely identified and contains the hash of the preceding block (Tönnissen & Teuteberg, 2020). The network's users document transactions in chronological order. Blockchain technology's authentication procedure controls the addition of new blocks to the system and ensures that their records are linked using encode values (Kumari et al., 2024). In order to protect the integrity of historical data through an authenticating process, an adversary must surmount the entire network in order to generate an extended chain in the event that they attempt to modify an existing block (Beck et al., 2017).

Applications that use blockchain technology are moving from being developed as experimental endeavors to being used on a larger scale in the real world. Grand View Research (GVR) projects that the worldwide market for blockchain technology will expand from $10,015.2 million in 2022 to $1,431,535.5 million in 2030, a compound annual growth rate (CAGR) of 85.9% from 2023 to 2030. Blockchain technology presents significant possibilities for the implementation of digital ledger systems, which holds particular relevance in supply chain networks (Cole et al., 2019; Xu et al., 2021; Kumari et al., 2024) (Fig. 8.1 shows the significant increase starting from 2025, with a sharp rise in 2029 and 2030 and reaching over 1,750,000 US$ by 2030).

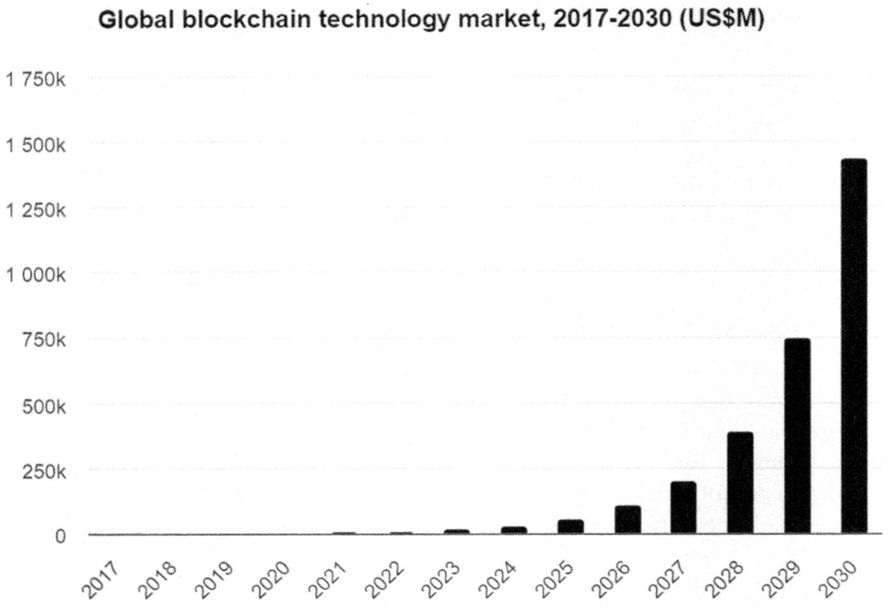

Fig. 8.1 Global blockchain technology market. (**Source:** Own Source, 2024)

8.3.1 Blockchain Technology in Supply Chain Governance

Supply chain management (SCM) encompasses a complex array of stakeholders, processes, and the flow of goods, services, and information. The rise of globalization has intensified these networks, demanding robust solutions for transparency, efficiency, and reliability. The ability of blockchain technology to streamline operations and save expenses brings a huge boon to supply chain network systems. Kamble et al. (2019) highlight that those smart contracts, which are self-executing contracts containing coded terms, have the potential to automate a variety of supply chain procedures. The potential for this technology to lessen the need for middlemen might speed up transactions and save expenses.

Similarly, Francisco and Swanson (2018) investigate how blockchain technology may simplify supply chain traceability. Traditional supply chains, as pointed out by the authors, are often reliant on clumsy and time-consuming paper paperwork. Digitizing these papers on the blockchain may help companies save a lot of money since it speeds up processing and makes it more accurate.

In addition, blockchain technology could significantly contribute to certifying the authenticity of products, particularly in sectors where counterfeiting poses substantial challenges. The study by Casino et al. (2019) illustrates that with the help of blockchain technology, secure records that trace a product's path across supply chain networks can be efficiently created. The processes engaged in the supply

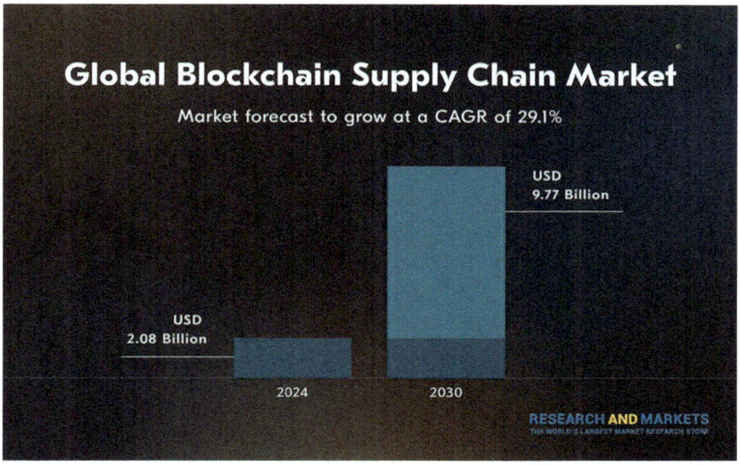

Fig. 8.2 Global blockchain supply chain market. (**Source:** Research and Markets, Acropolium)

chain are made more efficient overall by this integration, which also increases the efficacy of security measures.

According to Kouhizadeh et al. (2020), blockchain technology might additionally assist the climate and people by making it easier to see how actions in the supply chain affect those areas. Customers are able to make more informed selections and businesses are motivated to adopt more sustainable practices because of this transparency and trust. Therefore, blockchain technology is more accurately described as a confidence machine than a trustless technology (De Filippi et al., 2020). This is due to the fact that blockchain technology tackles trust problems by offering a method that decreases the need for trust in transactions (Cole et al., 2019). Consequently, within the realm of supply chain management on a worldwide scale, blockchain technology is being increasingly used. The below Fig 8.2 represents the blockchain supply chain market, currently valued at $2.08 billion in 2024, is projected to experience a compound annual growth rate of 29.14%, ultimately reaching $9.77 billion by the year 2030 (Research and Markets, Acropolium) (Fig. 8.2).

8.3.2 *Blockchain Technology and Information Transparency in Supply Chain*

Blockchain facilitates the sharing of data across a decentralized network, rendering it exceedingly challenging to modify or eliminate any type of recorded information. This inherent characteristic of blockchain renders it especially well-suited for applications that necessitate high levels of trust and security, such as supply chain management (Cole et al., 2019).

The manufacturing and distribution phases of supply chains are sometimes quite complex, involving many different parties located in different parts of the world. Because of its intrinsic complexity, attempts to assure ethical material sourcing, regulatory compliance, and product authenticity may be more difficult to achieve (Kouhizadeh & Sarkis, 2018). These problems may also relate to visibility and responsibility. According to Abeyratne and Monfared (2016), traditional supply chain data management methods are prone to fraud, inefficiencies, and inaccuracies as they often rely on manual processes and centralized databases (Herrgos et al., 2020). Consequently, there is a great need for solutions that increase supply chain transparency and accountability (Wang et al., 2019).

Its ability to provide complete traceability is one of blockchain's main advantages for enhancing supply chain transparency (Qian & Papadonikolaki, 2020). According to Casino et al. (2019), supply chain visibility is enhanced when every transaction is recorded on the blockchain, which makes it possible to monitor the origin, movement, and ultimate destination of items (Lohmer et al., 2020).

The pharmaceutical sector has also seen blockchain's promise in preventing the sale of fake medications by recording and verifying each step of the supply chain (Mettler, 2016). In addition to safeguarding customers' and pharmaceutical businesses' reputations, this degree of traceability may aid in the dissemination of counterfeit or inferior drugs (Uddin, 2021).

Blockchain technology has the capacity to significantly improve supply chains; however, its implementation is fraught with numerous potential impediments. Since blockchain technology is a relatively new technology but may not be able to handle the huge transaction volumes characteristic of global supply chains, scalability is one of the main areas of worry (Khan et al., 2021). Additionally, supply chain stakeholders must interact and communicate information in a fundamentally different manner when incorporating blockchain, which calls for a significant infrastructure investment (Rejeb et al., 2021). Furthermore, there are persistent worries about the legal and regulatory ramifications of blockchain adoption, particularly in sectors that are subject to stringent standards (Dehghani et al., 2022).

Blockchain technology offers significant opportunities to enrich accountability in supply chains, while offering a secure and decentralized framework for transaction documentation and verification, therefore alleviating the challenges inherent in traditional supply chain processes (Khan et al., 2024).

8.3.3 Blockchain Technology in Different Sectors

Blockchain technology has the potential to break through research barriers, open up new avenues for study, and propel research forward across a range of industrial domains. Additionally, blockchain powers innovation in new businesses, a slew of new apps, and the elimination of middlemen—all of which would contribute to blockchain's ability to promote innovation across sectors (Hughes et al., 2019). This literature looks at how blockchain technology is being used in several fields and the revolutionary potential it offers to each.

8.3.3.1 Blockchain in Finance

Blockchain technology, according to Tapscott and Tapscott (2016), might do away with middlemen entirely in financial transactions by enabling direct payments between users. This capacity is most noticeable in international money transfers, where blockchain technology drastically cuts down on fees and processing times (Guo & Liang, 2016). Also, smart contracts powered by the blockchain may automate the enforcement of contractual agreements, cutting down on the need for arbitrators (Peters & Panayi, 2016).

8.3.3.2 Blockchain in Healthcare

Blockchain may guarantee that medical records are safely transferred across approved healthcare professionals while protecting patient anonymity by offering a decentralized and unchangeable ledger (Mettler, 2016). This may result in less expensive healthcare, better patient outcomes, and more precise diagnosis.

8.3.3.3 Blockchain in Real Estate

According to Zhang et al. (2024), blockchain technology may help streamline and protect the transfer of property titles, cutting out middlemen like brokers. Additionally, smart contracts may simplify a number of procedures, which in turn reduces costs and delays (Bennett et al., 2021). These procedures include, but are not limited to, land administration and rental agreements (Saari et al., 2022). A further benefit of blockchain technology is that it creates a clear record of property ownership, reducing the possibility of dispute (Singh et al., 2024).

8.3.3.4 Blockchain in Government and Public Services

Blockchain technology is attracting a lot of interest from public sector organizations and governments because of its potential to increase trust, decrease corruption, and improve the quality of services provided (Cagigas et al., 2021). A study conducted by Ølnes et al. (2017) suggests that blockchain technology has the potential to be used in several domains, including identity verification and public records administration. One example is voting systems that use blockchain technology (Berryhill et al., 2018). These systems are tamper-proof, transparent, and safe, which means that voters may feel more confident in the process and the results (Yavuz et al., 2018).

8.3.3.5 Blockchain in Education

While blockchain technology is still in its development stage as a means of education, it shows great potential for streamlining the authentication and administration of credentials (Bhaskar et al., 2021). Blockchain may provide a safe and reliable record of academic accomplishments (Bhaskar et al., 2021; Chen et al., 2018). This lowers the risk of misconduct and makes it simpler for institutions and companies to validate credentials.

Blockchain technology has exhibited substantial potential in a variety of sectors, providing solutions to long-standing challenges related to transparency and security. In spite of the fact that the technology has a variety of applications, it continues to face widespread difficulties, including adaptability and regulatory compliance. To realize its full potential, blockchain will need constant study and cooperation between industry players and legislators as it develops.

8.4 Future Opportunities of Blockchain Technology in Supply Chain Management

The transformational potential of blockchain technology in supply chain management (SCM) is becoming more widely acknowledged. Even while its present uses are focused on enhancing traceability, there is even more potential for revolutionizing supply chains in the future (Singh et al., 2022).

8.4.1 Blockchain for Enhanced Sustainability

In supply chain management, blockchain's capacity to propel sustainability efforts is among its most attractive possibilities. Companies are looking for strategies to reduce their carbon impact and assure the ethical procurement of resources as global supply chains become more environmentally sensitive (Parmentola et al., 2022). The upside of blockchain technology lies in its ability to offer complete transparency, allowing businesses to track the ecological footprint of their products (Saberi et al., 2019; Kumari et al., 2024). In order to achieve their sustainability objectives, organizations may benefit from this openness by lowering the possibility of environmental propaganda and increasing compliance with environmental standards (Varriale et al., 2020).

8.4.2 Integration with Emerging Technologies

Blockchain's ability to integrate with cutting-edge technologies like IoT and AI presents SCM with yet another significant possibility. With this connection, the supply chain can be more data-driven and responsive, enabling better decision-making with real-time access (Zelbst et al., 2019). After this, blockchain-verified data may be analyzed by AI algorithms to improve inventory management, forecast demand, and plan routes (Saberi et al., 2019).

8.4.3 Automation through Smart Contracts

Smart contracts, made possible by blockchain technology, provide yet another chance to revolutionize supply chain management. According to Peters and Panayi (2016), smart contracts are agreements that are encoded into code and include provisions that, if certain criteria are satisfied, automatically enforce compliance. Using smart contracts, supply chains may streamline payment processing, order fulfilment, and dispute resolution, cutting out the middlemen and saving time (Casino et al., 2019). As a result of this automation, administrative burdens may be greatly reduced while accuracy and efficiency are greatly improved.

8.4.4 Enabling Resilient and Agile Supply Chains

Supply chains may be made more robust with the use of blockchain technology, which allows for speedier decision-making in the face of interruptions and real-time insight into supply chain operations (Kshetri, 2021; Wang et al., 2019). Chang et al. (2019) assert that blockchain's capacity to furnish precise, real-time information regarding inventory levels and supplier performance can facilitate the rapid response of organizations to unexpected events, including natural disasters.

8.4.5 Facilitating Circular Economies

Blockchain's ability to trace the life cycle of resources and goods may facilitate the shift to circular economies. According to Perboli et al. (2018), a circular economy reduces waste and its effect on the environment by designing items for reuse or recycling. It is possible that having the ability of blockchain technology to offer a transparent route that a product has taken could foster the recycling and reuse of materials. This would be accomplished by ensuring that all necessary parties or stakeholders have access to data about the product's provenance (Kumari et al.,

2024; Khan et al., 2022). This may support sustainable practices, reduce waste, and maximize resource utilization for manufacturers.

As businesses endeavor to establish resilient and transparent supply chains, blockchain will emerge as an indispensable instrument for confronting such challenges. But in order to make use of these possibilities, current impediments pertaining to scalability and regulatory compliance must be addressed.

8.5 Conclusion

Blockchain technology is becoming well-known for its ability to transform local and international supply chains by increasing transparency and operational efficiency. Corporations and governments across numerous sectors are granted a strategic advantage by the emergence of blockchain. Challenges such as standardization and the integration of multiple blockchain platforms are being addressed as the technology continues to expand into new domains. By incorporating recent advancements in blockchain, investigating its applications in a variety of industries, and assessing relevant studies, this research contributes to the field, providing valuable insights for academicians, scientists, researchers and professionals. In conclusion, this research suggests that businesses operating in economies of the third world should prioritize the implementation of blockchain technology applications in order to efficiently oversee trust and transparency.

References

Abeyratne, S. A., & Monfared, R. P. (2016). Blockchain ready manufacturing supply chain using distributed ledger. *International Journal of Research in Engineering and Technology, 5*(9), 1–10.

Agrawal, T. K., Kumar, V., Pal, R., Wang, L., & Chen, Y. (2021). Blockchain-based framework for supply chain traceability: A case example of textile and clothing industry. *Computers & Industrial Engineering, 154*, 107130. https://doi.org/10.1016/j.cie.2021.107130

Alcácer, V., & Cruz-Machado, V. (2019). Scanning the industry 4.0: A literature review on technologies for manufacturing systems. *Engineering Science and Technology, an International Journal, 22*(3), 899–919. https://doi.org/10.1016/j.jestch.2019.01.006

Beck, R., Avital, M., Rossi, M., & Thatcher, J. B. (2017). Blockchain technology in business and information systems research. *Business & Information Systems Engineering, 59*(6), 381–384. https://doi.org/10.1007/s12599-017-0505-1

Bennett, R., Miller, T., Pickering, M., & Kara, A.-K. (2021). Hybrid approaches for smart contracts in land administration: Lessons from three Blockchain proofs-of-concept. *Land, 10*(2), 220. https://doi.org/10.3390/land10020220

Berryhill, J., Bourgery, T., & Hanson, A. (2018). Blockchains unchained: Blockchain technology and its use in the public sector, *OECD Working Papers on Public Governance*, No. 28, OECD Publishing, Paris, https://doi.org/10.1787/3c32c429-en.

Bhaskar, P., Tiwari, C. K., & Joshi, A. (2021). Blockchain in education management: Present and future applications. *Interactive Technology and Smart Education, 18*(1), 1–17. https://doi.org/10.1108/ITSE-07-2020-0102

Cagigas, D., Clifton, J., Diaz-Fuentes, D., & Fernandez-Gutierrez, M. (2021). Blockchain for public services: A systematic literature review. *IEEE Access, 9*, 13904–13921. https://doi.org/10.1109/ACCESS.2021.3052019

Casino, F., Dasaklis, T. K., & Patsakis, C. (2019). A systematic literature review of blockchain-based applications: Current status, classification and open issues. *Telematics and Informatics, 36*, 55–81. https://doi.org/10.1016/j.tele.2018.11.006

Centobelli, P., Cerchione, R., Vecchio, P. D., Oropallo, E., & Secundo, G. (2022). Blockchain technology for bridging trust, traceability and transparency in circular supply chain. *Information & Management, 59*(7), 103508. https://doi.org/10.1016/j.im.2021.103508

Chang, S. E., Chen, Y.-C., & Lu, M.-F. (2019). Supply chain re-engineering using blockchain technology: A case of smart contract-based tracking process. *Technological Forecasting and Social Change, 144*, 1–11. https://doi.org/10.1016/j.techfore.2019.03.015

Chen, G., Xu, B., Lu, M., & Chen, N.-S. (2018). Exploring blockchain technology and its potential applications for education. *Smart Learning Environments, 5*(1), 1. https://doi.org/10.1186/s40561-017-0050-x

Cole, R., Stevenson, M., & Aitken, J. (2019). Blockchain technology: Implications for operations and supply chain management. *Supply Chain Management: An International Journal, 24*(4), 469–483. https://doi.org/10.1108/SCM-09-2018-0309

De Filippi, P., Mannan, M., & Reijers, W. (2020). Blockchain as a confidence machine: The problem of trust & challenges of governance. *Technology in Society, 62*, 101284.

Dehghani, M., Popova, A., & Gheitanchi, S. (2022). Factors impacting digital transformations of the food industry by adoption of blockchain technology. *Journal of Business & Industrial Marketing, 37*(9), 1818–1834. https://doi.org/10.1108/JBIM-12-2020-0540

Dutta, P., Choi, T.-M., Somani, S., & Butala, R. (2020). Blockchain technology in supply chain operations: Applications, challenges and research opportunities. *Transportation Research Part E: Logistics and Transportation Review, 142*, 102067. https://doi.org/10.1016/j.tre.2020.102067

Francisco, K., & Swanson, D. (2018). The supply chain has no clothes: Technology adoption of blockchain for supply chain transparency. *Logistics, 2*(1), 2.

Ghobakhloo, M., Iranmanesh, M., Mubarik, M. S., Mubarak, M. F., Amran, A., & Khanfar, A. A. A. (2024). Blockchain technology as an enabler for sustainable business ecosystems: A comprehensive roadmap for socioenvironmental and economic sustainability. *Business Strategy & Development, 7*(1), e319. https://doi.org/10.1002/bsd2.319

Guo, Y., & Liang, C. (2016). Blockchain application and outlook in the banking industry. *Financial Innovation, 2*(1), 24. https://doi.org/10.1186/s40854-016-0034-9

Herrgos, L., Lohmer, J., Schneider, G., & Lasch, R. (2020). Development and evaluation of a blockchain concept for production planning and control in the semiconductor industry. *2020 IEEE International Conference on Industrial Engineering and Engineering Management (IEEM)*, 440–444. https://doi.org/10.1109/IEEM45057.2020.9309979.

Hughes, L., Dwivedi, Y. K., Misra, S. K., Rana, N. P., Raghavan, V., & Akella, V. (2019). Blockchain research, practice and policy: Applications, benefits, limitations, emerging research themes and research agenda. *International Journal of Information Management, 49*, 114–129. https://doi.org/10.1016/j.ijinfomgt.2019.02.005

Kamble, S., Gunasekaran, A., & Arha, H. (2019). Understanding the Blockchain technology adoption in supply chains-Indian context. *International Journal of Production Research, 57*(7), 2009–2033. https://doi.org/10.1080/00207543.2018.1518610

Khan, D., Jung, L. T., & Hashmani, M. A. (2021). Systematic literature review of challenges in Blockchain scalability. *Applied Sciences, 11*(20), 9372. https://doi.org/10.3390/app11209372

Khan, S., Ali, S. S., & Singh, R. (2022). Determinants of remanufacturing adoption for circular economy: A causal relationship evaluation framework. *Applied System Innovation, 5*(4), 62.

Khan, S., Khan, M. I., & Singh, R. (2023). Modeling the barriers of Blockchain technology implementation in supply chain. *Journal of Industrial Integration and Management, 8*(03), 367–385.

Khan, S., Singh, R., Alnahas, J., Abbate, S., & Centobelli, P. (2024). Navigating the smart circular economy: A framework for manufacturing firms. *Journal of Cleaner Production, 480*, 144007.

Kouhizadeh, M., & Sarkis, J. (2018). Blockchain practices, potentials, and perspectives in greening supply chains. *Sustainability, 10*(10), 3652. https://doi.org/10.3390/su10103652

Kouhizadeh, M., Saberi, S., & Sarkis, J. (2020). Blockchain technology and the sustainable supply chain: Theoretically exploring adoption barriers. *International Journal of Production Economics, 231*, 107831. https://doi.org/10.1016/j.ijpe.2020.107831

Kshetri, N. (2021). Blockchain and sustainable supply chain management in developing countries. *International Journal of Information Management, 60*, 102376. https://doi.org/10.1016/j.ijinfomgt.2021.102376

Kumari, A., Kumar, P., & Golyan, A. (2024). Introduction of Blockchain for biomedical and healthcare system. In P. Kumar & A. Kumari (Eds.), *Blockchain for biomedical research and healthcare* (pp. 1–16). Springer Nature Singapore. https://doi.org/10.1007/978-981-97-4268-4_1

Lin, S.-Y., Zhang, L., Li, J., Ji, L., & Sun, Y. (2022). A survey of application research based on blockchain smart contract. *Wireless Networks, 28*(2), 635–690. https://doi.org/10.1007/s11276-021-02874-x

Lohmer, J., Bugert, N., & Lasch, R. (2020). Analysis of resilience strategies and ripple effect in blockchain-coordinated supply chains: An agent-based simulation study. *International Journal of Production Economics, 228*, 107882. https://doi.org/10.1016/j.ijpe.2020.107882

Mettler, M. (2016). Blockchain technology in healthcare: The revolution starts here. *2016 IEEE 18th International Conference on E-Health Networking, Applications and Services (Healthcom)*, 1–3. https://doi.org/10.1109/HealthCom.2016.7749510.

Nandi, S., Hervani, A., & Helms, M. M. (2020). Circular economy business models—Supply chain perspectives. *IEEE Engineering Management Review, 48*(2), 193–201. https://doi.org/10.1109/EMR.2020.2991388

Nurmilaakso, J.-M. (2008). Adoption of e-business functions and migration from EDI-based to XML-based e-business frameworks in supply chain integration. *International Journal of Production Economics, 113*(2), 721–733. https://doi.org/10.1016/j.ijpe.2007.11.001

Ølnes, S., Ubacht, J., & Janssen, M. (2017). Blockchain in government: Benefits and implications of distributed ledger technology for information sharing. *Government Information Quarterly, 34*(3), 355–364. https://doi.org/10.1016/j.giq.2017.09.007

Park, A., & Li, H. (2021). The effect of Blockchain technology on supply chain sustainability performances. *Sustainability, 13*(4), 1726. https://doi.org/10.3390/su13041726

Parmentola, A., Petrillo, A., Tutore, I., & De Felice, F. (2022). Is blockchain able to enhance environmental sustainability? A systematic review and research agenda from the perspective of sustainable development goals (SDGs). *Business Strategy and the Environment, 31*(1), 194–217. https://doi.org/10.1002/bse.2882

Pazaitis, A., De Filippi, P., & Kostakis, V. (2017). Blockchain and value systems in the sharing economy: The illustrative case of Backfeed. *Technological Forecasting and Social Change, 125*, 105–115. https://doi.org/10.1016/j.techfore.2017.05.025

Perboli, G., Musso, S., & Rosano, M. (2018). Blockchain in logistics and supply chain: A lean approach for designing real-world use cases. *IEEE Access, 6*, 62018–62028. https://doi.org/10.1109/ACCESS.2018.2875782

Peters, G. W., & Panayi, E. (2016). Understanding modern banking ledgers through Blockchain technologies: Future of transaction processing and smart contracts on the internet of money. In P. Tasca, T. Aste, L. Pelizzon, & N. Perony (Eds.), *Banking beyond banks and money* (pp. 239–278). Springer International Publishing. https://doi.org/10.1007/978-3-319-42448-4_13

Pradhan, P. P., Malia, A., Das, B., Bisoyi, B., & Satpathy, I. (2023). Greener economy for sustainable development through AI intervention: Demystifying critical factors. In *Handbook of research on sustainable consumption and production for greener economies* (pp. 327–343). IGI Global.

Qian, X. (Alice), & Papadonikolaki, E. (2020). Shifting trust in construction supply chains through blockchain technology. *Engineering, Construction and Architectural Management, 28*(2), 584–602. https://doi.org/10.1108/ECAM-12-2019-0676.

Queiroz, M. M., Telles, R., & Bonilla, S. H. (2020). Blockchain and supply chain management integration: A systematic review of the literature. *Supply chain management: An international journal, 25*(2), 241–254.

Rajput, S., & Singh, S. P. (2019). Identifying industry 4.0 IoT enablers by integrated PCA-ISM-DEMATEL approach. *Management Decision, 57*(8), 1784–1817. https://doi.org/10.1108/MD-04-2018-0378

Rejeb, A., Keogh, J. G., Simske, S. J., Stafford, T., & Treiblmaier, H. (2021). Potentials of blockchain technologies for supply chain collaboration: A conceptual framework. *The International Journal of Logistics Management, 32*(3), 973–994. https://doi.org/10.1108/IJLM-02-2020-0098

Risius, M., & Spohrer, K. (2017). A Blockchain research framework: What we (don't) know, where we go from here, and how we will get there. *Business & Information Systems Engineering, 59*(6), 385–409. https://doi.org/10.1007/s12599-017-0506-0

Saari, A., Vimpari, J., & Junnila, S. (2022). Blockchain in real estate: Recent developments and empirical applications. *Land Use Policy, 121*, 106334. https://doi.org/10.1016/j.landusepol.2022.106334

Saberi, S., Kouhizadeh, M., Sarkis, J., & Shen, L. (2019). Blockchain technology and its relationships to sustainable supply chain management. *International Journal of Production Research, 57*(7), 2117–2135. https://doi.org/10.1080/00207543.2018.1533261

Singh, R., Khan, S., & Centobelli, P. (2022). Investigating the interplay between social performance and organisational factors supporting circular economy practices. *Sustainability, 14*(24), 16781.

Singh, R., Khan, S., & Dsilva, J. (2023). A framework for assessment of critical factor for circular economy practice implementation. *Journal of Modelling in Management, 18*(5), 1476–1497.

Singh, R., Gupta, A., Bajpai, A., & Kandpal, V. (2024). Exploring challenges of circular economy initiatives for smart cities. In *Smart cities and circular economy: The future of sustainable urban development* (pp. 59–69). Emerald Publishing Limited.

Song, J. M., Sung, J., & Park, T. (2019). Applications of Blockchain to improve supply chain traceability. *Procedia Computer Science, 162*, 119–122. https://doi.org/10.1016/j.procs.2019.11.266

Sunmola, F., & Burgess, P. (2023). Transparency by design for Blockchain-based supply chains. *Procedia Computer Science, 217*, 1256–1265. https://doi.org/10.1016/j.procs.2022.12.324

Tapscott, D., & Tapscott, A. (2016). *Blockchain revolution: How the technology behind bitcoin is changing money, business, and the world*. Penguin.

Tönnissen, S., & Teuteberg, F. (2020). Analysing the impact of blockchain-technology for operations and supply chain management: An explanatory model drawn from multiple case studies. *International Journal of Information Management, 52*, 101953. https://doi.org/10.1016/j.ijinfomgt.2019.05.009

Tripathi, S., & Gupta, M. (2020). A framework for procurement process re-engineering in industry 4.0. *Business Process Management Journal, 27*(2), 439–458. https://doi.org/10.1108/BPMJ-07-2020-0321

Uddin, M. (2021). Blockchain Medledger: Hyperledger fabric enabled drug traceability system for counterfeit drugs in pharmaceutical industry. *International Journal of Pharmaceutics, 597*, 120235. https://doi.org/10.1016/j.ijpharm.2021.120235

Upadhyay, A., Mukhuty, S., Kumar, V., & Kazancoglu, Y. (2021). Blockchain technology and the circular economy: Implications for sustainability and social responsibility. *Journal of Cleaner Production, 293*, 126130.

Varriale, V., Cammarano, A., Michelino, F., & Caputo, M. (2020). The unknown potential of Blockchain for sustainable supply chains. *Sustainability, 12*(22), 9400. https://doi.org/10.3390/su12229400

Wang, Y., Singgih, M., Wang, J., & Rit, M. (2019). Making sense of blockchain technology: How will it transform supply chains? *International Journal of Production Economics, 211*, 221–236. https://doi.org/10.1016/j.ijpe.2019.02.002

Wong, L.-W., Leong, L.-Y., Hew, J.-J., Tan, G. W.-H., & Ooi, K.-B. (2020). Time to seize the digital evolution: Adoption of blockchain in operations and supply chain management among Malaysian SMEs. *International Journal of Information Management, 52*, 101997. https://doi.org/10.1016/j.ijinfomgt.2019.08.005

Xu, P., Lee, J., Barth, J. R., & Richey, R. G. (2021). Blockchain as supply chain technology: Considering transparency and security. *International Journal of Physical Distribution & Logistics Management, 51*(3), 305–324. https://doi.org/10.1108/IJPDLM-08-2019-0234

Yavuz, E., Koc, A. K., Cabuk, U. C., & Dalkilic, G. (2018). Towards secure e-voting using ethereum blockchain. *2018 6th International Symposium on Digital Forensic and Security (ISDFS)*, 1–7. https://doi.org/10.1109/ISDFS.2018.8355340.

Zelbst, P. J., Green, K. W., Sower, V. E., & Bond, P. L. (2019). The impact of RFID, IIoT, and Blockchain technologies on supply chain transparency. *Journal of Manufacturing Technology Management, 31*(3), 441–457. https://doi.org/10.1108/JMTM-03-2019-0118

Zhang, L., Ci, L., Wu, Y., & Wiwatanapataphee, B. (2024). The real estate time-stamping and registration system based on Ethereum blockchain. *Blockchain: Research and Applications, 5*(1), 100175. https://doi.org/10.1016/j.bcra.2023.100175

Chapter 9
Unveiling Blockchain Technology Dynamics: Empowering Future Generations to Build a Resilient UAE Circular Economy

Rekha Pillai, Jacinta Dsilva, and Godwin Francis

Abstract Blockchain technology has emerged as a transformative force in addressing global sustainability challenges, offering innovative solutions to build a resilient and efficient circular economy. This chapter aims to identify the challenges and potential opportunities in implementing blockchain technology within the UAE's existing circular economy frameworks. Further, it explores how blockchain technology can catalyse innovation in current and upcoming business models and generate new market opportunities for sustainable circular practices in the UAE. Through the lens of the UAE's strategic vision for sustainability, the discussion highlights how blockchain applications—ranging from smart contracts to supply chain optimization—can enhance resource efficiency, reduce waste, and promote ethical consumption patterns. Eventually, focusing on setting policy implications for effectively integrating blockchain technology into the UAE's circular economy framework. The chapter contextualises the discussion to develop an understanding of how blockchain can be implemented to achieve circular economy as well as examines the role of educational institutions and policymakers in equipping the young generation with skills and knowledge to leverage blockchain for sustainability goals. By offering strategic recommendations for addressing implementation barriers, this work contributes to the growing discourse on integrating emerging technologies into sustainable development frameworks.

Keywords Blockchain technology · Circular economy · UAE sustainability · Resource efficiency · Technological literacy

---

R. Pillai · G. Francis
The British University in Dubai, Dubai, UAE
e-mail: Rekha.pillai@buid.ac.ae; godwin.francis@buid.ac.ae

J. Dsilva (✉)
SEE Research Center, SEE Institute, The Sustainable City, Dubai, UAE
e-mail: jacinta@seeinstitute.ae

R. Singh et al. (eds.), *Blockchain Technologies for Smart Circular Economy and Organisational Sustainability*, Environmental Footprints and Eco-design of Products and Processes, https://doi.org/10.1007/978-3-031-89183-0_9

9.1 Introduction

The past few years have witnessed the emergence of circular economy as a core element of industrial and environmental policies in several parts of the world as well as a major concern for corporates and regional authorities (Upadhyay et al., 2021) as this regenerative system denigrates waste, emissions, and energy depletion activities by upholding sustainable practices such as encouraging durable designs, maintenance, repair, reuse, remanufacturing, refurbishing, and recycling (Geissdoerfer et al., 2017, p. 759). With the surge in prominence of a circular economy, there seems to be an ever-developing interest in multi-stakeholder technologies that might enable the persistence of circular economy practices, such as the Internet of Things, cloud computing, and blockchain technology. Blockchain technology is a new paradigm that enables the sharing and changing of information through a decentralized, peer-to-peer, open-access network by linking ledgers or databases, thus ensuring the data is stored and updated securely in a tamper-proof and irreversible way (Lansiti & Lakhani, 2017). Blockchain technology, being a disruptive innovation (Khan et al., 2023), can underpin the circular economy by ensuring a transparent, secure, and unalterable environment for tracking products and materials throughout their whole life cycle, hence allowing better recycling, reuse, remanufacturing, and management of resources (Khan et al., 2022a, b). This will ensure that the transparency in the movement of materials, thanks to blockchain, further strengthens the trust and accountability between the different stakeholders involved, hence facilitating the circular economy practices related to product repair, refurbishment, and responsible disposal. In an era marked by the rapid evolution of technologies and increased focus on sustainable approaches, the convergence of blockchain technology with principles of a Circular Economy proved to be one of the essential drivers of change across all major industries in the global scenario (Shou & Domenech, 2022; Khan et al., 2022a, b).

The principle focus of this chapter is to explore the potential of blockchain technology in expanding the circular economy principles within the UAE. The concept is of vital importance in the UAE as it surfaces as one of the world's highest per capita waste generation countries (Mawed et al., 2020), with policies still nascent to achieve the sustainable development goals set by the United Nations (Nie et al., 2024; Khan et al., 2024; Singh et al., 2024a, b, c, d). Therefore, the chapter is expected to discuss how the adoption of blockchain will play a significant role in advancing resource efficiency, transparency, and accountability, thereby driving sustainability practices across various industries within the country. Given the UAE's commitment to sustainability initiatives such as the UAE Vision 2021 and the National Climate Change Plan of the country, this chapter provides insight into how innovative technologies such as blockchain will enable the UAE to develop a resilient, circular economy in line with its environmental and economic objectives by addressing two research questions namely 'How can blockchain technology augment the efficiency and transparency of circular economy practices in the UAE? and what are the key impediments and opportunities for integrating blockchain technology into the UAE's circular economy initiatives?. To answer the former, the chapter aims to achieve threefold objectives. First, to identify the inherent challenges and potential opportunities in implementing blockchain technology within the UAE's existing circular economy frameworks. Secondly, to explore how blockchain technology can catalyse innovation in current/upcoming business models and generate new market opportunities for sustainable circular practices in the UAE and finally to set policy implications for effectively integrating blockchain technology into the UAE's circular economy framework (Kumar et al., 2024).

The chapter is of vital significance for future generations, innovators, and policymakers in deciphering the role of blockchain technology in advancing the UAE circular economy. For future generations, especially Gen Alpha, the chapter identifies the need to guarantee sustainability and provides the roadmap obligatory for the development of resource efficiency, transparency, and accountability in light of mounting environmental challenges. For innovators, the chapter will serve as a beacon for how blockchain can catalyse existing/new business models and uncover market opportunities for sustainable circular practices. For policymakers, the chapter will provide evidence-based recommendations and policy implications that may guide blockchain technology integration into the UAE's circular economy framework, ensuring the country's march toward its environmental and economic objectives.

9.2 Theoretical Review and Prior Literature

Extant research has perceived BCT as a socio-technical assemblage (Shin & Ibahrine, 2020), thereby tracing its roots from the socio-technical systems (STS) theory (Trist, 1981) which emphasizes the interdependency and joint optimization of social and technical factors for overall performance. Here the social components encompass human beings, organizational relationships, values, competencies, and attitudes (Singh et al., 2022, 2023), while the technical components consist of technologies, tools, and processes necessary to produce the intended output. These two systems are interdependent, meaning that improvements in one system necessitate simultaneous improvement in the other for optimizing and synergizing performance; an independent focus on one will not yield good results. The principles of Socio-technical Systems theory align with the goals of a circular economy, which requires a combination of social and technological innovations to achieve sustainable, resource-efficient use. This involves social systems comprised of organizations, consumers, and governments working together cooperatively in support of circular modes of recycling, reuse, and remanufacturing. Still, technical systems, such as blockchain for transparency, IoT for the follow-up on resources, and automation for production, are required for optimization and maintenance of resource flow through closed-loop systems. Here, STS theory meets mutual optimization of social and technical elements to ensure improvements by either side, new business models, or technological advancement with regard to circular economy strategy performance. Blockchain exemplifies how a technical system can enable social innovations. Its decentralized ledger enables traceability and accountability in resource management, giving stakeholders access to verified information regarding sustainability practices (Kamble et al., 2020). In that sense, the role of IoT further complements the use of blockchain. IoT devices enable real-time monitoring of resources, optimizing operations and reducing waste. The IoT embodies the need for a holistic view in circular economy interventions, where technology advancement meets social structures—for example, influencing consumers to adopt sustainable lifestyles. In addition, we need to be mindful that leaders must exist who can foster collaborative practices and embed sustainability in corporate strategies, since these are key enablers of socio-technical change (Geng et al., 2020; Singh et al., 2024a, b, c, d).

9.2.1 Blockchain Technology: A Transformative Force

Once perceived as the pivot of cryptocurrencies, blockchain technology has progressed into a resourceful platform that surpasses traditional financial services. It is important to note that blockchain technology refers to a decentralized, distributed ledger system that records and verifies transactions on many different computers. At its heart, these principles are immutability, transparency, and security. Immutability ensures no transaction can be erased once registered (Masadeh et al., 2024). Transparency also allows all parties to see and verify the data that is recorded. Furthermore, the safeguarded and unhackable features of the blockchain are another pillar of its security, ensured by cryptographic hashing and consensus mechanisms (Hulea et al., 2024). These features make blockchain well-suited for scenarios that demand both high security and transparency, like supply chain management. There are several types of blockchain with distinct features and applications, from public blockchains for Bitcoin to private or regulated blockchains used by organizations (Abid et al., 2024; Sharma et al., 2025). Depending on the scenario of the circular economy being charted, select the most suitable blockchain infrastructure. Moreover, the fundamental characteristics of what blockchain technology implies such as decentralization, transparency, immutability, and security make it one of the most potential mechanisms for managing complex systems and building trust among groups (Abid et al., 2024). Hence, it is necessary also to understand the potential advantages from blockchain. These attributes are crucial for tracking materials in the circular economy, enhancing product authenticity, and facilitating reverse logistics. Generating an open and verifiable ledger of a product's lifecycle enables circularity (Giovanni, 2022). That extra traceability helps people treat their waste that much better, "recycle better and more responsibly." However, challenges persist.

9.2.2 The Circular Economy Concept and Its Importance

The circular economy encompasses a general approach to separate economic development from resource consumption and the creation of waste (Steenmans et al., 2021; Chaudhari et al., 2022). It highlights the 4Rs: reduce, reuse, recycle, recover to reduce waste and increase the use efficiency of resources (Abid et al., 2024). The CE model is fundamentally different from the traditional economy, which predominantly follows a "take-make-dispose" paradigm that causes environmental degradation and resource scarcity with a transition to a circular economy requiring changes in production, consumption and waste management. Multiple frameworks have been developed for this transition focusing on different aspects such as product design, supply chain management and waste management (Kouhizadeh et al., 2022), and this is where the partnership is key for successful CE transition: government, business and consumers must be partners in a CE transition (Chaudhari et al., 2022).

9.2.3 Link Between Blockchain Tech and Circular Economy Practices

Integration of blockchain technology with circular economy practices holds great potential benefits (Giovanni, 2022). Transparency and traceability characteristics of blockchain increase products accountability and traceability during their lifecycle, from raw material sourcing to end-of-life management (Abid et al., 2024; Hristova et al., 2023). This enables increased visibility, which helps trace and treat supply chain inefficiencies, minimize fraud and counterfeiting, and improve resource recovery. An example of this application is tracking recycled materials and verifying their processing, thus avoiding diversion to a landfill (Hassan et al., 2023). Smart contracts, self-executing contracts that are recorded on the blockchain, can facilitate the automation of several processes (e.g., payments for recycled materials and incentives for consumers to participate in product take-back programs) (Giovanni, 2022). According to Bekrar et al. (2021) blockchain can enable decentralized marketplaces to trade used or recycled products.

However, the implementation of blockchain technology in the circular economy faces numerous limitations (Abid et al., 2024; Benabdellah et al., 2023). For example, scalability issues can limit blockchain networks' capacity to handle large data volumes (Benabdellah et al., 2023). Thus, interoperability is one of the biggest challenges in blockchain, for if disparate blockchain networks cannot communicate with each other, interoperability dilemmas subsequently arise (Benabdellah et al., 2023). As Steenmans et al. (2021) state, the introduction of new technology like blockchain always brings along data protection and privacy issues; regulations and legal uncertainty on the use of information on blockchain are considered a hindrance to the adoption of blockchain. Difficult integration of blockchain platforms into existing supply chains due to the high degree of non-standardization and lack of interoperability (Benabdellah et al., 2023). In essence, for technology based on blockchain to be applied, a large amount of money needs to be realized, and this becomes an obstacle for small and medium enterprises (Butera & Gatteschi, 2024). Overcoming these challenges calls for continued investigation, technology development, and policy regulations, which will provide the basis for the use of blockchain technology in a circular economy (Steenmans et al., 2021; Abid et al., 2024).

Numerous published research studies investigating specific use cases for blockchain technology in a circular economy. Key studies analyse blockchain applications towards traceability and transparency along agricultural supply chains (Chaudhari et al., 2022; Hristova et al., 2023), waste material management (Hassan et al., 2023) and closed-loop supply terms (Giovanni, 2022). Other studies address the potential of blockchain technology to support product stewardship programs, encourage recycling, and optimize reverse logistics (Bekrar et al., 2021). Also being investigated is the potential of blockchain to enable multiple circular economy approaches (Sharma et al., 2025) including the use of product passports to track the life of a product from the cradle to the grave (Govindan, 2022). Additionally, Kouhizadeh et al. (2022) provide initial evidence of the links between the application of blockchain and various dimensions of the circular economy but also systematically examine practice gaps and potential tensions. Data transparency, market competition, lack of infrastructure, standardization, and complicated protocols are consistently cited as barriers to wide-scale adoption (Benabdellah et al., 2023). Studies investigating barriers to blockchain adoption in such different settings as remanufacturing (Lin et al., 2024) and the automotive industry (Lee et al., 2023) provide useful insights about challenges and opportunities.

That said, a limiting factor is public awareness and understanding of blockchain technology and, more specifically, research on the role of blockchains for circular economies is still in its infancy (Steenmans et al., 2021), so a detailed overview of existing cases and their strengths and weaknesses is warranted. Although many of the potential benefits of blockchain to circular economy models have not yet been fully realized, as challenges such as scalability and interoperability remain (Abid et al., 2024), several promising applications have been noted.

We are experiencing the potential of blockchain solutions to improve aspects of circular economies through a range of international initiatives. Supply chain management is one such area of intense focus. As a distributed ledger technology, blockchain is characterized by its level of decentralization and system immutability; when properly configured, it allows for the establishment of transparent, secure systems that guarantee product accountability and traceability (Abid et al., 2024). This is still important in industries with complex supply chains where materials and products need to be tracked through multiple stages. The addition of blockchain-based applications in used products reverse logistics (Bekrar et al., 2021) as well as recycling or recovery facilitation. This means that the materials can be traced throughout their life, which makes it easier to manage the material and reduce waste, all of which are core principles of a circular economy.

Moreover, the transparency provided by blockchain increases the trust between stakeholders, which is key for successful cooperation in circular economy initiatives. In the agricultural sector, blockchain could improve product traceability and consumers can know the entire journey from farm to fork and how products or food was sourced and processed (Chaudhari et al., 2022; Hristova et al., 2023). This can boost consumer confidence and can also foster the development of sustainable agricultural practices. Blockchains can help in waste management as well as monitor waste streams so a better recycling and waste management system can be created (Hassan et al., 2023; Khan et al., 2021).

Another important aspect of blockchain technology, the deployment of smart contracts, enables automation and increases efficiency in circular economy processes (Ishak et al., 2024). Smart contracts can allow for automatic payments on recycled materials, incentivize recycling program participation, and automate product returns. As a result, this automation lowers administrative costs and uses process efficiency in circular economy. Evidence shows that blockchain can facilitate the creation of closed-loop supply chains (Giovanni, 2022), which reuse the same materials through repeated use and recycling, reducing waste and waste of resources. However, several barriers to general adoption remain, the prominent among them being the necessity for standardization and interoperability between distinct blockchain networks (Benabdellah et al., 2023).

9.2.4 Blockchain in the UAE

The UAE has promoted blockchain technology as central to economic growth and digital transformation. Initiatives like the Dubai Blockchain Strategy have been put in place by the government itself, with the aim of establishing Dubai as the global centre for blockchain technology by 2021. This approach represents a national effort to harness blockchain's promise in different industries, from finance and supply chain management to government services. Many government departments have already implemented blockchain solutions for better efficiency and transparency. That said, the adoption to date in the private sector varies; there are frontrunners, but also onlookers. While smart cities and digital infrastructure make the UAE an ideal recipient of blockchain technology, challenges remain for industry-wide take-up (Singh et al., 2024a, b, c, d).

9.2.4.1 A Economics of Blockchain for a Circular Economy in the UAE

Though the UAE is extremely progressive, there are some barriers to proactively mainstreaming blockchain into a circular economy.

1. **Regulatory and legal framework**: The UAE has been proactive in adopting new technologies, but a cohesive regulatory framework governing the use of blockchain within a circular economy context is under development. Outdated regulations governing data ownership, privacy, and liability have the potential to

add uncertainty to the minds of prospective businesses that are interested in implementing the technology, especially with regard to aspects that have nothing to do with payouts. In addition, the adoption of blockchain technology needs to be well thought out in the context of current legal and regulatory frameworks to prevent any potential clashes and to ensure legal obligations are met. The lack of standardized regulations among emirates can also cause fragmentation and pose an impediment to interoperability.

2. **Interoperability and standardization**: Among the existing challenges is the absence of interoperability between numerous blockchain platforms and systems (Abid et al., 2024; Benabdellah et al., 2023). Data must flow through various stakeholders over the product lifecycle for a circular economy to function. Without standardized protocols and data formats, there may be obstacles to the exchange and integration of data, which can hinder the effectiveness of blockchain implementations (Benabdellah et al., 2023). This issue, however, becomes even more significant as the UAE economy is broad and deep, with a number of sectors and organizations employing unique blockchain technologies.
3. **Scalability and infrastructure**: The scalability of blockchain technology remains a keystone concern, particularly if it is applied to the management of large volumes of data consistent with a circular economy (Abid et al., 2024). At the same time, the UAE is pursuing an ambitious vision of a holistic circular economy that depends on infrastructure that can consistently and securely handle massive volumes of data transactions. Faster and more efficient performance in power therapy and bandwidth. Further, the energy consumption of some blockchain networks presents a challenge to achieving sustainability targets (Masadeh et al., 2024).
4. **Data security and privacy challenges**: Blockchain possesses security features that are compatible with a circular economy and equally raises data security and privacy issues (Abid et al., 2024). The importance of protecting sensitive information regarding product provenance, formulation and lifecycle cannot be understated. As data breaches and unauthorized access to sensitive information remain major threats, maintaining robust cybersecurity protocols and establishing data governance frameworks are essential. Another factor to consider would be regulatory compliance with various data protection laws, like the UAE Personal Data Protection Law, and how they guarantee that the application of blockchain technology is being conducted in an ethical and responsible manner.
5. **Indifference and technical know-how**: For the mass adoption of blockchain, a large workforce is needed that has been trained in the development, deployment, and management of blockchain systems (Kouhizadeh et al., 2022). With an ever-growing list of startups and scaleups to hit the UAE, the country now also has a shortage of blockchain professionals, limiting the capacity to innovate and develop at blistering speeds. Moreover, there is a demand to inform organizations and citizens about the advantages and uses of blockchain to motivate more extensive utilization and acceptance. The educational efforts and programs that help bridge the skills gap and that show participants how to perceive the ramifications of blockchain technology in a circular economy need to be fast-tracked.

6. **Cost and implementation challenges**: Implementing blockchain solutions is costly and challenging (Benabdellah et al., 2023), especially for small and medium-sized enterprises (SMEs). The initial investments required in infrastructure, software and expertise can be overwhelming for most organizations, especially those with fewer resources. Furthermore, the ongoing maintenance and upgrades required to keep blockchain systems secure and efficient have also emerged as a primary operational expense. Addressing these financial constraints is vital to enable small businesses to take advantage of blockchain technology, and it will not happen without some degree of government support in terms of tax incentives or grants and collective efforts.

9.2.4.2 Blockchain Opportunities to Build Sustainability in the UAE

Despite those inherent obstacles in blockchain, several key opportunities exist in the UAE to tap into blockchain technology to improve sustainability and enter a more circular economy.

1. **Improved traceability and transparency**: The features of transparency and stability of blockchain technology can be leveraged to transform supply chain management by way of tracking the life cycle of a product (from the cradle to the grave) through an auditable and granular log that can be retrieved in real time (Abid et al., 2024; Chaudhari et al., 2022). This augmented visibility may solve problems related to counterfeit products, improve accountability from production to sale, and allow improved waste stream management. In the UAE's varied economy, particularly in the food and agriculture sectors, establishing product authenticity and origin verification has never been more critical.
2. **Enhanced waste control**: The tracking of waste materials from the point of generation to their eventual disposal or recycling can be improved by using blockchain technology (Steenmans et al., 2021; Hassan et al., 2023). In addition to better recycling rates, you can optimize waste collection routes and ensure hazardous materials are properly disposed of. Similar processes such as the payment for waste collection or rewarding active participation in recycling programs can be automated, thus achieving more efficient and sustainable waste management using smart contracts.
3. **Sustainable supply chain management**: Blockchain can enhance transparency, traceability, and collaboration between stakeholders, thus enabling more sustainable and resilient supply chains (Chaudhari et al., 2022). With construction materials tracking, minimizing counterfeiting and better sourcing of materials, businesses can shrink their environmental footprint and improve social responsibility too. It becomes more relevant, especially with the ardent commitment of the UAE towards the 17 Sustainable Development Goals (Singh et al., 2024a, b).
4. **Landings facilitating the product lifecycle**: Blockchain also aids in formulating digital product passports that provide detailed insights about a product's material composition, sourcing, and possible end-of-life alternatives (Sharma

et al., 2025). Necessary data can make highly intelligent recycling and remanufacturing methods possible, effectively prolonging the lifespan of products and decreasing their impact on the environment. The UAE can also benefit from the integration of blockchain-based product lifecycle management systems, which facilitates sustainable manufacturing and product design with cutting-edge technology.

5. **Encouraging green finance and investments**: Tokenized assets representing environmental services or sustainable projects enabled by blockchain could foster the establishment of green finance initiatives. This move can harness investments in renewable projects, waste management projects, and sustainable development projects. The growing fintech industry in the UAE is likely to be instrumental in shaping this green finance revolution, creating blockchain-based systems for sustainable investments.
6. **Empowering consumers**: The technology allows blockchain to introduce more transparency into consumption habits and empower consumers with greater control over it. The more information consumers have regarding the sustainability credentials of a product, the smarter their purchasing decisions will be, and that eventually will make sure businesses are doing the right thing in working towards more sustainable practices. In fact, such heightened awareness and engagement of consumers form the very basis of a circular economy.

9.3 Implications

9.3.1 Policy Implications

The interplay between blockchain technology and circular economy principles underscores the need for socio-technical integration. This allows collaboration on the part of all the concerned stakeholders with enhanced efficiency in resources, hence developing circular economy strategies that are both scalable and sustainable within the UAE. These findings emphasize the need for an integrated regulatory framework that must cater to scalability, interoperability, and data security challenges in adopting blockchain. Policymakers should come up with a very clear policy direction that coincides with UAE Vision 2021 and the National Climate Change Plan, creating an enabling environment that nurtures innovation and ensuring a level of legality and ethics within blockchain applications. The ability of blockchain technology to enhance aspects of transparency and traceability creates new business model innovation opportunities. Industries in the UAE, especially in the waste management, agriculture, and manufacturing sectors, stand to benefit by leveraging BCT in operational optimization, enhancing accountability among stakeholders, and developing decentralized marketplaces for recycled and reused products. These will contribute to catalysing economic diversification and meeting the sustainability objectives of the UAE. The inclusion of blockchain in educational curricula will ensure that a generation is built to innovate within the CE frameworks. Engagement of youth through hackathons, workshops, and training programs will drive awareness and innovation, hence long-term sustainability. This will position Gen Alpha as a proactive contributor to resource-efficient and sustainable economic systems. Barriers to the adoption of blockchain require public-private collaboration. The creative entrepreneurship can harness the blockchain-enabled CE practices through public-private partnerships in the regulatory oversight and thus spread widely across the UAE.

9.3.2 Practical Implications

It is of utmost importance that young people are involved in the creation of a sustainable and strong economy. Younger generations are well acquainted with technology and, thus, are equipped with the necessary skills and ingenuity to employ new technologies such as blockchain for the circular economy (Kouhizadeh et al., 2022). This type of engagement can take various forms such as hackathons, competitions, and workshops that target the development of blockchain applications for waste management, supply chain tracking and other circular economy challenges (Khan et al., 2021). Providing the youth with the means to learn about how blockchain technology applies to the circular economy gives them vital skills and knowledge they will need to enter a rapidly shifting job market (Llano et al., 2024). The involvement of youth in the creation and use of blockchain technology can help the UAE create a better version of the country which is technologically advanced and environment-friendly for our coming generation.

9.4 Conclusion

Blockchain technology has become a game-changing intervention in the advancements of CE principles in the UAE. The transparency, decentralization, and innate stability in blockchain address critical challenges to resource management, from material traceability to product authentication. These qualities build trust and accountability, crucial in sustainable practice. Success with blockchain implementation in circular economy practices will depend upon how existing challenges—such as interoperability issues, scalability, and regulatory uncertainty—are sorted out. Full potential realization requires standardization of protocols and heavy technological infrastructure investment. Evidence-based policy recommendations may include guiding blockchain integration into the UAE's CE framework by incentivizing its adoption, supporting research and development, and establishing frameworks that ensure seamless implementation across sectors. The UAE will lead in blockchain-enabled CE models by embracing this technology. This will be an enhancement towards sustainability and will also reflect the diversification of the UAE's economy, positioning its leadership in technological innovation and environmental stewardship.

Nevertheless, several limitations are inherent in the chapter. The lack of empirical evidence and the setting of an independent country reduces the generalizability of results. Future research should empirically investigate how blockchain integrates with other emerging technologies, such as artificial intelligence and IoT, in devising more advanced solutions for CE. Long-term socio-economic impact assessments of blockchain are required to frame its role in the building of sustainable and resilient economies. Cross-country studies are recommended to ensure a comprehensive outlook on the blockchain-circular economy nexus.

References

Abid, I., Fuad, S. M. Z. A., Chowdhury, M. J. M., Chowdhury, M. S., & Ferdous, M. (2024). A systematic literature review on the use of blockchain technology in transition to a circular economy. arXiv.org. https://doi.org/10.48550/arXiv.2408.11664

Bekrar, A., Cadi, A. A. E., Todosijevi, R., & Sarkis, J. (2021). Digitalizing the closing-of-the-loop for supply chains: A transportation and blockchain perspective. *Sustainability, 13*, 2895. https://doi.org/10.3390/SU13052895

Benabdellah, A. C., Zekhnini, K., Cherrafi, A., GarzaReyes, J., Kumar, A., & Baz, J. E. (2023). Blockchain technology for viable circular digital supplychains: An integrated approach for evaluating the implementation barriers. *Benchmarking: An International Journal, 30*(4), 4397–4424. https://doi.org/10.1108/bij-04-2022-0240

Butera, A., & Gatteschi, V. (2024). EU battery regulation and the role of blockchain in creating digital battery passports. In *Conference on Blockchain Research & Applications for Innovative Networks and Services*. https://doi.org/10.1109/BRAINS63024.2024.10732126

Chaudhari, R. S., Mahajan, S. K., Rane, S., & Agrawal, R. (2022). Modeling barriers in circular economy using TOPSIS: Perspective of environmental sustainability & blockchain-IoT technology. *International Journal of Mathematical, Engineering and Management Sciences, 7*(6), 820–843. https://doi.org/10.33889/ijmems.2022.7.6.052

Geissdoerfer, M., Savaget, P., Bocken, N. M. P., & Hultink, E. J. (2017). The circular economy—A new sustainability paradigm? *Journal of Cleaner Production, 143*, 757–768. https://doi.org/10.1016/J.JCLEPRO.2016.12.048

Geng, R., Mansouri, A., Aktas, E., & Yen, D. A. (2020). An empirical study of green supplier collaboration in the Chinese manufacturing sector: The double-edged sword effect of guanxi. *Supply Chain Management: An International Journal, 25*(3), 359–373.

Giovanni, P. D. (2022). Leveraging the circular economy with a closed-loop supply chain and a reverse omnichannel using blockchain technology and incentives. *International Journal of Operations and Production Management, 42*, 959. https://doi.org/10.1108/ijopm-07-2021-0445

Govindan, K. (2022). Tunneling the barriers of blockchain technology in remanufacturing for achieving sustainable development goals: A circular manufacturing perspective. *Business Strategy and the Environment, 31*, 3769. https://doi.org/10.1002/bse.3031

Hassan, N. M., Khan, S., Ashraf, M., & Sheikh, A. A. (2023). Interconnection between the role of blockchain technologies, supply chain integration, and circular economy: A case of small and medium-sized enterprises in Pakistan. *Science in Progress, 106*, 368504231186527. https://doi.org/10.1177/00368504231186527

Hristova, T., Savov, N., & Yanev, N. (2023). Municipal solid waste and non-hazardous waste processing for sustainable circular economy through blockchain. *Journal of Chemical Technology and Metallurgy, 58*, 513. https://doi.org/10.59957/jctm.v58i3.81

Hulea, M., Miron, R., & Murean, V. (2024). Digital product passport implementation based on multi-blockchain approach with decentralized identifier provider. *Applied Sciences, 14*, 4874. https://doi.org/10.3390/app14114874

Ishak, Z., Amran, A. S. A., Esfahani, M. D., Ayyasamy, R., Abdullah, M. O. B., & Ann, Y. J. (2024). Determining the readiness factors to adapt blockchain technology in the palm oil industry to facilitate circular economy transition. https://doi.org/10.1109/AiDAS63860.2024.10730330

Kamble, S. S., Gunasekaran, A., & Sharma, R. (2020). Modeling the blockchain-enabled traceability in agriculture supply chain. *International Journal of Information Management, 52*, 101967.

Khan, S. A. R., Yu, Z., Sarwat, S., Godil, D., Amin, S., & Shujaat, S. (2021). The role of blockchain technology in circular economy practices to improve organizational performance. *International Journal of Logistics Research and Applications, 25*(2), 1–18. https://doi.org/10.1080/13675567.2021.1872512

Khan, S., Ali, S. S., & Singh, R. (2022a). Determinants of remanufacturing adoption for circular economy: A causal relationship evaluation framework. *Applied System Innovation, 5*(4), 62. https://doi.org/10.3390/ASI5040062

Khan, S., Singh, R., Sá, J. C., Santos, G., & Ferreira, L. P. (2022b). Modelling of determinants of logistics 4.0 adoption: Insights from developing countries. *Machines, 10*(12), 1242. https://doi.org/10.3390/machines10121242

Khan, S., Khan, M. I., & Singh, R. (2023). Modeling the barriers of blockchain technology implementation in supply chain. *Journal of Industrial Integration and Management, 8*, 367. https://doi.org/10.1142/S2424862223500070

Khan, S., Singh, R., Alnahas, J., Abbate, S., & Centobelli, P. (2024). Navigating the smart circular economy: A framework for manufacturing firms. *Journal of Cleaner Production, 480*, 144007. https://doi.org/10.1016/j.jclepro.2024.144007

Kouhizadeh, M., Zhu, Q., & Sarkis, J. (2022). Circular economy performance measurements and blockchain technology: An examination of relationships. *International Journal of Logistics Management, 34*(3), 720–743. https://doi.org/10.1108/ijlm-04-2022-0145

Kumar, K., Singh, R., Sharma, A., & Lehri, B. P. S. (2024). Climate change adaptation strategies for achieving net-zero economy. In *Approaches to global sustainability, markets, and governance* (pp. 163–183). Springer. https://doi.org/10.1007/978-981-97-5756-5_8

Lansiti, M., & Lakhani, K. R. (2017). The truth about blockchain. *Harvard Business Review, 95*(1), 118–127. https://enterprisersproject.com/sites/default/files/the_truth_about_blockchain.pdf

Lee, C., Xiao-Ai, H., Ho, X., Eu, Y., Tan, J., Xiao-Lin, J., Husni, M. H., Chai, W., & Aye, J. T. (2023). Innovation helixes, blockchain, open innovation, and phased rewards in promoting circular design in waste management: A conceptual HCI study. In *International Conference on Software, Knowledge, Information Management and Applications*. https://doi.org/10.1109/SKIMA59232.2023.10387357

Lin, W. L., Leow, N. X., Lim, W. M., Ho, M. K., Yong, J. Y., & Pek, C. (2024). The influence of blockchain technology on circular economy implementation in the automotive sector: From a GMM model to a new machine learning algorithm. *Business Strategy and the Environment, 34*, 1043. https://doi.org/10.1002/bse.4003

Llano, E. A., Hurmelinna-Laukkanen, P., & Haapanen, L. (2024). Blockchain for the circular economy: Implications for public governance. *International Journal of Public Sector Management*. https://doi.org/10.1108/ijpsm-12-2023-0365

Masadeh, R., Jaber, M., Sharabati, A., Nasereddin, A., & Marei, A. (2024). The blockchain effect on courier supply chains digitalization and its contribution to industry 4.0 within the circular economy. *Sustainability, 16*(6), 7218. https://doi.org/10.3390/su16167218

Mawed, M., Al Nuaimi, M. S., & Kashawni, G. (2020). Construction and demolition waste management in the UAE: Application and obstacles. *GEOMATE Journal, 18*(70), 235–245. https://geomatejournal.com/geomate/article/view/645

Nie, P., Dahanayake, K. C., & Sumanarathna, N. (2024). Exploring UAE's transition towards circular economy through construction and demolition waste management in the pre-construction stage—A case study approach. *Smart and Sustainable Built Environment, 13*(2), 246–266. https://doi.org/10.1108/SASBE-06-2022-0115

Sharma, R., Samad, T. A., Jabbour, C. J. C., & Queiroz, M. J. D. (2025). Leveraging blockchain technology for circularity in agricultural supply chains: Evidence from a fast-growing economy. *Journal of Enterprise Information Management, 38*(1), 32–67. https://doi.org/10.1108/jeim-02-2021-0094

Shin, D., & Ibahrine, M. (2020). The socio-technical assemblages of blockchain systems: How blockchains are framed and how the framing reflects societal contexts. *Digital Policy, Regulation and Governance, 22*(3), 245–263. https://doi.org/10.1108/DPRG-11-2019-0095

Shou, M., & Domenech, T. (2022). Integrating LCA and blockchain technology to promote circular fashion—A case study of leather handbags. *Journal of Cleaner Production, 373*, 133557. https://doi.org/10.1016/J.JCLEPRO.2022.133557

Singh, R., Khan, S., & Centobelli, P. (2022). Investigating the interplay between social performance and organizational factors supporting circular economy practices. *Sustainability, 14*(24), 16781. https://doi.org/10.3390/SU142416781

Singh, R., Khan, S., & Dsilva, J. (2023). A framework for assessment of critical factor for circular economy practice implementation. *Journal of Modelling in Management, 18*(5), 1476–1497. https://doi.org/10.1108/JM2-06-2021-0145

Singh, R., Khan, S., Dsilva, J., Akram, U., & Haleem, A. (2024a). Modelling the organisational factors for implementation of corporate social responsibility: A modified TISM approach. *Global Journal of Flexible Systems Management, 25*(2), 283–301. https://doi.org/10.1007/s40171-024-00388-x

Singh, R., Gupta, A., Bajpai, A., & Kandpal, V. (2024b). Exploring challenges of circular economy initiatives for smart cities. In V. Kandpal, E. D. Santibanez-Gonzalez, P. Chatterjee, & M. K. Nallapaneni (Eds.), *Smart cities and circular economy* (pp. 59–69). Emerald Publishing Limited. https://doi.org/10.1108/978-1-83797-957-820241005

Singh, R., Kumar, K., & Khan, S. (2024c). A comprehensive views of artificial intelligence (AI)-based technologies for sustainable development goals (SDGs). In A. Kumar, R. Singh, S. Khan, & V. Kumar (Eds.), *Artificial intelligence enabled management: An emerging economy perspective* (pp. 183–195). De Gruyter.

Singh, R., Crowther, D., & Ravesangar, K. (2024d). Harnessing human capital for a sustainable future: HR's pivotal role toward net zero policies. In R. Singh & D. Crowther (Eds.), *Transition towards a sustainable future. Approaches to global sustainability, markets, and governance*. Springer. https://doi.org/10.1007/978-981-97-5756-5_2

Steenmans, K., Taylor, P., & Steenmans, I. (2021). Regulatory opportunities and challenges for blockchain adoption for circular economies. In *International Congress on Blockchain and Applications*. https://doi.org/10.1109/Blockchain53845.2021.00086

Trist, E. (1981). *Evolution of socio-technical systems*. https://sistemas-humano-computacionais.wdfiles.com/local%2D%2Dfiles/capitulo%3Aredes-socio-tecnicas/Evolution_of_socio_technical_systems.pdf

Upadhyay, A., Mukhuty, S., Kumar, V., & Kazancoglu, Y. (2021). Blockchain technology and the circular economy: Implications for sustainability and social responsibility. *Journal of Cleaner Production, 293*, 126130. https://doi.org/10.1016/J.JCLEPRO.2021.126130

Part III
Issues and Challenges in Blockchain Technologies

Chapter 10
Blockchain Technology and Its Impact on Circular Economy: Real-World Use Cases and Future Potential

Amit Kumar Singh, Vikas Garg, and Pooja Kaushik

Abstract The contemporary integrated economy and sustainable business are gaining value due to blockchain technology as the circular economy pivots through decentralized, transparent, and immutable advanced approaches. This paper aims to discuss the application of blockchain and AI, IoT and 5g in relation to efficient use of resources, efficiency in waste management and sustainable development. It goes over the applications in vertical sectors, recycling, energy markets, and supply chain traceability to highlight the relevant ethical implications of blockchain for responsible resource consumption and pollution reduction. Also, the research acknowledges global projects such as the EU Blockchain Partnership and the Sustainable Development Goals of the United Nations, which state that maximum prospects of blockchain applications can be achieved through cooperation at the international level and in the development of global legislation. Limitations of the technology are also examined, including the issues of scalability, power use, and environmental factors such as governmental restraints, are also examined in order to furnish a fair determination of the shortfalls of the offering. The study's conclusion is focused on presenting future research directions in the achievement of innovation and collaboration to advance the circular economy.

Keywords Circular economy · Blockchain technology · Sustainable development · Supply chain transparency · Smart contracts

A. K. Singh
Department of Commerce, Delhi School of Economics, New Delhi, India
e-mail: aksingh1@commerce.du.ac.in

V. Garg (✉)
Symbiosis Institute of Business Management, Noida, India

Symbiosis International (Deemed University), Pune, India

P. Kaushik
Department of Management Studies, Galgotias College of Engineering and Technology, Greater Noida, Uttar Pradesh, India
e-mail: pooja.kaushik@galgotiacollege.edu

R. Singh et al. (eds.), *Blockchain Technologies for Smart Circular Economy and Organisational Sustainability*, Environmental Footprints and Eco-design of Products and Processes, https://doi.org/10.1007/978-3-031-89183-0_10

10.1 Introduction

10.1.1 Global Context of Blockchain Technology

Blockchain technology has made a massive shift in data storage and data sharing, and in the verification process around the world. Originally developed as a backbone for Bitcoin in 2008 by an unknown person or a group under the pseudonym of Satoshi Nakamoto, blockchain has grown from being a simple infrastructural technology for cryptocurrencies to Industry 4.0 technology applicable in the health sector, supply chain, finance, and sustainability, among others.

In developed countries the rate of blockchain implementation has increased significantly due to the demand in the market for openness, speed and reliability. For example, in North America and Europe, industries are using blockchain for secure information management and optimizing supply chain visibility (Fosso Wamba et al., 2020). The governments of Estonia and Sweden have been employing blockchain technology to address the issues of digital identities and registers of land ownership, respectively (Treiblmaier, 2019).

On the other hand, developing countries are adopting blockchain technology to address infrastructure dilemmas and efficiency in resource utilization. Blockchain is already implemented in Africa, where the DeFi platforms have improved the access to banking services for those who do not have a bank account (Nassiry, 2018a, b).

10.1.2 Circular Economy: A Global Perspective

The circular economy is an economic model that is planned and implemented systematically with the aim of creating value for organizations, society and the earth. Compared to the conventional linear model of 'take, make, dispose,' circular economy was designed to minimize negative impact on the environment and operation, to reintroduce returned resources back into use, and to renew ecological systems (Ellen MacArthur Foundation, 2013).

Across the world, there are signs that both governments and organizations of different maturity levels have started to realize the need for change from the linear economy to the circular economy to meet the climate change challenges and the scarcity of resources (Kumar et al., 2024). It looks like the European Union (EU) has been leading the charge with a specific circular economy action plan that sets the ground policy to encourage sustainable resource use, enhance the management of waste, and support recycling efforts (European Commission, 2020a, b). Likewise, China's Circular Economy Promotion Law targets the efficient use of resources and waste reduction in the country's growing industrialisation economy (Zeng et al., 2017).

10.1.3 The Role of Blockchain in Addressing Global Circular Economy Challenges

The applicability of a circular economy on a global level presents important issues with regard to supply chain transparency, materials circulation and the ability to gain the necessary trust from all stakeholders. These issues have presented themselves in various forms and are quite fitting since blockchain infrastructure's inherent characteristics, such as decentralization, transparency, and immutability, make it easy to address them (Saberi et al., 2019).

Those gaps may be bridged by blockchain, which ensures full supply chain management to reveal the state of resources at any stage of the process. Such a level of traceability is especially important in sectors such as fashion and electronics, where products and or raw materials pass through several borders before reaching the final consumer (Bai et al., 2022).

Besides traceability, another facet that blockchain brings about is trust together with other interested parties and actors. For instance, blockchain can help to ensure that every claim to use recycled material in a particular product matches the set international standards on sustainability (Behnke & Janssen, 2020). In addition, smart contracts, part of the blockchain system, handle issues like payment and resource dispatch, thereby eliminating delays and promoting international cooperation (Treiblmaier, 2019).

10.2 Blockchain Technology: Key Features for Circular Economy Applications

Based on the characteristics of blockchain technology, namely, transparency, decentralization, smart contracts, and data security, it is most suitable for solving global challenges in the circular economy.

10.2.1 Transparency and Traceability in Global Supply Chains

Their values are high, but their transparency and ability to track products throughout their lives make blockchain a leading technology of the circular economy. This is because, with its ability to keep an incorruptible record of everything, blockchain brings a sense of transparency in the record of the products and other materials' journey and change throughout the transformation process, giving the stakeholders the relevant data to ensure that they can trust the sustainability claims that is being forwarded to them and to ensure that proper ethical standards have been followed (Singh et al., 2022).

Food industries are among the most recognizable use cases in this field: Walmart, Carrefour, and other giants already apply blockchain. For instance, while sourcing mangoes, Walmart uses IBM's Food Trust blockchain platform (Kamath, 2020). Tracking the journey in seconds instead of days enables a level of traceability that is crucial for promoting sustainability and reducing food waste across the globe. Such a degree of traceability is vital to setting up sustainability and minimizing food waste worldwide.

10.2.2 Decentralization for Cross-border Collaboration

Blockchain also supports involving initiatives from more than one country through a B2B platform in the circular economy. For example, the Plastic Bank project is a digital business platform for recycling plastic waste from developing countries, which applies blockchain technology. Sorted plastic wastes are termed "Social Plastic," which has a physical attribute labelled and documented to be tradable on the blockchain market. Though structured to support international cooperation, this effort induces global consumption of recycled products (Andrady, 2022).

In Australia, rooms such as Power Ledger offer consumers a strategy to trade excess solar electricity, thereby consuming such energy instead of relying on the wheels of traditional fossil energy (Mengelkamp et al., 2018).

Distributed in the same way as the ledgers over which they are created, blockchain also embraces international relations on resources. For example, the Basel Convention on the Control of Transboundary Movements of Hazardous Wastes and Their Disposal could use blockchain as an instrument to monitor the flows of hazardous waste crossed boundaries and compliance with the legal framework (Behnke & Janssen, 2020).

10.2.3 Smart Contracts for Global Trade in Circular Economy

The distribution of smart contracts across the globe presents tangible opportunities for automating global trade, especially in circular economy. Through them, middlemen and third-party enforcements are removed, and agreement is ensured through stipulated parameters, hence lowering the transaction costs while increasing trust.

In waste management, payments for waste collection can be distributed as soon as the required amount of recycled material is brought in and checked. It has the following benefits: This has become a system free from concealment and encourages recycling involving the borders (Saberi et al., 2019).

In resource sharing, i.e., usage of community resources, smart contracts help allocate and manage resources in the best possible ways. For instance, firms can employ blockchain technology to implement automatic leasing of equipment or other materials (Singh et al., 2024b). The things returned under the contract must be in usable condition and do not require anyone to remind the users since it is on an automated system where penalties are imposed once an item is late. Such systems have been known to support the circular economy since they enhance resource productivity and reduce waste (Bai et al., 2022).

Smart contracts are used broadly in international recycling projects. For instance, ventures such as Loop Industries employ blockchain technology to issue smart contracts that ensure that the companies agree to do what they promised in terms of sustainability. This level of integration minimizes fraud and validates the implementation of circular economy principles (Shen, 2022).

10.2.4 Data Security and Immutability for Global Stakeholders

There is a significant use of blockchain in recording data regarding e-waste. Blockchain is applied to e-waste in the same way and is being trialled by organizations, including Dell Technologies, which records the recycling of used electronics through blockchain (Hsu, 2021). Blockchain technologies such as VeChain ensure the documentation of carbon credits for global stakeholders to assess the verity of sustainability claims and monitor ecological injury (Treiblmaier, 2019). Using cryptographic features, blockchain guarantees the privacy of data and, at the same time, opens only the necessary data to specific subjects. It is useful for industries such as healthcare and agriculture because the security of data is important for generating trust among the customers (Fosso Wamba et al., 2020).

10.3 Use Cases of Blockchain in the Global Smart Circular Economy

The application of blockchain in the circular economy has created many innovations in various industrial fields. It is for these reasons that through encouraging transparency, trust and automation, challenges related to waste management, supply chain, sharing of resources, energy trading and agriculture can be addressed through blockchain (Singh et al., 2023). This section covers important cases according to the geographical regions of the world (Singh, 2024a).

10.3.1 Waste Management and Recycling Across Borders

Waste management is one of the key elements of the circular economy, especially during its application to the production of e-waste and hazardous waste.

The Basel Convention makes provisions for the control of the transboundary movement of hazardous wastes and their disposal, and the European Union has applied blockchain to track hazardous wastes as envisaged under the Basel Convention. Blockchain solutions like Circularise allow the company to report and share information on waste streams while keeping it as private and secure as possible (Behnke & Janssen, 2020).

In Asia, for instance, both China and South Korea engage in using the blockchain system to oversee waste handling. For example, VeChain cooperates with the municipal government of Shanghai to build a blockchain life cycle platform for medical and industrial waste. Likewise, India has identified blockchain for integrating and identifying electronic waste recycling through startups, including TraceX, that trace the flow of used electronics to authorized recyclers (Gupta et al., 2021).

Blockchain is being utilised to solve problems with informal waste collection in Africa. The Egypt- and South Africa-based initiative known as the Plastic Bank pays people in plastic waste, which is then recorded over the blockchain and sold further to manufacturers under Social Plastic (Andrady, 2022). This model not only encourages recycling but also opens job opportunities for vulnerable groups of society.

10.3.2 Circular Supply Chain Management

Circular supply chain management involves identifying and tracking materials throughout their lifecycle to prepare them for reuse, recycling, and other uses. Blockchain offers the framework that is necessary for global giants to put such systems in place efficiently.

In the electronics industry, blockchain is currently applied to track the journey of items and materials through their lifecycle. HP and Dell are examples of corporations that leverage blockchain to track the reprocessing of printer cartridges and computers, respectively, to check compliance with EPR rules (Hsu, 2021).

When it comes to the automotive industry, one of the four sections where circularity is enhanced through blockchain is through components such as batteries and tyres. For instance, the BMW company applies blockchain technology to manage the cobalt sourcing for electric vehicles and batteries and support its responsible mining and recycling (Bai et al., 2022).

An important topic that is growing in importance in circular supply chains is digital product passports. These passports consist of key information about a product, where its components were sourced, its design and how it should be recycled in order to help global actors make better decisions and to facilitate circular economy (Saberi et al., 2019).

10.3.3 The Sharing Economy: A Global Perspective

Blockchain has once again disrupted the sharing economy through decentralized, trustless transactions. This technology enables the global sharing of resources in different sectors: machinery, transportation, and shelter. GoShare is an example of a platform that was developed based on the application of blockchain in the Asia-Pacific region and aims to unite users who want to rent tools and equipment. Smart contracts guarantee the proper execution of the transactions without much controversy among users (Mengelkamp et al., 2018).

In Europe Share &Charge has, for instance, taken the role of a broker to allow for decentralized sharing of EV charging stations via the use of blockchain technology. The above initiative will help in reducing its reliance on central-owned structures and encourage the use of environmentally friendly means of transport (Treiblmaier, 2019).

In Latin America, through decentralized housing platforms, the deficit of housing is solved since individuals are connected by resource-sharing agreements on housing usage. They help to make the rental business clear and eliminate hurdles to housing (Fosso Wamba et al., 2020).

10.3.4 Renewable Energy and Resource Sharing

Their decentralised models of renewable energy trading and global resource sharing have seen blockchain technology become a fundamental pillar in the deployment of blockchain enablement in the renewable energy sector.

10.3.4.1 Peer-to-Peer Energy Trading

In Germany, for instance, there are blockchain firms such as Power Peers, which help households sell excess solar energy to their neighbours. These systems rely on smart contracts that make the transactions fair with the right price and safe exchange (Mengelkamp et al., 2018).

Power Ledger has achieved the same outcomes in Australia; community-scale energy resources can be managed autonomously. This model not only helps in using distributed energy resources and minimise the dependency on both the central and residential energy networks but also contributes to the use of renewable energy (Behnke, & Janssen, 2020).

10.3.4.2 Cross-national Models of Resource Distribution

Classically, the application of blockchain can also facilitate international cooperation in resource sharing. For example, the Energy Web Foundation utilise Saidu to manage renewable energy credits worldwide; it makes sure that countries adhere to carbon emission-cutting deals such as the Paris Accord (Treiblmaier, 2019).

10.3.5 E-Waste and Rare Earth Metal Management

The disposal of e-waste and rare earth metals is a concern of contemporary global business since improper disposal leads to environmental pollution and depletion of resources. Blockchain provides solutions for monitoring and conducting a proper disposal of these materials.

In the Asia region, blockchain is being used to oversee e-waste centres in China and Malaysia, amongst others. In the VeChain platform, the lifecycle of discarded electronics is recorded on the blockchain; accordingly, precious metals, including gold and palladium, are properly and safely recycled.

Blockchain also supports the mobilisation of international cooperation for e-waste management. For example, the United Nations' Step Initiative has utilised blockchain to register global e-waste shipments and dispose of records that must follow International Waste Management standards (Behnke & Janssen, 2020).

10.3.6 Sustainable Agriculture and Food Systems

Agriculture is also one of the fields where blockchain is used as a driver for circularity. As a consequence, blockchain promotes global cooperation in agriculture by increasing the level of supply chain transparency while supporting sustainable farming practices (Singh, 2024a, b).

This chapter examines the issue of traceability in agricultural supply chains for the following reasons:

- To identify and evaluate the various approaches to traceability used in agricultural supply chains.
- To assess how companies operating in agricultural supply chains can implement this concept.
- To understand the benefits that might be derived from the use of traceability in agricultural supply chains.
- To determine how companies in agricultural supply chains can maximize the returns from their investments in trace.

10.4 Challenges and Limitations: A Global View

10.4.1 Scalability and Energy Consumption

The two main fears people have regarding the technology are its scalability issue and the energy problem, especially where public ledgers utilizing the PoW consensus algorithm are involved. While more blockchain networks are developed, efforts grow towards consuming energy to process and manage transactions as well as update the distributed ledger.

Presumably, the problems of using blockchain in China would merge with global challenges if the rate of adoption of such technologies were sustained: when it comes to the impact of blockchain on the environment, the situation in China is critical. The country is home to many mining farms due to the availability of cheap power from coal-fired power stations. This has stirred the horn of debate on the sustainability of blockchain technology. The report conducted by the Cambridge Centre for Alternative Finance in 2021 revealed that over 65% of Bitcoin's computational power is in China, according to the two sources cited by the authors Zohar and Dhanraj (2021).

Europe, on the other hand, has maintained an environmentally friendly approach to the adoption of blockchain technology. Recently, the European Union (EU) has acknowledged the environmental impacts that blockchain presents and has developed legislation to ensure that new technologies being adopted follow environmentalism standards. For instance, the EU passed the Digital Finance Package in 2020 and the Green Deal to promote the creation of environmentally friendly blockchain solutions. Further, there is encouragement to seek other energy-elaborate consensus mechanisms such as PoS, which the EU supports, and such mechanisms are far more efficient than PoW (Bai & Yung, 2022).

10.4.2 Regulatory and Legal Barriers

The consumer protection issue is one more important and evident challenge, rendering global blockchain adoption critically reliant upon the steady establishment of frameworks. The approaches to the regulation of blockchain systems have significantly differed in various countries, which has led to the formation of a fragmented system that hampers the overall creation of international applications of blockchain in the sphere of circular economy.

In the European Union, GDPR was enacted in May 2018 to advance the ways in which data privacy is handled in blockchain architectures. The GDPR limits conditions under which data can be processed and shared, which is problematic for blockchain because it cannot be changed or erased once written. Therefore, while implementing GDPR, there is a conflict between the GDPR regulation and some inherent properties of blockchain, such as decentralisation and transparency (Elzain & Nasr, 2021).

On the other hand, some countries accept blockchain technology with minimal legal restraints, some of which are India and Brazil, while others, like Nigeria and Vietnam, have approached the technology with precautions. This external structure raises challenges when it comes to the implementation of blockchain across some of these nations since there is still ambiguity on how blockchain will be fitted into these laws, especially in the regions of intellectual property, contract law, and possibly taxation (Rajagopal et al., 2020).

10.4.3 Economic and Technological Disparities

Although there have been tremendous developments in the blockchain industry in recent years, the high costs required for attaining blockchain facilities in the existing business structure still make it hard to popularize blockchain solutions, especially in less developed countries. Blockchain technology entails massive physical and human capital investment in hardware, software, and manpower and receives regular updates and maintenance, which many small companies or governments might not afford (Singh et al., 2024a).

Global adoption of blockchain is also influenced by disparities in technology. Digital literacy, access to the internet and relevant support infrastructure for blockchain systems are still issues of concern in many developing countries. Currently, about 40% of the world population does not have effective access to an internet connection, which is a primary requirement for deploying all blockchain solutions that are anchored on the provision of real-time data inputs.

However, it can be noted that the cost of energy in diverse regions further affects the usage of blockchain in the circular economy. Whereas some nations have affordable RE, others use fossil energy, which greatly increases the cost of adopting blockchain solutions in areas such as energy markets and supply chain tracking. For example, Norway, with favourable and sufficient hydroelectric power supplies, is the best place to implement blockchain solutions for energy management, while developing nations such as India and South Africa are disadvantaged because they use coal and other non-renewable sources of energy power (Gupta et al., 2021).

10.4.4 Data Privacy and Ethical Concerns

They point to the intrinsic character of data in a blockchain, as well as the data ownership and control questions that it provokes. Another critical area involves the question of data ownership globally since, in blockchain systems, the data involved normally constitutes personal or business-sensitive data which is hosted on decentralized networks (Rubee Singh, 2024).

In the EU, for example, due to their privacy policies, the question of who owns the data stored in a blockchain is highly debatable. One of the principle processing operations, according to the GDPR, is deletion or modification, which cannot be accomplished on the blockchain itself because of its permanent and unalterable records. This has resulted in discussions on whether blockchain can meet data protection laws and how businesses can meet the privacy standards and transparency brought about by blockchain (Elzain & Nasr, 2021).

On the other hand, in developing countries where privacy laws are not as elaborate, the work of the blockchain is twofold; it is as beneficial as it is harmful. Although it can be useful for striving for belief systems like voting or financial transactions, it can present critical information, especially in scenarios where cyberspace security is nascent (Nassiry, 2018a, b).

The ethical issues arising from the application of blockchain technology are also anchored to the environmental issues of the block as explained under scalability and energy use, as noted earlier in this block. First, the case indicates that countries with high blockchain adoption have a challenge of balancing between technological development and the ethical importance of a non-harming environment. The topic of blockchain sustainability is still one of the most contentious discussions to the present day because a majority of applications that use the framework cannot rely on sustainable consensus algorithms such as proof-of-work (Zohar & Dhanraj, 2021).

10.5 Global Collaboration for Scaling Use Cases

The World Economic Forum has been at the forefront of pushing for blockchain in circular economy uses, such as promoting sustainable mining and e-waste recycling. For example, WEF's Global Blockchain Council project was established to assemble blockchain specialists from various industries with the goal of creating blockchain solutions that can support the implementation of SDGs (World Economic Forum, 2022).

10.5.1 Key Global Use Cases and Learnings from Diverse Regions.

1. **Waste Management and Recycling**: Currently, there is an application of blockchain technology to follow the waste management streams and the levels of compliance with recycling regulations all over Europe. For instance, the Circularise project applies blockchain to manage material flows, particularly in the automotive and electronics sectors, identifying which valuable material is being wasted instead of being reclaimed. This project is a clear pointer to the need for collaborative governance since various players, including companies,

governments, and consumers, are responsible for making the system effective (Circularise, 2021).
2. **Renewable Energy Trading**: In Germany and Australia, blockchain is being applied to establish a P2P energy market where consumers and producers of energy from renewable sources are allowed to sell and buy energy to and from each other. This decentralized model enhances energy utilization efficiency, brings down carbon emissions and increases renewable energy technology uptake. These include the Power Ledger, which focuses on the use of blockchain in energy in Australia, and the Energy Web Foundation, which focuses on the same in Europe. These cases show how decentralized technology can lower the cost of entry for SPPs in bringing electricity access to remote areas (Energy Web Foundation, 2022).
3. **Circular Supply Chains in Fashion**: Some examples from fashion brands that currently implement blockchain solutions are Patagonia and Levi's, both of which utilise blockchain to track the entire life cycle of a garment, from the acquisition of materials to recycling. This application of blockchain in traceability makes it possible to guarantee that ordered textiles and garments are produced in the right way and that their impact on the environment is brought to a minimum. In the Fashion Industry, still an area infamous for numerous sustainability issues, blockchain is assisting in activities like The Fashion Industry Charter for Climate Action, which aims at cutting on waste, enhancing the product lifespan, and getting better use of resources (Fashion Revolution, 2021).
4. **Agriculture and Food Systems**: Blockchain is another industry that is benefitting sustainable agriculture and food systems. In areas such as India and the South American region, blockchain solutions are being applied to enhance the supply chain information for commodities such as coffee, cocoa, and palm oil, amongst others. These blockchain-based platforms help consumers track the origin of their meals and confirm that they are received as a result of appropriate means (Khan et al., 2023). In Africa, a mobile app called AgUnity uses blockchain to improve the transparency of agriculture markets and support farmers in finding a fair buyer and organising their work adequately (AgUnity, 2021).

10.6 Implications

The incorporation of blockchain technology into circular economy opens up the potential for innovation in many industries with the potential of bringing in more effectiveness, efficiency as well as improved transparency. In supply chains, blockchain works to ensure that proper ethical standards are followed and that resources are made the best use of. This is made possible through smart contracts that enable people to collaborate by automating transactions, lowering costs, eliminating middlemen, and currently recycling in the energy trading and sharing of resources. These decentralized models include blockchain in supporting P2P energy markets and waste management, which results in lowering environmental impacts and promoting RE. Also, blockchain's ability to provide secure peer-to-peer data exchange creates confidence in the various stakeholders who, in turn, carry out efficient international collaborations in an environmentally conscious manner. Such implications point to blockchain as a crucial enabler of system transformation toward a sustainable, resilient global economy (Anil Kumar, 2024).

10.7 Conclusion

There is great promise in the blockchain approach in the future development of the circular economy since it can bring the benefits of increased supply chain transparency, better tracking, and improved cooperation. It also presents its revolutionary uses in fields including manufacturing, supply, waste management, energy, and trading, particularly renewable energy. However, there are crises such as high energy consumption consensus algorithms, the ability to scale the networks and disjointed regulatory environments that hamper the efficient adoption of blockchain systems. The use of AI and IoT as complementary technologies is crucial in the realization of such gaps for efficient and effective implementation. The utilisation of the EU Blockchain Partnership, the UN SDGs, and cooperative forums on an international level is necessary to reconcile regulations, encourage innovation, and create best practices for blockchain applications of circular economy systems. Understanding the weak points that currently define the technology and overlaying them with unique strengths can enable blockchain to become a model for resourceful and sustainable economies of the future.

10.8 Limitations

There are certain challenges that blockchain comes with: scalability and high energy consumption, particularly for blockchains implementing the proof-of-work protocol. The varying degrees of regulation on the use of blockchain throughout the world prevent harmonization between countries, making cross-border relations a challenge. Since blockchain is permanent and linear, privacy and ethical issues are bound to surface with regard to data protection laws such as GDPR, as well as rights of data usage and ownership. Lack of funding and technological advancement are other barriers to the use of blockchain solutions in developing nations since the use comes with high costs, and the nations lack requisite digital infrastructure. More to the point, the carbon footprint of blockchain mining in areas that depend on fossil energy is still a concern of discussion. These limitations justify the incapability of blockchain to optimise circular economy to its fullest without more enhanced technology, legislative measures, and structures.

10.9 Future Scope of Research

The actual development research on blockchain in the circular economy is envisioned to be focused on optimizing the application of artificial intelligence, the Internet of Things, and 5G technology to form a series of perfect circles on resource utilization and waste reduction (Khan et al., 2024). Future research could extend to improve the application of smart contracts in supporting circular supply chain operations, such as monitoring the circular supply chain and the recycling processes more effectively (Tuzhilin, 2022). Studying the energy-efficient approach of blockchain, how Ethereum recently shifted from PoW to PoS, is essential in reducing environmental costs (Ethereum Foundation, 2022). I tentatively pioneered the analysis of the standardized guidelines and harmonized regulations of the EU Blockchain Partnership and the UN SDGs from the perspectives of the worldwide launching base (United Nations, 2021). Also, a case of a similar kind of application like Circularise has the potential to provide real-world insight into how the tracking of material could be enhanced with the help of IoT sensors and blockchain (Circularise, 2021). In addition, AI forecast applications can also help with the identification of the spots within the supply chain where a particular resource has not been optimally deployed, which a business can use the insights obtained from the analysis to improve the loops in the field.

References

AgUnity. (2021). Blockchain for sustainable agriculture. *AgUnity*. https://www.agunity.com

Andrady, A. L. (2022). Microplastics in the environment: A global problem. *Science, 376*(6593), 449–451. https://doi.org/10.1126/science.abc8901

Anil Kumar, R. S. (2024). *Artificial intelligence enabled management an emerging economy perspective*. De Gruyter.

Bai, C., & Yung, R. (2022). Blockchain and sustainability: Enabling circular supply chains. *Journal of Cleaner Production, 321*, 128–135. https://doi.org/10.1016/j.jclepro.2021.128998

Bai, C., Sarkis, J., & Dou, Y. (2022). Blockchain technology for enhancing circular economy in supply chains. *International Journal of Production Research, 60*(11), 3276–3290. https://doi.org/10.1080/00207543.2021.1988024

Behnke, K., & Janssen, M. F. W. H. A. (2020). Boundary conditions for traceability in circular supply chains using blockchain technology. *International Journal of Information Management, 52*, 101969. https://doi.org/10.1016/j.ijinfomgt.2019.05.025

Circularise. (2021). Blockchain for transparency in the circular economy. *Circularise*. https://www.circularise.com

Ellen MacArthur Foundation. (2013). *Towards the circular economy: Economic and business rationale for an accelerated transition*. https://www.ellenmacarthurfoundation.org/publications

Elzain, M. A., & Nasr, M. (2021). Blockchain and GDPR: Compatibility and challenges. *International Journal of Law and Information Technology, 29*(1), 1–20. https://doi.org/10.1093/ijlit/eaaa011

Energy Web Foundation. (2022). *The future of energy trading: Blockchain-based peer-to-peer energy markets*. Energy Web Foundation. https://www.energyweb.org

Ethereum Foundation. (2022). *Ethereum 2.0: Energy efficiency and the transition to proof of stake*. Ethereum Foundation. https://ethereum.org/en/upgrades/
European Commission. (2020a). *A new Circular Economy Action Plan for a cleaner and more competitive Europe*. https://ec.europa.eu/environment/circular-economy/
European Commission. (2020b). *Digital finance package*. European Commission. https://ec.europa.eu/info/business-economy-euro/banking-and-finance/digital-finance_en
Fashion Revolution. (2021). *Blockchain in the fashion industry: Transparency and traceability*. Fashion Revolution. https://www.fashionrevolution.org
Fosso Wamba, S., Queiroz, M. M., & Trkman, P. (2020). Blockchain in operations and supply chain management: Benefits, challenges, and future research opportunities. *International Journal of Information Management, 52*, 102064. https://doi.org/10.1016/j.ijinfomgt.2019.102064
Gupta, A., Kumar, V., & Verma, R. (2021). Energy consumption and sustainability: Blockchain as a solution. *Energy Policy, 153*, 112204. https://doi.org/10.1016/j.enpol.2021.112204
Hsu, C. (2021). Blockchain for e-waste management. *Journal of Cleaner Production, 281*, 125259. https://doi.org/10.1016/j.jclepro.2020.125259
Kamath, R. (2020). Food traceability on blockchain: Walmart's pork and mango pilots with IBM. *The Journal of the British Blockchain Association, 1*(2), 371–377. https://doi.org/10.31273/fbba.v1i1.34
Khan, S., Khan, M. I., & Singh, R. (2023). Modelling the barriers of blockchain technology implementation in supply chain. *Journal of Industrial Integration and Management, 8*(3), 367–385. https://doi.org/10.1142/S2424862223500070
Khan, S., Singh, R., Alnahas, J., Abbate, S., & Centobelli, P. (2024). Navigating the smart circular economy: A framework for manufacturing firms. *Journal of Cleaner Production, 480*, 144007. https://doi.org/10.1016/j.jclepro.2024.144007
Kumar, K., Singh, R., Sharma, A., & Lehri, B. P. S. (2024). Climate change adaptation strategies for achieving net-zero economy. In *Transition towards a sustainable future. Approaches to global sustainability, markets, and governance*. Springer. https://doi.org/10.1007/978-981-97-5756-5_8
Mengelkamp, E., Gärttner, J., Rock, K., Kessler, S., Orsini, L., & Weinhardt, C. (2018). Designing microgrid energy markets: A case study: The Brooklyn Microgrid. *Applied Energy, 210*, 870–880. https://doi.org/10.1016/j.apenergy.2017.06.054
Nassiry, D. (2018a). *The role of fintech in unlocking green finance: Policy insights for developing countries* (ADB Institute Working Paper). https://doi.org/10.2139/ssrn.3104188
Nassiry, S. (2018b). Blockchain in emerging markets: Challenges and opportunities. *Technology in Society, 56*, 64–71. https://doi.org/10.1016/j.techsoc.2018.03.005
Rajagopal, R., Dandekar, P., & Sharma, P. (2020). Blockchain in developing countries: A barrier or opportunity? *Global Business Review, 21*(4), 1023–1042. https://doi.org/10.1177/0972150919878381
Rubee Singh, S. K. (2024). Modelling the organisational factors for implementation of corporate social responsibility: A modified TISM approach. *Global Journal of Flexible Systems Management, 25*, 283–301.
Saberi, S., Kouhizadeh, M., Sarkis, J., & Shen, L. (2019). Blockchain technology and its relationships to sustainable supply chain management. *International Journal of Production Research, 57*(7), 2117–2135. https://doi.org/10.1080/00207543.2018.1533261
Shen, B. (2022). Blockchain-enabled supply chain transparency in fashion. *Clothing and Textiles Research Journal, 40*(1), 31–45. https://doi.org/10.1177/0887302X211045674
Singh, M. B. (2024a). Talent management facilitates net-zero transition through employee green behavior. In R. Singh, S. Khan, A. Kumar, S. Luthra, & H. Chokshi (Eds.), *Net zero economy, corporate social responsibility and sustainable value creation*. Springer.
Singh, R. (2024b). A comprehensive view of AI based technologies for sustainable development goals. In R. Singh, S. Khan, A. Kumar, & V. Kumar (Eds.), *Artificial intelligence enabled management an emerging economy perspective*. De Gruyter.

Singh, R., Khan, S., & Centobelli, P. (2022). Investigating the interplay between social performance and organisational factors supporting circular economy practices. *Sustainability, 14*(24), 16781. https://doi.org/10.3390/su142416781

Singh, R., Khan, S., & Dsilva, J. (2023). A framework for assessment of critical factor for circular economy practice implementation. *Journal of Modelling in Management, 18*(5), 1476–1497. https://doi.org/10.1108/JM2-06-2021-0145

Singh, R., Crowther, D., & Ravesangar, K. (2024a). Harnessing human capital for a sustainable future: HR's pivotal role toward net zero policies. In *Transition towards a sustainable future*. Springer. https://doi.org/10.1007/978-981-97-5756-5_2

Singh, R., Gupta, A., Bajpai, A., & Kandpal, V. (2024b). Exploring challenges of circular economy initiatives for smart cities. In *Smart cities and circular economy*. Emerald Publishing Limited. https://doi.org/10.1108/978-1-83797-957-820241005

Treiblmaier, H. (2019). Combining blockchain technology and the physical internet to achieve triple bottom line sustainability: A comprehensive research agenda for modern logistics and supply chain management. *Logistics, 3*(10), 1–13. https://doi.org/10.3390/logistics3010001

Tuzhilin, A. (2022). AI-powered blockchain for smart circular economy. *Journal of Artificial Intelligence in Business, 7*(3), 87–98. https://doi.org/10.1016/j.artifintell.2022.04.003

United Nations. (2021). *Blockchain for sustainable development*. United Nations. https://www.un.org/blockchain-sdg

World Economic Forum. (2022). *The role of blockchain in sustainable development*. World Economic Forum. https://www.weforum.org/reports/blockchain-sdgs

Zeng, W., Dai, L., & Liu, M. (2017). Circular economy: From theoretical discussion to policy practice. *Journal of Environmental Management, 197*, 109–116. https://doi.org/10.1016/j.jenvman.2017.03.001

Zohar, S., & Dhanraj, R. (2021). Blockchain's energy consumption: A global perspective. *Journal of Energy Management, 22*(2), 97–110. https://doi.org/10.1016/j.jenman.2021.01.004

Chapter 11
Blockchain-Enabled Smart Circular Economy: Strategies for Sustainable Organizational Practices

Rubee Singh, **Akshay Singh**, **Aastha Jain**, and **Vikas Kumar**

Abstract The shift from linear to circular economic models is critical for addressing resource depletion and environmental challenges. Blockchain technology, with its attributes of transparency, traceability, and decentralized data management, provides innovative solutions to enhance circular economy (CE) practices. This study develops a conceptual framework to integrate blockchain into CE, focusing on optimizing environmental, economic, and social sustainability. Through an extensive review, the study identifies blockchain's role in tracking material lifecycles, reducing waste, and automating processes like recycling through smart contracts. Findings reveal that blockchain-enabled CE practices minimize greenhouse gas emissions, reduce operational costs, and enhance resource efficiency. Additionally, these practices create green jobs, promote fair labor standards, and foster community engagement.

The study concludes that blockchain-powered CE strategies significantly support sustainability goals but require addressing challenges like energy consumption and IoT integration. This research offers actionable strategies for implementing blockchain-enabled CE systems.

Keywords Blockchain technology · Circular economy · Organizational practices · Sustainability · Efficient management

---

R. Singh (✉)
Institute of Business Management, GLA University, Mathura, India

A. Singh
School of Management Studies, Indira Gandhi National Open University, New Delhi, India

A. Jain
Department of Management Studies, Vaish College of Engineering, Rohtak, Haryana, India

V. Kumar
University of Portsmouth, Portsmouth, UK
e-mail: vikas.kumar@port.ac.uk

R. Singh et al. (eds.), *Blockchain Technologies for Smart Circular Economy and Organisational Sustainability*, Environmental Footprints and Eco-design of Products and Processes, https://doi.org/10.1007/978-3-031-89183-0_11

11.1 Introduction

Since the twentieth-century industrial revolution, globalization has driven nations to compete by producing more goods and services to meet market demand and generate more profits (Abad-Segura et al., 2020). As economies and nations become more interconnected, regional conflict like the Ukraine-Russia war has significant and wide-ranging consequences. Rapid globalization, along with fast-paced economic and industrial development, has triggered significant challenges, such as resource scarcity and socioenvironmental problems worldwide (Sachs, 2015). Industrial production is a significant source of waste generation, which continues to be a major factor contributing to environmental pollution (Abdel-Shafy & Mansour, 2018). The world predominantly follows a linear model based on the "take, make, and dispose" approach (Shojaei et al., 2021). While this conventional model has brought economic growth, it has also led to severe environmental degradation (Khan et al., 2021). This situation highlights the urgent need for a new system, like the circular economy (CE), to improve environmental sustainability (Geissdoerfer et al., 2017).

Nowadays, the CE has gained attention as a resolution to problems such as limited resources, environmental harm, and unsustainable production and consumption habits. CE focuses on reducing waste by reusing, repairing, recycling, and restoring materials while promoting cooperation between industries (Khan et al., 2021, 2022a). Implementing CE effectively requires a comprehensive, 360-degree strategy encompassing all stages of production, including planning, designing, operations, and evaluation. Researchers assert that CE has the potential to drastically contribute to various "UN Sustainable Development Goals" (SDGs), predominantly SDG-12, which focuses on "responsible consumption and production" (Singh et al., 2024c; Schroeder et al., 2019). Consequently, exploring how authorities and organizations can trace product life cycles to encourage reutilization and successfully integrate CE systems into sustainable organizational practices has become a critical area of research (Schroeder et al., 2018).

This study emphasizes the potential of blockchain technology (BCT) to support the CE paradigm with its inherent attributes of transparency, traceability, and tamper-proof record-keeping. Blockchain can address key challenges in the implementation of CE systems (Friedman & Ormiston, 2022; Khan et al., 2021; Kouhizadeh et al., 2020; Leng et al., 2020). BCT enables the integration and analysis of data, which is essential for optimizing resource flows in a circular economy. For example, blockchain-powered digital networks can improve information sharing across the supply chain, reducing wastes and ensuring the efficient reuse of resources (Böckel et al., 2021). These networks facilitate the flow of accurate and real-time data, which can improve decision-making processes and help businesses implement sustainable practices that line up with circular economy principles.

While existing research highlights the individual benefits of blockchain and circular economy practices, there is a limited understanding of their integrated application and resultant impacts on performance metrics. This paper aims to address this gap by presenting a conceptual framework that links blockchain technology, CE practices, and sustainable performance.

11.2 Literature Review

11.2.1 Blockchain-Technology-and-Circular Economy

The emergence of blockchain technology plays a key role in making over "economic," "environmental," and "organizational performance" (Swan, 2015). Blockchain operates by managing uniquely identified blocks that are linked-together in a chain (Treleaven et al., 2017). The main attributes of blockchain technology (BT) comprise its decentralized ledger system, traceability, transparency, the permanence of transactions, and accessibility and credibility (Xu et al., 2019; Iansiti & Lakhani, 2017). Its decentralized nature ensures that all accomplices in a supply-chain have access to a shared, unchallengeable ledger, facilitating the tracking of material lifecycles and ensuring adherence to sustainability standards (Böckel et al., 2021; Di Vaio & Varriale, 2020). For instance, in recycling systems, blockchain enables efficient tracking of waste materials, ensuring proper recovery and reuse while preventing leakage into the environment. Blockchain fosters such collaboration by reducing information asymmetry and building through verifiable data (Rejeb et al., 2021). A crucial factor in building customer trust and optimizing supply chain operations is the system's traceability, which enables smoother movement of products and materials (Khan et al., 2022b; Alexandris et al., 2018). Moreover, blockchain technology offers strong security against data breaches, protecting organizational confidentiality (Kouhizadeh et al., 2020). It ensures continuous data integrity, preventing issues like double-spending, data corruption, and false ownership claims (Khan et al., 2021). Specifically, fraud prevention in the supply chain reduces costs and enhances efficiency (Upadhyay et al., 2021). These factors suggest that BCT adoption offers significant benefits that are crucial for supporting and enabling the circular economy (CE) model.

11.2.2 Circular Economy as a Sustainable Economic Model

The concept of a circular economy (CE) has emerged as a transformative approach to achieving sustainability by redefining traditional linear economic models. Unlike the traditional linear economic approach of "take, make, and dispose," the CE model prioritizes resource recovery, reuse, refurbishment, and recycling to ensure a more sustainable future (Singh et al., 2024a; Khan et al., 2021). By adopting CE-oriented production systems, organizations can ensure the optimal performance of materials and products. This approach enhances resource efficiency, which, in turn, contributes to better operational outcomes for firms (Sehnem et al., 2019). In their study, Zhu and Sarkis (2004) demonstrated that organizations can reduce costs and eliminate waste from their processes by integrating green practices. Key CE practices include *design for longevity*, where products are made durable and repairable to extend their lifecycle, and *remanufacturing and refurbishing*, which recover value from end-of-life products. Another significant practice is *industrial symbiosis*, where byproducts from one industry are utilized as raw materials for another, thereby reducing waste (Khan et al., 2021). Implementing circular economy (CE) practices proposes substantial assistances across "environmental," "economic," and "social" dimensions. From an environmental perspective, these practices help cut down greenhouse gas emissions, reduce waste production, and conserve natural resources by lowering their consumption (Khan et al., 2021; Geissdoerfer et al., 2017). Economically, they create new business opportunities, enhance resource productivity, and reduce costs related to the procurement of raw materials and disposal of wastes (Khan et al., 2021, 2024). Socially, CE practices foster innovations, create green jobs, and promote sustainable lifestyles among consumers (Upadhyay et al., 2021). Arguably, as organizations increasingly align their operations with sustainability goals, CE practices are anticipated to play a central role in accomplishing long-term economic and environmental resilience.

11.2.3 Sustainable Performance Metrics

Sustainable performance metrics are essential for evaluating the "environmental," "economic," "social," and "organizational" dimensions of sustainability, enabling businesses and stakeholders to align their operations with long-term sustainability goals (Chishty, 2023; Ghobakhloo et al., 2023). These metrics serve as tools to assess and track progress, identify gaps, and implement improvements across various aspects of performance.

Environmental performance involves evaluating a company's effect on natural resources, such as waste management, greenhouse gas emissions, and energy usage (Geissdoerfer et al., 2017). This enables organizations to monitor their ecological footprint and adopt strategies for minimizing environmental degradation. *Economic performance*, on the other hand, focuses on financial sustainability by assessing factors such as cost efficiency, profitability, and return on investment, ensuring that businesses maintain economic viability while adopting sustainable practices (Khan et al., 2021; Yang et al., 2019).

Besides environmental and economic aspects, *social performance* is essential for measuring an organization's contribution to societal well-being (Singh et al., 2022, 2024b; Geng et al., 2016). This typically includes aspects like employee well-being, diversity and inclusion, community involvement, and compliance with ethical labor standards (Upadhyay et al., 2021). They highlight the importance of fostering positive relationships with employees, communities, and other stakeholders to drive long-term social sustainability. Lastly, *organizational performance* evaluates internal efficiency, innovation, and governance frameworks that facilitate sustainable operations. This assessment includes reviewing corporate governance practices, ensuring supply chain transparency, and incorporating sustainability into core business strategies (Chishty, 2023; Khan et al., 2021).

11.3 Role of Blockchain in Smart Circular Economy

Blockchain technology plays an imperative role in endorsing the principles of a smart circular economy. It facilitates enhanced transparency and accountability by maintaining absolute, decentralized records of resource flows, product life cycles, and management of wastes (Böckel et al., 2021; Rejeb et al., 2021; Di Vaio & Varriale, 2020). With the integration of blockchain, organizations can effectively monitor the origins, usage, and disposal of materials, ensuring efficient reuse, remanufacturing, or recycling (Lu, 2019). This traceability helps businesses ensure that they adhere to sustainability goals, comply with environmental regulations, and demonstrate their commitment to reducing waste and minimizing environmental impact (Hastig & Sodhi, 2020; Alexandris et al., 2018). By monitoring the entire product life cycle, organizations can make informed decisions about reusing materials, enhancing resource efficiency, and minimizing their reliance on raw materials, which is a key principle of a circular economy.

Moreover, blockchain boosts collaboration among various stakeholders like suppliers, manufacturers, consumers, and recyclers, by offering a transparent platform for exchanging data and ensuring the accuracy of sustainability claims (Rejeb et al., 2021). For example, blockchain can support "product-as-a-service models," where products are leased, maintained, and returned for recycling, dropping the requirement of new production and minimizing waste. Additionally, blockchain's use of smart contracts can automate processes like recycling, remanufacturing, and the return of products, ensuring that these steps are carried out efficiently and promptly (Leng et al., 2020; Liu et al., 2020). By eliminating intermediaries and reducing transaction costs, blockchain also makes the entire process more cost-effective, enabling organizations to achieve their sustainability objectives without compromising on profitability (Kshetri, 2021).

Furthermore, blockchain technology can incentivize sustainable practices through tokenization (Zhao et al., 2023). Organizations can reward consumers and businesses for engaging in eco-friendly behaviors such as recycling or returning used products for refurbishment. These tokens can be exchanged for discounts, rewards, or other incentives, thus driving greater participation in circular economy practices. Blockchain also supports organizations in meeting their sustainability reporting requirements by providing real-time, verifiable data on their environmental impact, including waste reduction, carbon emissions, and resource utilization (Alnogaidan, 2024; Alotaibi et al., 2024). This not only helps in achieving compliance but also strengthens the organization's reputation among consumers, investors, and other stakeholders concerned with sustainability.

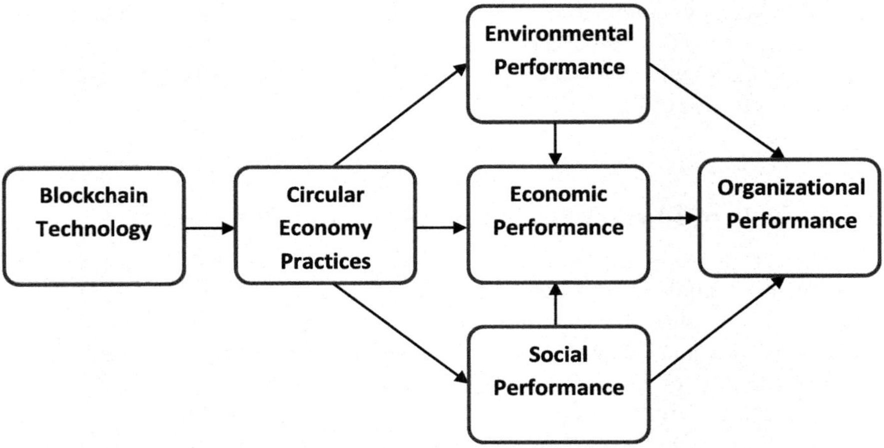

Fig. 11.1 Conceptual framework

Despite its potential, the integration of blockchain into organizational sustainability practices requires careful consideration of challenges, i.e., scalability, energy consumption, and the need for integration with Internet of Things (IoT) devices for real-time tracking. However, with continued advancements in blockchain technology, particularly in energy-efficient consensus mechanisms, these challenges can be addressed.

11.4 Conceptual Framework

The conceptual framework (refer to Fig. 11.1) illustrates the interconnections between blockchain technology, circular economy practices, and sustainable performance metrics.

Based on the above framework, the following propositions can be made:

Proposition 1 *Blockchain technology positively influences the efficiency of circular economy practices* (Lin et al., 2024; Khan et al., 2021, 2022a, b; Upadhyay et al., 2021; Mubarik et al., 2021).

Argument Blockchain technology significantly influences circular economy practices by offering a secure, transparent, and efficient platform to track materials, products, and transactions throughout their lifecycle. By utilizing blockchain, organizations can augment traceability and transparency in supply chains, guaranteeing that products and materials are reused, remanufactured, or recycled (Khan et al., 2023), which is central to circular economy principles. Blockchain allows for the creation of immutable records, ensuring that data on the origin, use, and disposal of products is accurate and cannot be tampered with, promoting trust among stakeholders. This transparency supports more effective recycling, waste management, and resource recovery practices.

Additionally, blockchain facilitates decentralized systems where consumers, manufacturers, and recyclers can directly interact, reducing reliance on centralized systems and permitting peer-to-peer transactions for shared resources or services. Smart contracts, motorized by blockchain, can automate processes such as product returns, recycling incentives, or the leasing of products, streamlining operations, and reducing transaction costs.

By enabling tokenized rewards for recycling or product returns, blockchain encourages sustainable consumer behavior, further promoting circularity. Overall, blockchain strengthens circular economy practices by improving data integrity, reducing waste, and fostering collaboration across all stakeholders, leading to a more sustainable and efficient economy.

Proposition 2 *Circular economy practices positively impact environmental performance* (Amin et al., 2024; Sahoo et al., 2023; Yin et al., 2023; Marrucci et al., 2022; Rehman Khan et al., 2022).

Argument One of the most important advantages of adopting CE practices is the improvement of environmental performance. Traditional production systems are resource-intensive and generate substantial waste, contributing to environmental degradation. By transitioning to circular models, organizations can mitigate these negative impacts. Circular economy practices, like resource efficiency, minimizing waste, and managing product lifecycle, reduce the reliance on raw materials and limit the waste directed to landfills. This leads to lower environmental footprints, including reductions in excess water consumption, greenhouse gas emissions, and energy usage (Silvestre & Ţîrcă, 2019).

The concept of closing the loop in CE practices focuses on extending the life cycle of products through reuse, remanufacturing, and recycling, which can significantly reduce resource extraction. For instance, rather than producing new goods from virgin materials, organizations can reclaim and reuse materials from old products, thereby decreasing environmental degradation associated with mining, deforestation, and other harmful extraction practices. By embracing circular principles, businesses not only reduce their carbon footprint but also contribute to preserving ecosystems and biodiversity.

The principles of the CE promote the creation of products that are durable, repairable, and recyclable, thereby reducing waste and extending the products' useful life. These practices lessen the demand for new resources, thereby alleviating the pressures on natural resources and contributing to sustainability goals. By using renewable materials and focusing on product stewardship, organizations can adopt a more environmentally responsible approach, creating a sustainable future for generations to come.

Proposition 3 *Blockchain enhances economic performance through cost reduction and new revenue streams* (Sahoo et al., 2023; Yin et al., 2023; Marrucci et al., 2022; Rehman Khan et al., 2022).

Argument CE practices also have a thoughtful impact on the economic performance of organizations. It focuses on generating value by optimizing resource use and eliminating waste. By adopting this approach, businesses can lower operational expenses, streamline supply chains, and generate additional sources of revenue. For example, organizations that implement product-as-a-service models can generate continuous income through leasing, renting, or refurbishing products rather than selling them outright. This not only prolongs the lifecycle of products but also fosters recurring revenue and strengthens customer loyalty.

Resource efficiency is another key economic benefit of circular economy practices. By using fewer raw materials and reducing waste, organizations can lower production costs, improving profitability. Moreover, circular practices reduce the financial risks associated with raw material price volatility and supply chain disruptions. With fewer dependencies on virgin materials, organizations are less susceptible to fluctuations in global commodity markets, which can lead to significant cost savings.

Circular economy practices also create opportunities for new business models and innovation. Organizations that embrace these practices often find new ways to create value, including the development of sustainable products, services, or technologies. For example, the growth of the second-hand market, repair services, and recycling technologies has created new business opportunities, contributing to job creation and economic growth. These practices enable businesses to diversify their income streams and tap into emerging markets focused on sustainability and environmental responsibility.

Additionally, cost savings associated with waste management and energy efficiency also contribute to enhanced economic performance. Businesses that implement circular practices can reduce waste disposal costs, optimize energy usage, and decrease the need for landfill space. This leads to direct financial savings, which improve profitability and allow organizations to reinvest in more sustainable initiatives.

Proposition 4 *Circular economy practices contribute to improved social performance* (Ghobakhloo et al., 2023; Upadhyay et al., 2021).

Argument The execution of CE practices not only yields environmental and economic advantages but also contributes to enhanced social performance. Circular economy initiatives contribute in the creation of jobs, the promotion of fair labor practices, and the enhancement of community engagement. By fostering industries centered around recycling, remanufacturing, and sustainable product design, organizations contribute to job creation in green sectors, which supports local economies and improves livelihoods. For example, the growing demand for recycling and waste management services has created jobs in sorting, processing, and reusing materials, benefiting both individuals and communities.

Circular economy practices also promote social equity by fostering inclusive growth and tackling inequality. In many cases, circular initiatives help underserved populations by providing opportunities for employment, skills development, and fostering entrepreneurship. For instance, communities involved in the informal recycling sector can benefit from more structured and sustainable recycling systems, which offer safer working conditions and fair wages. By empowering local communities and promoting fair labor standards, circular economy practices can help improve social well-being and reduce social disparities.

Moreover, organizations that prioritize consumer engagement and collaboration with stakeholders in circular economy initiatives promote a culture of sustainability. These practices foster trust and loyalty among consumers who value environmental and social responsibility. By communicating their circular practices and commitment to sustainability, organizations can build stronger relationships with customers and other stakeholders, including employees, suppliers, and investors.

Proposition 5 *Blockchain adoption in circular economy models leads to improved organizational performance* (Chishty, 2023; Samadhiya et al., 2023; Khan et al., 2021; Sehnem et al., 2019).

Argument The influence of circular economy practices on organizational performance is evident when considering the cumulative effects on "economic," "environmental," and "social" dimensions. Organizations that integrate circular economy principles into their core operations typically experience a range of benefits that contribute to long-term sustainability and growth. The environmental benefits, such as reduced waste and resource consumption, improve the company's image as a "responsible corporate entity," which can augment customer loyalty and attract environmentally conscious consumers.

Economically, adopting circular economy practices can lead to greater profitability, reduced costs, and increased revenue opportunities through new business models and markets. The cost savings associated with waste reduction, resource efficiency, and energy optimization directly contribute to improving the bottom line. Moreover, the circular economy can impel innovation, leading to new products and services that meet the rising demand for sustainable solutions.

From a social perspective, organizations that embrace circular economy practices improve their reputation and build stronger relationships with their customers, employees, and communities. By promoting sustainability and social equity, organizations contribute to positive societal impact and align with broader goals related to corporate social responsibility (CSR). This, in turn, can attract investment, strengthen brand value, and improve stakeholder engagement.

The adoption of circular economy practices within organizational operations significantly bolsters resilience. Companies that emphasize sustainability and resource efficiency are better equipped to navigate supply chain disruptions, adapt to regulatory shifts, and manage market volatility. By minimizing reliance on finite resources, these organizations foster greater stability and predictability, ultimately enhancing their overall performance and ensuring long-term viability.

11.5 Strategies for Implementing Blockchain-Enabled Smart Circular Economy

Implementing blockchain-enabled smart circular economy strategies in organizational practice involves integrating sustainability at every level of operations, from product design to supply chain management and waste reduction. The first key strategy is to adopt *transparent and traceable supply chains.* Organizations can leverage blockchain to track the movement of products and materials throughout their life cycle, from raw material extraction to disposal recycling. By creating immutable, real-time records, blockchain ensures complete visibility and accountability, making it easier to monitor sustainability practices and minimize environmental impact (Rani et al., 2024). Organizations should prioritize integrating blockchain with existing supply chain technologies and involve key stakeholders, such as suppliers and recyclers, to enable seamless data sharing across the network.

The second strategy is to incorporate *circular business models* into organizational practices. Blockchain facilitates the shift from a linear to a circular economy by enabling product-as-a-service models, where products are leased rather than sold. This approach encourages reuse, remanufacturing, and recycling, ensuring that resources stay in circulation longer and reducing waste (Aithal & Aithal, 2023; Singh et al., 2023). Smart contracts on the blockchain can automate returns, repairs, and remanufacturing processes, ensuring efficient and timely product life cycle management. Companies should design products with modularity and recyclability in mind, using blockchain to track the condition and lifecycle stage of each item.

A third strategy focuses on *rewarding sustainable consumer behavior*. Blockchain can introduce tokenized incentive systems that reward consumers for returning used products or participating in recycling programs. By offering digital tokens or benefits tied to blockchain-based rewards, organizations can encourage sustainable consumer habits and ensure that products are properly disposed of or recycled (Esmaeilian et al., 2020). This not only drives sustainability efforts but also helps build customer loyalty and enhances the brand reputation as an environmentally responsible business.

To support these strategies, organizations must *integrate blockchain with IoT and data analytics* to enable real-time monitoring and decision-making. Using IoT devices, businesses can track resource consumption, waste production, and product conditions, feeding this data into blockchain systems for accurate, real-time reporting (Bressanelli et al., 2022). Analytics can then help businesses identify inefficiencies and areas for improvement, allowing for more effective resource allocation and waste management (Esmaeilian et al., 2020). This data-driven approach ensures continuous improvement in sustainability performance.

Finally, *collaboration and partnership with industry stakeholders* are crucial for the successful implementation of blockchain in the smart circular economy. Organizations should work with suppliers, recycling companies, and industry partners to develop shared blockchain platforms, ensuring that sustainability practices are consistently applied across the value chain. This collaboration can help streamline logistics, improve recycling processes, and enhance the efficiency of the overall circular economy ecosystem (Awan et al., 2022; Veleva & Bodkin, 2018). By creating a unified, blockchain-enabled network, organizations can drive collective action toward sustainability and foster a circular economy that benefits all stakeholders.

11.6 Conclusion

Blockchain-enabled smart circular economy strategies transform how organizations approach sustainability and resource efficiency. Blockchain enhances transparency, traceability, and accountability, enabling businesses to monitor and optimize resource use throughout the entire product lifecycle. This integration ensures efficient reuse, remanufacturing, and recycling, reducing waste and promoting long-term environmental sustainability. By adopting blockchain, businesses establish robust systems that trace the flow of materials and provide immutable, tamper-proof records. These records not only help organizations comply with environmental regulations but also build trust among stakeholders by indicating a clear dedication to sustainability. Additionally, blockchain-driven solutions improve decision-making by providing real-time data on resource utilization, helping organizations align their operations with circular economy objectives.

Blockchain technology fosters collaboration among diverse stakeholders including suppliers, manufacturers, consumers, and recyclers by enabling transparent and decentralized platforms for data exchange. The integration of smart contracts enhances operational efficiency by automating critical processes, such as product returns, recycling, and remanufacturing. These advancements not only lower operational costs and streamline workflows but also promote the adoption of circular business models, i.e., product-as-a-service. This approach emphasizes leasing, reusing, and recycling goods, thereby reducing waste and optimizing resource efficiency.

Economically, blockchain-powered circular practices provide organizations with significant cost-saving opportunities. Businesses decrease their reliance on raw materials, minimize risks linked to resource shortages, and create new sources of revenue by adopting sustainable business models. Tokenized incentive systems reward eco-friendly consumer behavior, building customer loyalty and strengthening brand reputation. Moreover, businesses adopting these practices enhance their resilience against market volatility and regulatory challenges, ensuring long-term growth and stability.

Socially, circular economy practices powered by blockchain contribute to job creation, promote fair labor practices, and empower communities. Industries focused on recycling, remanufacturing, and sustainable product design generate green jobs, improving livelihoods and fostering inclusive growth. By engaging consumers and local communities in sustainability efforts, organizations enhance their relationships with stakeholders and foster a culture of environmental and social responsibility.

Despite its potential, organizations must address challenges associated with blockchain adoption, such as energy consumption and integration with IoT technologies. Continued advancements in energy-efficient blockchain solutions and real-time data analytics will play a critical role in overcoming these barriers. Additionally, collaboration with industry stakeholders to develop unified blockchain networks will amplify the impact of these technologies.

References

Abad-Segura, E., Fuente, A. B. D. L., González-Zamar, M. D., & Belmonte-Ureña, L. J. (2020). Effects of circular economy policies on the environment and sustainable growth: Worldwide research. *Sustainability, 12*(14), 5792. https://doi.org/10.3390/su12145792

Abdel-Shafy, H. I., & Mansour, M. S. (2018). Solid waste issue: Sources, composition, disposal, recycling, and valorization. *Egyptian Journal of Petroleum, 27*(4), 1275–1290. https://doi.org/10.1016/j.ejpe.2018.07.003

Aithal, S., & Aithal, P. S. (2023). Importance of circular economy for resource optimization in various industry sectors—A review-based opportunity analysis. *International Journal of Applied Engineering and Management Letters (IJAEML), 7*(2), 191–215.

Alexandris, G., Katos, V., Alexaki, S., & Hatzivasilis, G. (2018). Blockchains as enablers for auditing cooperative circular economy networks. In *2018 IEEE 23rd International Workshop on Computer Aided Modeling and Design of Communication Links and Networks (CAMAD)* (pp. 1–7). IEEE.

Alnogaidan, M. (2024). Blockchain for sustainability: Tracing environmental impacts in supply chains. *International Journal of Green Skills and Disruptive Technology, 1*(2), 12–22.

Alotaibi, E. M., Khallaf, A., Abdallah, A. A. N., Zoubi, T., & Alnesafi, A. (2024). Blockchain-driven carbon accountability in supply chains. *Sustainability, 16*(24), 10872. https://doi.org/10.3390/su162410872

Amin, M. B., Asaduzzaman, M., Debnath, G. C., Rahaman, M. A., & Oláh, J. (2024). Effects of circular economy practices on sustainable firm performance of green garments. *Oeconomia Copernicana, 15*(2), 637–682.

Awan, U., Sroufe, R., & Bozan, K. (2022). Designing value chains for industry 4.0 and a circular economy: A review of the literature. *Sustainability, 14*(12), 7084. https://doi.org/10.3390/su14127084

Böckel, A., Nuzum, A. K., & Weissbrod, I. (2021). Blockchain for the circular economy: Analysis of the research-practice gap. *Sustainable Production and Consumption, 25*, 525–539. https://doi.org/10.1016/j.spc.2020.12.006

Bressanelli, G., Visintin, F., & Saccani, N. (2022). Circular economy and the evolution of industrial districts: A supply chain perspective. *International Journal of Production Economics, 243*, 108348. https://doi.org/10.1016/j.ijpe.2021.108348

Chishty, S. K. (2023). Economic and environmental efficiencies of organizations: Role of technological advancements and circular economy practices. *Sustainability, 15*(22), 15935. https://doi.org/10.3390/su152215935

Di Vaio, A., & Varriale, L. (2020). Blockchain technology in supply chain management for sustainable performance: Evidence from the airport industry. *International Journal of Information Management, 52*, 102014. https://doi.org/10.1016/j.ijinfomgt.2019.09.010

Esmaeilian, B., Sarkis, J., Lewis, K., & Behdad, S. (2020). Blockchain for the future of sustainable supply chain management in industry 4.0. *Resources, Conservation and Recycling, 163*, 105064. https://doi.org/10.1016/j.resconrec.2020.105064

Friedman, N., & Ormiston, J. (2022). Blockchain as a sustainability-oriented innovation?: Opportunities for and resistance to blockchain technology as a driver of sustainability in global food supply chains. *Technological Forecasting and Social Change, 175*, 121403. https://doi.org/10.1016/j.techfore.2021.121403

Geissdoerfer, M., Savaget, P., Bocken, N. M., & Hultink, E. J. (2017). The circular economy—A new sustainability paradigm? *Journal of Cleaner Production, 143*, 757–768. https://doi.org/10.1016/j.jclepro.2016.12.048

Geng, Y., Sarkis, J., & Ulgiati, S. (2016). Sustainability, wellbeing, and the circular economy in China and worldwide. *Science, 351*, 76–79.

Ghobakhloo, M., Iranmanesh, M., Foroughi, B., Tirkolaee, E. B., Asadi, S., & Amran, A. (2023). Industry 5.0 implications for inclusive sustainable manufacturing: An evidence-knowledge-based strategic roadmap. *Journal of Cleaner Production, 417*, 138023. https://doi.org/10.1016/j.jclepro.2023.138023

Hastig, G. M., & Sodhi, M. S. (2020). Blockchain for supply chain traceability: Business requirements and critical success factors. *Production and Operations Management, 29*(4), 935–954. https://doi.org/10.1111/poms.1314

Iansiti, M., & Lakhani, K. R. (2017). The truth about blockchain. *Harvard Business Review, 95*(1), 118–127.

Khan, S. A. R., Zia-ul-haq, H. M., Umar, M., & Yu, Z. (2021). Digital technology and circular economy practices: An strategy to improve organizational performance. *Business Strategy & Development, 4*(4), 482–490. https://doi.org/10.1002/bsd2.176

Khan, S., Ali, S. S., & Singh, R. (2022a). Determinants of remanufacturing adoption for circular economy: A causal relationship evaluation framework. *Applied System Innovation, 5*(4), 62. https://doi.org/10.3390/asi5040062

Khan, S., Singh, R., Sá, J. C., Santos, G., & Ferreira, L. P. (2022b). Modelling of determinants of logistics 4.0 adoption: Insights from developing countries. *Machines, 10*(12), 1242. https://doi.org/10.3390/machines10121242

Khan, S., Khan, M. I., & Singh, R. (2023). Modeling the barriers of blockchain technology implementation in supply chain. *Journal of Industrial Integration and Management, 8*(3), 367–385. https://doi.org/10.1142/S2424862223500070

Khan, S., Singh, R., Alnahas, J., Abbate, S., & Centobelli, P. (2024). Navigating the smart circular economy: A framework for manufacturing firms. *Journal of Cleaner Production, 480*, 144007. https://doi.org/10.1016/j.jclepro.2024.144007

Kouhizadeh, M., Zhu, Q., & Sarkis, J. (2020). Blockchain and the circular economy: Potential tensions and critical reflections from practice. *Production Planning & Control, 31*(11–12), 950–966. https://doi.org/10.1080/09537287.2019.1695925

Kshetri, N. (2021). Blockchain and sustainable supply chain management in developing countries. *International Journal of Information Management, 60*, 102376. https://doi.org/10.1016/j.ijinfomgt.2021.102376

Leng, J., Ruan, G., Jiang, P., Xu, K., Liu, Q., Zhou, X., & Liu, C. (2020). Blockchain-empowered sustainable manufacturing and product lifecycle management in industry 4.0: A survey. *Renewable and Sustainable Energy Reviews, 132*, 110112. https://doi.org/10.1016/j.rser.2020.110112

Lin, W. L., Leow, N. X. C., Lim, W. M., Ho, M. K., Yong, J. Y., & Pek, C. K. (2024). The influence of blockchain technology on circular economy implementation in the automotive sector: From a GMM model to a new machine learning algorithm. *Business Strategy and the Environment, 34*, 1043. https://doi.org/10.1002/bse.4003

Liu, X. L., Wang, W. M., Guo, H., Barenji, A. V., Li, Z., & Huang, G. Q. (2020). Industrial blockchain based framework for product lifecycle management in industry 4.0. *Robotics and Computer-Integrated Manufacturing, 63*, 101897. https://doi.org/10.1016/j.rcim.2019.101897

Lu, Y. (2019). The blockchain: State-of-the-art and research challenges. *Journal of Industrial Information Integration, 15*, 80–90. https://doi.org/10.1016/j.jii.2019.04.002

Marrucci, L., Daddi, T., & Iraldo, F. (2022). The circular economy, environmental performance and environmental management systems: The role of absorptive capacity. *Journal of Knowledge Management, 26*(8), 2107–2132. https://doi.org/10.1108/JKM-06-2021-0437

Mubarik, M., Raja Mohd Rasi, R. Z., Mubarak, M. F., & Ashraf, R. (2021). Impact of blockchain technology on green supply chain practices: Evidence from emerging economy. *Management of Environmental Quality: An International Journal, 32*(5), 1023–1039. https://doi.org/10.1108/MEQ-11-2020-0277

Rani, P., Sharma, P., & Gupta, I. (2024). Toward a greener future: A survey on sustainable blockchain applications and impact. *Journal of Environmental Management, 354*, 120273. https://doi.org/10.1016/j.jenvman.2024.120273

Rehman Khan, S. A., Yu, Z., Sarwat, S., Godil, D. I., Amin, S., & Shujaat, S. (2022). The role of blockchain technology in circular economy practices to improve organisational performance. *International Journal of Logistics Research and Applications, 25*(4–5), 605–622. https://doi.org/10.1080/13675567.2021.1872512

Rejeb, A., Keogh, J. G., Simske, S. J., Stafford, T., & Treiblmaier, H. (2021). Potentials of blockchain technologies for supply chain collaboration: A conceptual framework. *The International Journal of Logistics Management, 32*(3), 973–994. https://doi.org/10.1108/IJLM-02-2020-0098

Sachs, J. D. (2015). *The age of sustainable development*. Columbia University Press.

Sahoo, S., Upadhyay, A., & Kumar, A. (2023). Circular economy practices and environmental performance: Analyzing the role of big data analytics capability and responsible research and innovation. *Business Strategy and the Environment, 32*(8), 6029–6046. https://doi.org/10.1002/bse.3471

Samadhiya, A., Agrawal, R., Kumar, A., & Garza-Reyes, J. A. (2023). Blockchain technology and circular economy in the environment of total productive maintenance: A natural resource-based view perspective. *Journal of Manufacturing Technology Management, 34*(2), 293–314. https://doi.org/10.1108/JMTM-08-2022-0299

Schroeder, P., Dewick, P., Kusi-Sarpong, S., & Hofstetter, J. S. (2018). Circular economy and power relations in global value chains: Tensions and trade-offs for lower income countries. *Resources, Conservation and Recycling, 136*, 77–78. https://doi.org/10.1016/j.resconrec.2018.04.003

Schroeder, P., Anggraeni, K., & Weber, U. (2019). The relevance of circular economy practices to the sustainable development goals. *Journal of Industrial Ecology, 23*(1), 77–95. https://doi.org/10.1111/jiec.12732

Sehnem, S., Jabbour, C. J. C., Pereira, S. C. F., & de Sousa Jabbour, A. B. L. (2019). Improving sustainable supply chains performance through operational excellence: Circular economy approach. *Resources, Conservation and Recycling, 149*, 236–248. https://doi.org/10.1016/j.resconrec.2019.05.021

Shojaei, A., Ketabi, R., Razkenari, M., Hakim, H., & Wang, J. (2021). Enabling a circular economy in the built environment sector through blockchain technology. *Journal of Cleaner Production, 294*, 126352. https://doi.org/10.1016/j.jclepro.2021.126352

Silvestre, B. S., & Ţîrcă, D. M. (2019). Innovations for sustainable development: Moving toward a sustainable future. *Journal of Cleaner Production, 208*, 325–332. https://doi.org/10.1016/j.jclepro.2018.09.244

Singh, R., Khan, S., & Centobelli, P. (2022). Investigating the interplay between social performance and organisational factors supporting circular economy practices. *Sustainability, 14*(24), 16781. https://doi.org/10.3390/su142416781

Singh, R., Khan, S., & Dsilva, J. (2023). A framework for assessment of critical factor for circular economy practice implementation. *Journal of Modelling in Management, 18*(5), 1476–1497. https://doi.org/10.1108/JM2-06-2021-0145

Singh, R., Gupta, A., Bajpai, A., & Kandpal, V. (2024a). Exploring challenges of circular economy initiatives for smart cities. In *Smart cities and circular economy: The future of sustainable urban development* (pp. 59–69). Emerald Publishing Limited. https://doi.org/10.1108/978-1-83797-957-820241005

Singh, R., Khan, S., Dsilva, J., Akram, U., & Haleem, A. (2024b). Modelling the organisational factors for implementation of corporate social responsibility: A modified TISM approach. *Global Journal of Flexible Systems Management, 25*(2), 283–301. https://doi.org/10.1007/s40171-024-00388-x

Singh, R., Kumar, K., & Khan, S. (2024c). A comprehensive view of artificial intelligence (AI)–based technologies for sustainable development goals (SDGs). In *Artificial intelligence enabled management: An emerging economy perspective* (p. 183). Walter de Gruyter GmbH.

Swan, M. (2015). *Blockchain: Blueprint for a new economy*. O'Reilly Media.

Treleaven, P., Brown, R. G., & Yang, D. (2017). Blockchain technology in finance. *Computer, 50*(9), 14–17.

Upadhyay, A., Mukhuty, S., Kumar, V., & Kazancoglu, Y. (2021). Blockchain technology and the circular economy: Implications for sustainability and social responsibility. *Journal of Cleaner Production, 293*, 126130. https://doi.org/10.1016/j.jclepro.2021.126130

Veleva, V., & Bodkin, G. (2018). Corporate-entrepreneur collaborations to advance a circular economy. *Journal of Cleaner Production, 188*, 20–37. https://doi.org/10.1016/j.jclepro.2018.03.196

Xu, M., Chen, X., & Kou, G. (2019). A systematic review of blockchain. *Financial Innovations, 5*(1), 27. https://doi.org/10.1186/s40854-019-0147-z

Yang, Y., Chen, L., Jia, F., & Xu, Z. (2019). Complementarity of circular economy practices: An empirical analysis of Chinese manufacturers. *International Journal of Production Research, 57*, 6369–6384. https://doi.org/10.1080/00207543.2019.1566664

Yin, S., Jia, F., Chen, L., & Wang, Q. (2023). Circular economy practices and sustainable performance: A meta-analysis. *Resources, Conservation and Recycling, 190*, 106838. https://doi.org/10.1016/j.resconrec.2022.106838

Zhao, R., Wang, J., & Xue, F. (2023, July). A blockchain-based token economic model for incentivizing ESG in the construction industry. In *EC3 Conference 2023* (Vol. 4). European Council on Computing in Construction.

Zhu, Q., & Sarkis, J. (2004). Relationships between operational practices and performance among early adopters of green supply chain management practices in Chinese manufacturing enterprises. *Journal of Operations Management, 22*(3), 265–289. https://doi.org/10.1016/j.jom.2004.01.005

Chapter 12
Enhancing Supply Chain Transparency and Traceability in Waste Management Practices Through Blockchain Technologies and Digital Innovations

Hiranya Dissanayake, Otilia Manta, and Shahbaz Khan

Abstract The globalization of economies has expanded supply chains into complex, interconnected networks, posing challenges to transparency, traceability, and sustainability. Simultaneously, the dominance of the linear economic model has exacerbated waste management issues, necessitating a transition to circular economy principles. This paper examines the transformative potential of digital technologies—blockchain, the Internet of Things (IoT), and digital product passports (DPPs)—in addressing these challenges. The study highlights how these technologies enhance supply chain visibility, promote ethical sourcing, reduce waste, and enable circular economy practices. Case studies from diverse industries demonstrate the practical applications and benefits of digital innovations, such as real-time monitoring, immutable records, and life cycle tracking of products. However, barriers, including high implementation costs, data silos, and regulatory disparities, impede widespread adoption. To address these challenges, the current study proposes actionable strategies for stakeholders. Industry leaders are encouraged to invest in digital tools and redesign operations to align with circular economy principles. Policymakers are urged to develop standardized frameworks, provide financial incentives, and foster public-private partnerships. Researchers are called to explore interdisciplinary approaches, evaluate scalability solutions, and assess socioeconomic impacts. The findings underscore the synergies between transpar-

---

H. Dissanayake (✉)
Department of Accountancy, Faculty of Business Studies and Finance, Wayamba University of Sri Lanka, Kuliyapitiya, Sri Lanka
e-mail: hiranya@wyb.ac.lk

O. Manta
Romanian Academy, "Victor Slavescu" Centre for Financial and Monetary Research, Bucharest, Romania
e-mail: otilia.manta@rgic.ro

S. Khan
Department of Industrial Engineering, University of Tabuk, Tabuk, Saudi Arabia

R. Singh et al. (eds.), *Blockchain Technologies for Smart Circular Economy and Organisational Sustainability*, Environmental Footprints and Eco-design of Products and Processes, https://doi.org/10.1007/978-3-031-89183-0_12

ency and waste management, advocating for collaborative efforts to build sustainable, resilient, and ethical supply chains. By leveraging digital technologies, stakeholders can overcome implementation barriers, enhance stakeholder collaboration, and contribute to a sustainable future. This paper provides a roadmap for integrating transparency and waste management strategies, aligning global supply chains with sustainability goals.

Keywords Sustainable waste management · Blockchain · The Internet of Things (IoT) · Digital product passports (DPPs)

12.1 Introduction

Economic globalization has witnessed the enlargement and complex supply chains much more than before. This has made them a mammoth of networks that cross countries and industries. These changes have posed heightened issues of opacity and control that have caused a key barrier to achieving the much-needed transparency and traceability that promotes ethical sourcing, improved environmental sustainability, and sustainable practices. Thus, transparency has transitioned from accountability compliance to a business imperative for trust, shockproofing, and supply chain differentiation (Kumar, 2024; Modgil et al., 2024). Nonetheless, auditable supply chain systems do not provide adequate insight, especially where operations in supply chain tiers are challenging to assess and control (Kashmanian, 2017).

At the same time, waste management emerges as a critical issue for global concern due to the prevalence of the linear economy model known as "take, make, dispose." This model has resulted in increased generation of waste, depletion of resources, and more environmental pollution. There is a fast-growing concern to move to a more circular economy approach that seeks to reuse, recycle, and reinstate resources (Singh et al., 2022; Zhang & Seuring, 2024). Nevertheless, they state that this change is challenging to implement due to the lack of integrated datasets, poor information infrastructure, and organizational opposition (Jasrotia et al., 2024).

Of course, digital technologies bring transformative possibilities to address these connected problems (Singh et al., 2022). Blockchain, for instance, creates indelible and open transactions that enhance responsibility and credibility among all chain members (Modgil et al., 2024). The Internet of Things (IoT) entails the actual time tracking of supply chain activities with better control (Kumar et al., 2023). On the other hand, the digital product passports, part of the European Union's Circular Economy Action Plan, based on a product's lifecycle, allow transparency and circularity (Zhang & Seuring, 2024). However, the broad uptake of these technologies is not universally seamless, and the key challenges they present include high implementation costs, lack of a standard format/structure, and engagements by different entities (Kashmanian, 2017).

Nonetheless, several research gaps are still apparent in the continually expanding literature on digital innovation in supply chains. Nevertheless, there is a lack of research based on multiple technologies used in or for the supply chain and their application in multiple supply chain stages and waste management (Zhang & Seuring, 2024), such as blockchain and IoT. Secondly, there is a limited understanding of modes that support integrated stakeholder collectives to enact a circular economy (Jasrotia et al., 2024). Finally, there is a lack of studies on how the cost and scalability hitch prevent those applications from being adopted in various settings, specifically the developing world (Modgil et al., 2024; Kumar, 2023).

To fill these gaps, this paper reviews the theoretical literature and proposes low-level practical recommendations for using digital technologies to increase transparency and manage waste in supply chains. This paper serves as a qualitative contribution to extend the knowledge concerning the combination of transparency, traceability, and waste minimizations that these innovations might begin in support of significantly sustainable and resilient supply chains.

12.1.1 Objectives

1. To understand how blockchain, IoT and DPPs can be used to improve supply chain visibility and traceability.
2. This paper attempts to assess the effectiveness of these technologies in implementing circular economy strategies, especially in waste management.
3. To analyze key issues and define recommendations for further development of these innovations and the obstacles in their further scaling.

This paper is divided into six parts. The first section explains the challenges and existing gaps, what the paper will offer, and what the study aims to achieve. The following section, Sect. 12.2 discusses the meanings of transparency and the technologies supporting traceability, and it has case studies. Section 12.3 looks at conventional problems and how technology disrupts waste management. Section 12.4 examine the combined effects or benefits, existing policies and system innovations, and the overall impact. Based on these underpinnings, the following sections are presented: Practical Recommendations provide specific measures that stakeholders may consider, and Sects. 12.6 and 12.5.3 outline the study's findings and suggest avenues for future research.

To address these crucial matters, this paper endeavors to add to the existing literature and actual practice of sustainable supply chain management. As a digitalized value chain, it provides a blueprint for how circularity can be achieved and how transparency can be facilitated to solve the difficulties and complexity of the current supply chain and disposal system.

12.2 Supply Chain Transparency and Traceability

Supply chain visibility and accountability are among the keys to creating sustainable and ethical supply chain management. As supply chains become elongated and large, there is a competing need to know, keep track of, and address issues related to visibility, responsibility, and compliance most appropriately and effectively. This section addresses questions about transparency in new media literacy, the role of new digital skills, and practical applications that show the possibilities of using new media tools.

12.2.1 Importance of Transparency

Traceability is, therefore, the extent to which organizations can deliver timely information regarding the sourcing and flow, as well as the ecological and social footprints of products and services in supply chains. Of pivotal relevance is establishing and rebuilding trust, moderating risk, and promoting effectiveness in operation (Kumar, 2023). In transparent supply chain structures, organizations can see the wrong things, such as forced labor or overexploitation of natural resources and work to prevent them. In addition, legal requirements, like the European Union's Circular Economy Action Plan, which promotes transparency among its objectives, demand sustainability (Zhang & Seuring, 2024).

Transparency colors also make businesses vulnerable to reputational risks and risks associated with disrupting supply chains. For example, when firms lack specific knowledge of suppliers' activities, they could fund detrimental environmental processes or experience latency due to unclear processes of supply chain tiers further upstream (Modgil et al., 2024). They also increase consumer rights and capacity to choose products that are ethical and environmentally friendly (Kashmanian, 2017).

12.2.2 Enhancing Traceability with Digital Tools

Supply chain traceability has been identified as a key area that hugely benefits from digital technologies. This means that the supply chain tools mentioned above offer real-time information and increase the chances of having correct and accurate information about the supply chain.

12.2.2.1 Blockchain Technology

Previously, traceability of the supply chain was a big issue. Hence, now, with the help of blockchain, many organizations are providing a distributed ledger which maintains every transaction or event in the supply chain. This makes the data tamper-proof, restoring confidence with the stakeholders (Modgil et al., 2024). For example, blockchain systems have been implemented to monitor the supply chain origin of raw materials across various sectors, including the food chain, garment, and textile and electronics and appliances, by meeting ethical sourcing standards (Jasrotia et al., 2024).

12.2.2.2 Internet of Things (IoT)

Internet of Things, like sensors and RFID tags, tracks the shipment of goods within the supply chain in real-time. These technologies build visibility by getting certain information such as temperature, position, and handling situation, which can be valuable for perishable items and medicines (Kumar, 2023). IoT also makes it possible to predict future risks and enhance logistics optimization, as analyzed by Zhang and Seuring (2024).

12.2.2.3 Digital Product Passports (DPPs)

DPPs refer to digital documents containing full details of every aspect of a product, from the materials used in the manufacturing process to how to dispose of it. Following the EU's Circular Economy Action Plan, Extended Producer Responsibilities or DPPs help achieve circular economy objectives by enabling easier product recycling, repair, and reuse (Zhang & Seuring, 2024). For instance, electric vehicle batteries must contain DPPs to guarantee correct recycling and disposal.

12.2.3 Case Studies Applications

12.2.3.1 The Food Industry

Blockchain technology simplifies food traceability systems in the food industry to improve food safety. For instance, through blockchain, Walmart could follow the journey of mangoes and pork. This system brought the time needed to track products down from days to seconds, and that increased safety and effectiveness (Modgil et al., 2024).

12.2.3.2 The Fashion Industry

The fashion industry has adopted blockchain and IoT technologies to counter problems of unethical labor practices and environmental degradation. Systems like Everledger provide data on the origin of diamonds and other valuable products, including minimizing the spread of fakes (Jasrotia et al., 2024).

12.2.3.3 Automotive and Electronics

DPPs are influential in the automotive and electronics industries; the responsible sourcing of materials such as cobalt and lithium is vital. These tools also contribute to the recyclability of components to enhance circularity (Zhang & Seuring, 2024).

12.2.4 Causes and Consequences

At the same time, important obstacles still exist. Although these technologies promise a lot, the following problems are still present. Implementation costs are often high, normativity is lacking, and stakeholders are usually resistant to change, which is the case with digital tools (Kumar, 2023). Furthermore, data governance on decentralized systems like blockchain for accuracy and security remains a topic analyzed in Modgil et al. (2024).

However, the prospects are just as promising. When applied in an organization, these technologies promote organizational resilience and compliance with legal requirements, as well as carry out activities with high levels of transparency and accountability due to emerging advanced technologies. These and similar challenges call for integrating efforts by governments, industries, and consumers to unlock the full value of digital innovations that address supply chain traceability.

Supply chain visibility and accountability are not choices but imperatives in the supply chain. Blockchain, IoT, and DPPs as digital technologies are the primary and effective helpers in increasing transparency and effectiveness of actions. By exploring these challenges and building on these opportunities, organizations can deploy their business activities and strategic management in ways that support common sustainability goals while building customer trust and sustainable future value.

12.3 Waste Management

Waste management is part of sustainable development, which defines programs and measures for handling waste and using resources effectively and ecologically. This paper examines how far digital innovations enhance the revolution of traditional waste management measures in the shift towards a circular economy. This section looks at the prospects of distinct conventional approaches, the possibilities of digital transformation, and case studies in progressing sustainable waste management.

12.3.1 Problems in Conventional Solid Waste Disposal

Most conventional waste disposal and management structures still only function within the framework of the linear economy model, working with extraction, manufacturing, use, and disposal of material resources. This approach has led to increased environmental decay, such as full landfills, pollution, and depletion of natural resources. Key challenges include the following:

1. Inefficiencies in Waste Sorting and Recycling: There are often no established measures for executing these processes, which can cause contamination of recyclable materials and minimize recycling efficiency (Jasrotia et al., 2024).
2. Insufficient Data for Decision-Making: As Zhang and Seuring (2024) observed, the lack of integrated information hampers the waste collection and processing improvement for the agencies in charge of waste management.
3. Limited Public Awareness and Participation: This paper attributes the inefficiencies to consumer behaviors and a lack of knowledge about correctly sorting waste bins (Kumar, 2023).

12.3.2 The Use of Technology

Waste management continues to be revolutionized by digital technologies in the sense that more efficient tools are being made available.

12.3.2.1 Internet of Things (IoT)

The Internet of Things refers to the application of IoT in waste management. Internet-enabled smart bins with sensors help monitor waste levels to improve collection schedules, thereby cutting costs. For instance, smart bins in city squirrels have assisted the cities in reducing energy circulating in the city for wastes, curtailing emissions (Kumar, 2023).

12.3.2.2 Blockchain Technology

Blockchain can provide transparency and traceability in the waste streams depending on the reliability of the recording and transaction of materials. Projects like the "Plastic Bank" apply blockchain in tracking the physical collection of plastic waste and even offering the cycles recycling; they turn the plastic waste into an economical and worthwhile resource (Modgil et al., 2024).

12.3.2.3 Artificial Intelligence (AI)

It is one of the most famous subdisciplines of computer science. This content studies how to design intelligent agents to create beings that can perform tasks that need to be done in society but are dull or dangerous for human beings.

Applying natural resource management, sorting systems powered by artificial intelligence enable visual recognition of the kind of waste for sorting and minimizing contamination of recyclable products. These systems have been installed at the waste processing plants to increase material recovery and reduce human interaction (Zhang & Seuring, 2024).

12.3.2.4 Digital Product Passports: 2.4

DPPs contain the complete record of the product's cycle, make-up, and disposal procedures. By enhancing the sorting and recycling of products at the end of their life, DPPs complement circular economy objectives and the minimization of waste (Jasrotia et al., 2024).

12.3.3 Case Studies and Applications

12.3.3.1 Smart Waste Collection in Urban Centers

Smart waste collection is IoT-enabled and implemented in cities such as Amsterdam. These systems rely on sensors to help determine the bins' fill level so that collection can be tailored to be efficient and cost-effective without incurring an unnecessary carbon footprint (Kumar, 2023).

12.3.3.2 Blockchain-Enabled Recycling Programs

Plastic Bank has implemented blockchain technology to incentivize waste collection in developing countries. Collectors earn digital tokens for the plastic they collect, which can be redeemed for goods or services. This initiative has diverted thousands of tons of plastic from oceans while providing economic opportunities for marginalized communities (Modgil et al., 2024).

12.3.3.3 AI in Material Recovery Facilities

AI-driven sorting technologies have been implemented in material recovery facilities to enhance efficiency and accuracy in separating recyclables. These systems reduce contamination and increase the quality of recycled materials, supporting higher reuse rates in manufacturing (Zhang & Seuring, 2024).

12.3.4 Integration of Circular Economy Principles

Waste management practices are increasingly aligned with circular economy principles. Adopting technologies like IoT, blockchain, and AI facilitates the reusing, recycling, and remanufacturing of materials, reducing reliance on virgin resources (Khan et al., 2024; Khan et al., 2022a). Circular practices such as closed-loop systems are being implemented across industries to extend product lifecycles and minimize waste (Jasrotia et al., 2024).

12.3.5 Challenges and Opportunities

12.3.5.1 Challenges

1. **High Initial Costs**: Implementing digital technologies requires significant investment in infrastructure and training (Kumar, 2024; Singh et al., 2024b).
2. **Data Integration**: Fragmented data sources across stakeholders complicate the development of cohesive waste management systems (Modgil et al., 2024).
3. **Regulatory Barriers**: Varying regulations across regions hinder the scalability of innovative waste management practices (Zhang & Seuring, 2024).
4. **Lack of Trust among Supply Chain Partners** (Khan et al., 2022b).

12.3.5.2 Opportunities

1. **Enhanced Collaboration**: Digital platforms enable better stakeholder coordination, from consumers to recyclers and policymakers (Kumar, 2023).
2. **Consumer Engagement**: Digital apps and platforms incentivize recycling behaviors through gamification and rewards, increasing public participation (Modgil et al., 2024).
3. **Sustainability Metrics**: Advanced analytics enable detailed tracking of waste streams, supporting regulatory compliance and sustainability reporting (Zhang & Seuring, 2024).

Waste management is at the core of achieving sustainability goals and transitioning to a circular economy. Digital technologies like IoT, blockchain, AI, and DPPs provide innovative solutions to traditional waste management challenges, enhancing efficiency, transparency, and public engagement. By addressing implementation barriers and leveraging these technologies, stakeholders can create a sustainable waste management ecosystem that benefits the environment and society.

12.4 Intersections of Transparency and Waste Management

Transparent supply chain and waste management are two factors of sustainable development. Greater transparency makes material and environmental costs and effects, as well as ethical compliance opportunities, apparent at all supply chain tiers, making waste management possible. This section examines the relationships between transparency and waste management, as well as the internal connections and the role of digital developments, further reflecting on the integration issues.

12.4.1 Synergies in Circular Economy Principles

In this case, the disposal of waste and transparency are features that define the circular economy. Transparent supply chains allow the ability to identify the products and those materials used within the product throughout their usable life and disposal. This visibility reduces waste by extending use, reconditioning, and recycling (Zhang & Seuring, 2024). For instance, extended tracking mechanisms help organizations pinpoint underutilized materials, redirecting them to different sections of the value line or return them to manufacturing lines for conversion.

Digital product passports (DPPs) are a good example of such integration. They are informative documents that offer detailed information on the materials used in product manufacturing, the manufacturing process, and recycling options at the product's end of life. They implement waste education and enhance resource efficiency and welfare, which is in direct consonance with a circular economy (Modgil et al., 2024).

Transparency mediates between enthusiasm for digital technologies and efficiency in waste management. Such applications help stakeholders to base their decision-making on information, promote cooperation, and improve performance.

12.4.1.1 Blockchain for Trackable Waste Flows

Blockchain technology offers unalterable details of material flow, which ultimately guarantees some waste tracking from its origins until disposal. This brings an aspect of transparency that eliminates instances of either dumping or misusing these materials. For instance, blockchain-based solutions, such as Plastic Bank, structure the collection and recycling of waste while documenting the material's collection accurately (Kumar, 2023).

12.4.1.2 IoT for Real-Time Monitoring

Smart waste bins covered with IoT devices enable actual-time tracking of the waste level and its content. Such data can improve waste collection schedules, minimize avoidable costs, and increase focus on recycling activities (Zhang & Seuring, 2024).

12.4.1.3 AI and Predictive Analytics

IoT integrates data from IoT devices and other sources to understand waste generation forecasts and assign competent resources (Dissanayake et al., 2024). Trends learnt by AI systems allow industries to predict future waste streams and devise solutions that will lessen the adverse effects on the environment (Jasrotia et al., 2024; Singh et al., 2024d).

12.4.2 Policies and Regulatory Measures

Measures that make the supply chain more transparent relate to waste practices (Bhushan & Singh 2024; Singh et al., 2024a). For instance, the Circularity Gap report of the European Union's Circular Economy Action Plan mandates that reporting on the life cycle of a product should be revealed by industries to enhance waste management strategies (Kumar et al., 2024; Singh et al., 2023; Zhang & Seuring, 2024). Governments can also promote these practices by making traceability technologies compulsory and promoting circular economy strategies. The EU Circular Economy Action Plan establishes a common framework across the European Union, but different rules may apply in each area or country due to variations in national legislation, regional policies, and levels of implementation. A single set of rules applied worldwide lacks co-partnership, leads to inefficiencies, and hinders progress, whereas diversified risk sharing fosters development. Enhanced transparency ensures that material flows, environmental impacts, and compliance with ethical standards are visible throughout the supply chain, creating opportunities to optimize waste management practices. This section explores how transparency and waste management intersect, focusing on their synergies, the role of digital innovations, and the integration challenges.

12.4.3 Challenges in Integration

12.4.3.1 Data Silos

One of the biggest challenges in integrating transparency and waste management into the process is the disconnection of information between the parties involved. Integration issues hamper the ability of the connected systems, making a high degree of coordinated effort in waste management using installed systems a challenge (Modgil et al., 2024).

12.4.3.2 Cost and Scalability

Applying new technologies in organizations is capital-intensive since it involves purchasing equipment to support new technology and training employees in new ways of performing their tasks. Unfortunately, small and medium enterprises cannot afford to implement such systems, limiting integration across organizations (Jasrotia et al., 2024).

12.4.3.3 Regulatory Disparities

The differing and diverse existing regulations present difficulties for gigantic international organizations applying uniformity in transparency and waste control. The continued difficulty in synchronizing policies at an international level is another challenge (Zhang & Seuring, 2024).

12.4.4 Partnership Areas

12.4.4.1 Multi-stakeholder Engagement

Partnership engagements between autocratic governments, industries, and nongovernmental organizations can help create workable solutions to transparency and waste management (Singh et al., 2024c). That is why the circular economy programs promoted by the Ellen MacArthur Foundation suggest the effectiveness of integrated activities on a significantly larger scale (Kumar, 2023).

12.4.4.2 Consumer Involvement

The quality of using transparency tools in engaging consumers will help achieve better waste management. For example, social change through technological methods that deliver product and recycling information ensures consumers benefit from the knowledge and become active participants in recycling (Zhang & Seuring, 2024).

The area of focus is a conceptual intersection between transparency and waste management in sustainable and circular supply chains. To this end, digital technologies encompassed in this paper, including blockchain, IoT, and AI, are critical to addressing this gap, enhancing resource sharing and management, and reducing wastage. Taking on issues like data silos, high cost, and improper regulation may call for cooperation among all the actors. Transparency uptake in the waste approach can boost sustainability, compliance, and resilience in the long term.

12.5 Practical Implications

This section provides specific measures that industry practitioners, policymakers, and scholars can take to improve supply chain visibility and reduce waste through embracing digital technologies. These recommendations seek to solve technological, regulatory, and operational barriers that can enhance sustainability, supply chain security, and ethical conduct in global supply chains.

12.5.1 Practical Implications for Industry

12.5.1.1 Promote Digital Solution for Transparency

Blockchain and sensors IoT and digital product passports must be pursued to enhance measurable supply chain transparency. Blockchain can then help maintain an unalterable record of transactions, while IoT enables tracking goods in real time. Digital product passports (DPPs) can help identify the location and movement of materials over their life cycle and enhance traceability (Modgil et al., 2024; Zhang & Seuring, 2024).

12.5.1.2 Implementation of Circular Economy Principles

Many entities encourage clients to redesign their supply chain operations to reflect the circular economy. This entails practices such as the use of a closed-loop system in which waste products can be recycled and used again to construct new products. Recyclers and waste management firms must be appropriately engaged to provide proper recoverable resources (Jasrotia et al., 2024).

12.5.1.3 Promote Enhancement of Multitier Collaboration

Interorganizational transparency entails the unhampered flow of information across the supply chain, from upstream players to downstream resellers. Collaborate with appropriate digital tools to exchange data and team communication and ensure that required sustainability benchmarks are met (Kumar, 2023).

12.5.1.4 Engage Consumers

Bring aware and active consumers into focus through transparency activities. With the help of QR codes or additional applications, information about the availability of product life cycle information can be given to consumers to make them choose products that are safe to dispose of and reduce waste (Zhang et al., 2023, p. 87).

12.5.2 *Practical Implications for Policymakers*

12.5.2.1 Develop Standardized Frameworks

Create global standards for transparency and waste management technologies, ensuring interoperability and consistency. This includes guidelines for blockchain deployment, IoT integration, and the adoption of DPPs to facilitate cross-border compliance (Modgil et al., 2024).

12.5.2.2 Provide Financial Incentives

Offer subsidies, tax breaks, or grants to businesses adopting digital technologies for transparency and waste management. Such incentives can help small and medium enterprises overcome financial barriers to technology adoption (Kumar, 2023).

12.5.2.3 Promote Extended Producer Responsibility (EPR)

EPR policies must be set to make manufacturers accountable for their products from inception to disposal. With the aid of digital tools, it is possible to control noncompliance since the method offers precise tracking and reporting (Jasrotia et al., 2024).

12.5.2.4 Provide Support on Public-Private Partnerships

Promote collaboration between governments and industrial players, as well as academia, for the development of unique solutions. Er: Collaboration efforts can enhance the process of digitalization of sustainability innovations (Zhang & Seuring, 2024).

12.5.3 *Future Research Directions*

12.5.3.1 Interdisciplinary Research

Research the connection between supply chain transparency, waste management, and digital technologies. To fill these gaps, research should explore whether any theoretical frameworks link these domains and work towards creating models that present their interdependence (Modgil et al., 2024).

12.5.3.2 Analyze Applications

It is important to research about typical use cases of blockchain, IoT, and DPPs in various industries, their issues, and standard solutions. Cross-sectional studies can, therefore, identify the challenges and opportunities of a given sector from a distinct culture (Kumar, 2023).

12.5.3.3 Assess Socioeconomic Effects

Examine the extension of the uses of digital technologies on factors including employment and job markets, consumer trends, and citizenship concerns. These suggestions can help policymakers and firms frame inclusive initiatives (Zhang & Seuring, 2024).

12.5.3.4 Scalability Solutions Research Lesson

Design approaches that will help redress the marginal cost and scale issues raised by digital initiatives so that they are affordable and utilizable by organizations of all sizes. These are, for instance, impact investment and crowdfunding to fund technology uptake (Jasrotia et al., 2024).

12.6 Conclusion

This paper identified supply chain transparency and waste management as important factors towards sustainable development. Rising supply chain complexity at the global level and growing environmental factors, therefore, mean that there is a need for better solutions to global supply chain problems, including better visibility, better accountability, and higher efficiency. Blockchain, IoT, and DPPs are emerging in these domains as the new, most effective tools that help track, monitor, and improve the overall performance of supply chain processes, minimize waste, and promote circular economy. The use of these technologies presents a number of advantages, such as better stakeholder engagement, meeting legal and regulatory demands, and improving consumer confidence. But challenges still exist: high implementation costs, data isolation, imperative differences in the norms of the regions, etc. These issues must be solved at the instance where professionals representing the industry, government, and scholars come together.

Tackling the question of how transparency supports the generation of sustainable, resilient, ethical supply chains, this paper demonstrates how pressurization enabled by digital tools empowers waste management. Regardless of the degree of integration, interdisciplinary collaboration in achieving sustainable business outcomes creates an opportunity for change. Current technologies such as AI, big data, blockchain, and IoT when implemented together can greatly enhance interoperability and promote the adoption of global standards in digital connectivity. These technologies also serve as powerful marketing tools and can be applied to provide detailed information about a product's lifecycle, offering incentives such as recycling bonuses for consumers and enabling new business models. Digital tools can support compliance by providing detailed tracking and reporting capabilities. The opportunity to improve supply chain transparency and waste management will arise from developing new digital technologies, collaborating with different actors, and solving existing problems.

References

Bhushan, M., & Singh, R. (2024). Talent management facilitates net-zero transition through employee green behavior. In *Net zero economy, corporate social responsibility and sustainable value creation. CSR, sustainability, ethics & governance, Part F2880* (pp. 117–129). Springer. https://doi.org/10.1007/978-3-031-55779-8_7

Dissanayake, H., Manta, O., Iddagoda, A., & Palazzo, M. (2024). AI applications in business: Trends and insights using bibliometric analysis. *The International Journal of Management Education, 22*(3), 101075. https://doi.org/10.1016/J.IJME.2024.101075

Jasrotia, S. S., Rai, S. S., Rai, S., & Giri, S. (2024). Stage-wise green supply chain management and environmental performance: Impact of blockchain technology. *International Journal of Information Management Data Insights, 4*(2), 100241. https://doi.org/10.1016/J.JJIMEI.2024.100241

Kashmanian, R. M. (2017). Building greater transparency in supply chains to advance sustainability. *Environmental Quality Management, 26*(3), 73–104. https://doi.org/10.1002/TQEM.21495

Khan, S., Ali, S. S., & Singh, R. (2022a). Determinants of remanufacturing adoption for circular economy: A causal relationship evaluation framework. *Applied System Innovation, 5*(4), 62. https://doi.org/10.3390/ASI5040062

Khan, S., Singh, R., Sá, J. C., Santos, G., & Ferreira, L. P. (2022b). Modelling of determinants of logistics 4.0 adoption: Insights from developing countries. *Machines, 10*(12), 1242. https://doi.org/10.3390/MACHINES10121242

Khan, S., Khan, M. I., & Singh, R. (2023). Modelling the barriers of blockchain technology implementation in supply chain. *Journal of Industrial Integration and Management, 8*(3), 367–385. https://doi.org/10.1142/S2424862223500070

Khan, S., Singh, R., Alnahas, J., Abbate, S., & Centobelli, P. (2024). Navigating the smart circular economy: A framework for manufacturing firms. *Journal of Cleaner Production, 480*, 144007. https://doi.org/10.1016/J.JCLEPRO.2024.144007

Kumar, M. (2024). Using fuzzy ISM approach to develop a framework of sustainable technology footprint enablers in inter- and intra-organisational integrated supply chain transparency and traceability. *International Journal of Business Performance and Supply Chain Modelling.* https://www.academia.edu/114291434/Developing_a_framework_of_sustainable_technology_footprint_enablers_in_inter_and_intra_organisational_integrated_supply_chain_transparency_and_traceability_using_fuzzy_ISM_approach

Kumar, D., Singh, R. K., Mishra, R., & Daim, T. U. (2023). Roadmap for integrating blockchain with Internet of Things (IoT) for sustainable and secured operations in logistics and supply chains: Decision making framework with case illustration. *Technological Forecasting and Social Change, 196*, 122837.

Kumar, K., Singh, R., Sharma, A., & Lehri, B. P. S. (2024). Climate change adaptation strategies for achieving net-zero economy. In *Transition towards a sustainable future. Approaches to global sustainability, markets, and governance* (pp. 163–183). Springer. https://doi.org/10.1007/978-981-97-5756-5_8

Modgil, S., Gupta, S., Zhang, B., & Bag, S. (2024). Blockchain-enabled supply chain finance strategies for a circular economy revolution. *Business Strategy and the Environment, 33*, 7006. https://doi.org/10.1002/BSE.3853

Singh, R., Khan, S., & Centobelli, P. (2022). Investigating the interplay between social performance and organisational factors supporting circular economy practices. *Sustainability, 14*(24), 16781. https://doi.org/10.3390/SU142416781

Singh, R., Khan, S., & Dsilva, J. (2023). A framework for assessment of critical factors for circular economy practice implementation. *Journal of Modelling in Management, 18*(5), 1476–1497. https://doi.org/10.1108/JM2-06-2021-0145/FULL/XML

Singh, R., Crowther, D., & Ravesangar, K. (2024a). Harnessing human capital for a sustainable future: HR's pivotal role toward net zero policies. In *Transition towards a sustainable future. Approaches to global sustainability, markets, and governance* (pp. 27–47). Springer. https://doi.org/10.1007/978-981-97-5756-5_2

Singh, R., Gupta, A., Bajpai, A., & Kandpal, V. (2024b). Exploring challenges of circular economy initiatives for smart cities. In *Smart cities and circular economy: The future of sustainable urban development* (pp. 59–69). Emerald. https://doi.org/10.1108/978-1-83797-957-820241005/FULL/XML

Singh, R., Khan, S., Dsilva, J., Akram, U., & Haleem, A. (2024c). Modelling the organisational factors for implementing corporate social responsibility: A modified TISM approach. *Global Journal of Flexible Systems Management, 25*(2), 283–301. https://doi.org/10.1007/S40171-024-00388-X/METRICS

Singh, R., Kumar, K., & Khan, S. (2024d). Chapter 12 A comprehensive view of artificial intelligence (AI)-based technologies for sustainable development goals (SDGs). In *Artificial intelligence enabled management* (pp. 183–196). De Gruyter. https://doi.org/10.1515/9783111172408-012

Zhang, A., & Seuring, S. (2024). Digital product passport for sustainable and circular supply chain management: A structured review of use cases. *International Journal of Logistics Research and Applications, 27*, 2513–2540. https://doi.org/10.1080/13675567.2024.2374256

Zhang, A., Duong, L., Seuring, S., & Hartley, J. L. (2023). Circular supply chain management: A bibliometric analysis-based literature review. *The International Journal of Logistics Management, 34*(3), 847–872.

Chapter 13
Blockchain and the Smart Circular Economy: Real-World Applications and Insight

Labaran Isiaku and Awwal Muhammad Dawud

Abstract This chapter explores how blockchain significantly promotes the smart circular economy by improving transparency, traceability, and efficiency via its integration with technologies such as IoT and AI. Essential characteristics of blockchain, including decentralized governance, immutability, and smart contracts, tackle issues in waste management, tracking product life cycles, and ensuring supply chain transparency. Practical examples demonstrate blockchain's uses, such as promoting recycling, ensuring ethical sourcing, and managing circular supply chains. Although it has potential, issues like technical restrictions, regulatory obstacles, and financial limitations are examined, along with approaches to address these challenges. Future pathways emphasize the incorporation of blockchain with new technologies, promoting worldwide acceptance, and developing creative business frameworks to enhance sustainability. This chapter provides guidance for companies, government officials, and academics to utilize blockchain for an efficient and sustainable future, highlighting the importance of teamwork and creativity to speed up the shift to a circular economy.

Keywords Blockchain · Industry 4.0 · Circular economy · Sustainability · Carbon emissions

L. Isiaku (✉)
School of Applied Sciences, Department of Management Information Systems, Cyprus International University, Mersin, Turkey
e-mail: lisiaku@ciu.edu.tr

A. M. Dawud
Department of Computer Engineering, Cyprus International University, Mersin, Turkey

R. Singh et al. (eds.), *Blockchain Technologies for Smart Circular Economy and Organisational Sustainability*, Environmental Footprints and Eco-design of Products and Processes, https://doi.org/10.1007/978-3-031-89183-0_13

13.1 Introduction

13.1.1 Overview of the Circular Economy (CE)

The circular economy (CE) serves as an alternative economic model that highlights sustainability through the ongoing utilization of resources, minimization of waste, and restoration of natural systems. In contrast to the conventional linear economy that operates on a "take, make, discard" model, the circular economy aims to reduce resource usage and ecological damage by encouraging the reuse, recycling, and reprocessing of resources (Singh et al., 2022; Huynh, 2022). At its essence, CE seeks to establish a restorative and regenerative framework where products, materials, and resources remain in use for the maximum duration. Some of the essential concepts of the circular economy consist of creating for durability; preserving the worth of products, materials, and resources within the economy for as long as feasible; and guaranteeing the restoration and renewal of natural systems (Schroeder et al., 2019). This necessitates a transition from conventional consumption models, where products are utilized and thrown away, to approaches that emphasize sustainability, preserve value, and reduce waste. CE encourages the implementation of methods such as remanufacturing, recycling, and refurbishing to prolong product life cycles and lessen reliance on raw materials.

The significance of CE has become more apparent amid global environmental issues like resource depletion, pollution, and climate change. The circular economy offers a practical route to attain sustainability and tackle these urgent problems by minimizing waste and maximizing resource utilization. This transition also provides significant economic advantages, including fresh business prospects, employment generation, and lowered raw material expenses (Khan et al., 2022).

In practice, the implementation of CE principles has been seen throughout different industries, including manufacturing, fashion, electronics, and agriculture. Businesses are progressively acknowledging that a circular strategy not only aids the environment but also promotes innovation and improves their competitive advantage (Rodríguez-Espíndola et al., 2022). Nonetheless, despite its promise, the shift to a completely circular economy faces considerable hurdles, such as complicated supply chain management, insufficient traceability, and the necessity for increased cooperation among stakeholders.

13.1.2 Emergence of Blockchain Technology (BCT)

BCT is a decentralized, distributed ledger system that securely logs and validates transactions across various computers (Isiaku et al., 2024). It is created to be resistant to tampering and transparent, allowing peer-to-peer transactions without requiring intermediaries. The most recognized use of BCT is in digital currencies like Bitcoin and Ethereum. Nonetheless, the possible applications of blockchain reach well beyond digital currencies and are acknowledged as revolutionary for multiple industries, such as supply chain management, finance, healthcare, and logistics.

At its core, BCT is based on the principles of transparency, immutability, and decentralization. Every "block" in the chain contains a set of transactions, and once information is added to the blockchain, it cannot be modified or removed (Isiaku & Adalier, 2024). This unchangeability guarantees data integrity, while decentralization removes the need for a central authority to verify transactions, making the system more robust and secure.

The importance of BCT in modern industries is due to its ability to offer real-time transparency, streamline operations, and boost trust. In a progressively digital and interconnected world, marked by cross-border transactions and vast amounts of data, blockchain provides answers to numerous issues encountered by sectors, such as data privacy concerns, fraud, and inefficiencies in conventional systems. For example, BCT can improve supply chain visibility by offering a secure and unchangeable log of the source and transit of products, guaranteeing that each transaction is traceable and can be verified (Isiaku & Adalier, 2024; Agarwal et al., 2022). As a result, blockchain stands out as a perfect technology to tackle issues in multiple industries, particularly the circular economy, where transparency, traceability, and resource efficiency hold significant value.

13.1.3 Rationale for Combining Blockchain with the Circular Economy (CE)

Blockchain and the circular economy have a mutual goal of enhancing resource use, guaranteeing transparency, and developing systems that are efficient and sustainable. Although the circular economy provides a thorough framework for minimizing waste and optimizing the lifespan of products and materials, it faces challenges that BCT might be able to tackle. A major challenge is the absence of transparency and traceability in intricate supply chains, hindering the ability to track material flow and guarantee their proper reuse or recycling (Kouhizadeh et al., 2020).

Additionally, blockchain technology can improve the effectiveness of circular economy systems by leveraging smart contracts and tokenization (Souza et al., 2024). Smart contracts are self-implementing agreements that have the terms directly programmed, allowing for automated and secure transactions without the need for intermediaries (Souza et al., 2024). This technology can enhance processes like waste management, recycling, and product return programs, ensuring compliance with circular economy principles in a more effective and economical way. On the other hand, tokenization may encourage eco-friendly actions, like recycling or minimizing waste, by offering digital tokens or cryptocurrency incentives (Kahya et al., 2021).

A further challenge that blockchain technology can tackle is the problem of data fragmentation and absence of standardization in the circular economy. Blockchain facilitates the creation of a cohesive, decentralized database that can safely store and share information among parties involved (Bodkhe et al., 2020; Nawari & Ravindran, 2019). This ability enhances cooperation between producers, consumers, waste managers, and other participants in the circular economy, guaranteeing the accessibility of precise and uniform data for decision-making and optimization activities. The combination of blockchain and the circular economy can revolutionize industries by establishing more transparent, efficient, and sustainable systems. Therefore, the aim of this chapter is to explore the use of blockchain technology in different elements of the circular economy, showcasing actual use cases and assessing the advantages, obstacles, and prospects that emerge from this combination.

This chapter aims to provide a comprehensive analysis of the relationship between blockchain technology and the circular economy. It will start by examining the fundamental principles of the circular economy and the challenges that need to be addressed for its successful execution. The chapter will subsequently explore the features of blockchain technology, examining its ability to support and promote the goals of the circular economy. The primary emphasis will be on real-world applications, highlighting particular examples where blockchain is driving innovation and improving sustainability in circular economy initiatives.

13.2 Understanding the Smart Circular Economy (SCE)

The smart circular economy (SCE) builds upon the foundational principles of the traditional circular economy, enhancing them through the integration of digital technologies. The SCE is a digitally driven, data-rich model that leverages advanced technologies like the Internet of Things (IoT), artificial intelligence (AI), blockchain, and big data analytics to optimize the use of resources, minimize waste, and create sustainable economic systems. By embedding smart technologies into circular processes, the SCE enables real-time monitoring, intelligent decision-making, and dynamic system optimization (Alcayaga et al., 2019). Digital technologies play a critical role in enabling the SCE. For example:

IoT: IoT devices, such as sensors and trackers, facilitate real-time data collection on the condition, location, and usage of products and materials (Nižetić et al., 2020). This capability is vital for tracking product life cycles, monitoring resource flows, and ensuring efficient waste management.

Artificial Intelligence (AI): AI algorithms analyze large volumes of data to optimize resource allocation, predict maintenance needs, and identify areas for efficiency improvement (Nama et al., 2023). AI also aids in designing products and systems that align with circular economy principles (Sánchez-García et al., 2024).

Big Data Analytics: By processing and interpreting data from multiple sources, big data analytics uncover patterns and insights that guide sustainable decision-making (Mageto, 2021). For instance, analytics can predict material demand or identify inefficiencies in recycling processes.

The integration of these technologies enables dynamic systems where materials and resources are constantly monitored and managed, ensuring their optimal use. For example, smart technologies can help identify waste streams that are suitable for repurposing or recycling, reducing resource consumption and environmental impact.

In practice, the SCE operates across several dimensions, including product life cycle management, supply chain optimization, and waste recovery. It encourages businesses and governments to transition from static, manual systems to adaptive, automated ones that can respond to real-time information. This transformation ensures that the principles of the circular economy reduce, reuse, and recycle are implemented more effectively and at scale.

13.2.1 Key Characteristics of the Smart Circular Economy

The SCE is distinguished by several key characteristics that differentiate it from traditional approaches to the circular economy: Fig. 13.1 illustrates the key characteristics of the smart circular economy.

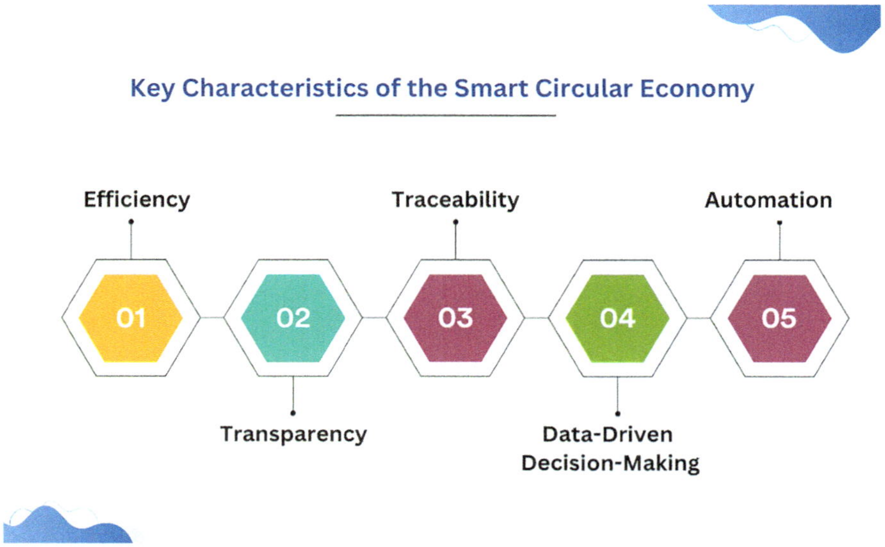

Fig. 13.1 Key characteristics of smart circular economy

13.2.1.1 Efficiency

Efficiency is central to the SCE, which seeks to maximize resource utilization while minimizing waste and energy consumption. Digital tools like AI and IoT optimize operations by providing real-time data and predictive analytics (Parida et al., 2024). For example, sensors in manufacturing systems can detect inefficiencies or waste, enabling immediate corrective actions that reduce material losses.

13.2.1.2 Transparency

Transparency is achieved through the digital tracking of products and materials across their life cycles. Blockchain technology plays a critical role here, creating tamper-proof records of transactions and product origins (Tyagi et al., 2024). This transparency builds trust among stakeholders, ensures compliance with sustainability standards, and enhances accountability in resource flows.

13.2.1.3 Traceability

Traceability refers to the ability to track the movement and transformation of resources throughout the value chain (Anastasiadis et al., 2022). IoT devices, which are integrated with blockchain, enable detailed records of product journeys, from raw material sourcing to end-of-life management. This capability supports practices such as responsible sourcing, ethical production, and extended producer responsibility (EPR).

13.2.1.4 Data-Driven Decision-Making

The SCE relies heavily on data to inform and optimize processes. Big data analytics provides actionable insights into resource usage patterns, supply chain bottlenecks, and waste management opportunities (Mageto, 2021). Data-driven decision-making ensures that actions align with sustainability goals, reducing inefficiencies and promoting innovation.

13.2.1.5 Automation

Automation streamlines circular economy processes through technologies such as robotics and AI-driven systems. Smart contracts, a feature of blockchain technology, automate transactions and enforce agreements without human intervention (Nzuva, 2019). For instance, in waste management, smart contracts can automatically trigger payments or rewards based on verified recycling actions (Alnuaimi et al., 2023).

By embedding smart technologies into circular economy practices, the SCE addresses many of the challenges faced by traditional circular approaches, such as lack of data, inefficient processes, and limited collaboration among stakeholders. It transforms circular economy systems into dynamic, interconnected networks capable of achieving greater levels of sustainability and economic efficiency.

As industries increasingly adopt the principles of the SCE, they are not only reducing their environmental footprint but also enhancing competitiveness and resilience in a rapidly changing world. This synergy between technology and sustainability positions the SCE as a critical driver of global efforts to achieve long-term resource efficiency and environmental stewardship (Singh et al., 2023).

13.3 Blockchain in the Circular Economy

Blockchain is a distributed ledger technology designed to securely record and verify transactions across a decentralized network (Komalavalli et al., 2020). Unlike traditional centralized systems, blockchain relies on a network of nodes to validate transactions, ensuring that no single entity has control over the entire system. This decentralized nature enhances security, transparency, and trust among participants.

Three key features of blockchain are central to its functionality:

- ***Decentralization***
- Blockchain operates without a central authority. Instead, transactions are validated by a network of nodes, making the system more resilient to failures or cyberattacks (Singh et al., 2021). This decentralization is crucial for fostering trust among stakeholders, particularly in circular economy ecosystems, where multiple parties with varying interests collaborate.
- ***Immutability***
- Once a transaction is recorded on the blockchain, it cannot be altered or deleted. This immutability ensures the integrity of data, providing a tamper-proof record of all transactions (Politou et al., 2019). In the context of the circular economy, this guarantees reliable tracking of materials and products throughout their life cycle.
- ***Transparency***
- Blockchain provides a shared, transparent ledger accessible to all authorized participants. This transparency builds trust by allowing stakeholders to verify transactions independently (Sedlmeir et al., 2022). For example, consumers can verify the sustainability credentials of products they purchase, while businesses can ensure compliance with circular economy principles.

Blockchain comes in two main types:

- ***Public Blockchains***

- The public or permissionless blockchains are open networks where anyone and any devices can participate and validate transactions without any kind of permission, the ledger is shared and transparent (Junejo et al., 2020). Public blockchains, such as Ethereum and Bitcoin, offer maximum transparency but may have scalability and energy consumption challenges (Andoni et al., 2019). The figure below shows an illustration of a public blockchain (Fig. 13.2).
- ***Private Blockchains***
- The private or permissioned blockchains are restricted networks whereby only authorized participants can join (Peng et al., 2021). Furthermore, such blockchains are often used in enterprise settings, offering greater control and efficiency and at the same time maintaining some level of transparency (Fig. 13.3).

Two additional features of blockchain that are particularly relevant to the circular economy are as follows:

- ***Smart Contracts***
- Smart contracts are self-executing contracts with predefined terms encoded directly into the blockchain (Nzuva, 2019). Smart contracts automate processes such as payments, waste recovery, and compliance checks, reducing the need for intermediaries.

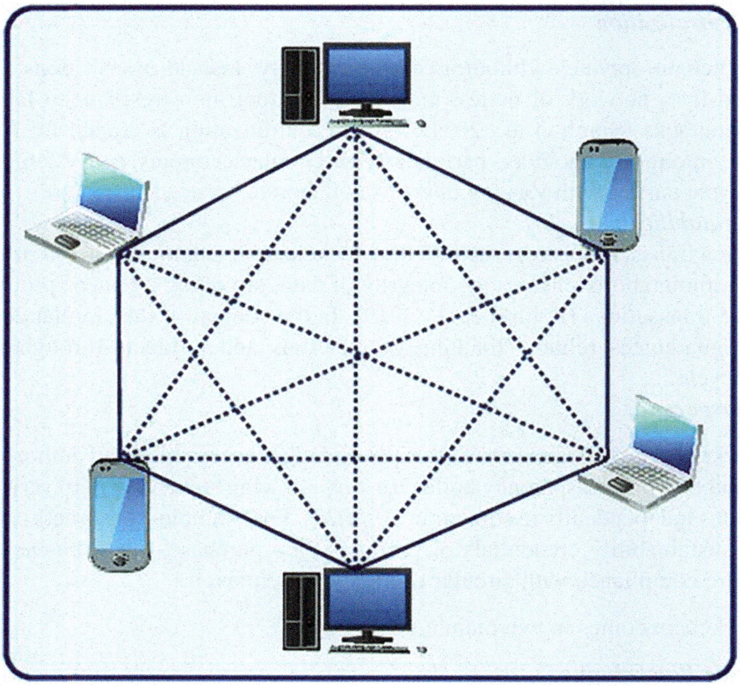

Fig. 13.2 Representation of a public or permissionless blockchain technology

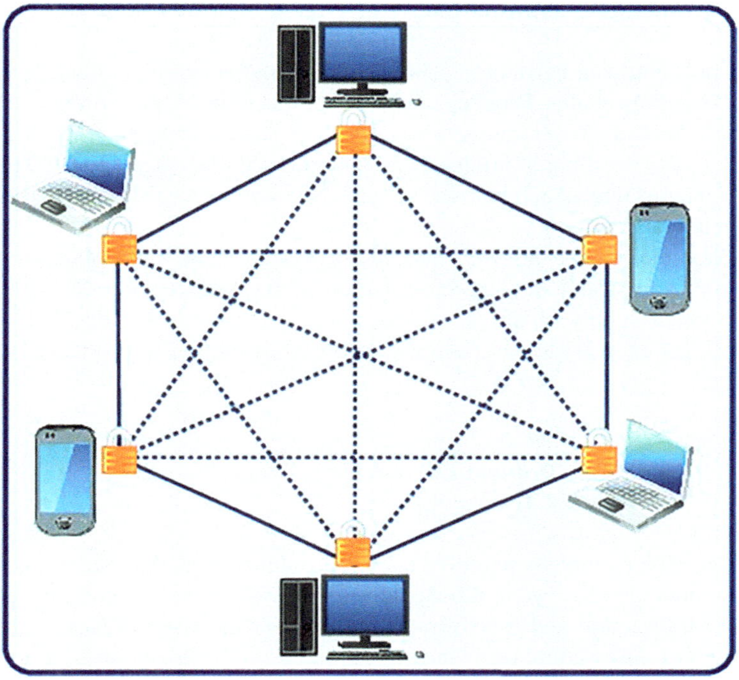

Fig. 13.3 Representation of a private or permissioned blockchain technology

- ***Tokenization***
- This is a representation of real-world assets or incentives as digital tokens on the blockchain (Sunyaev et al., 2021). Tokenization enables innovative business models, such as incentivizing recycling through token-based rewards.

13.3.1 Blockchain's Role in the Circular Economy

The implementation of circular economy principles faces significant hurdles that can be addressed through the innovative use of blockchain technology. This cutting-edge approach provides groundbreaking solutions by improving transparency, enabling the tracking of materials, and verifying product authenticity. As a result, blockchain empowers various stakeholders to develop more sustainable and effective systems.

13.3.1.1 Enhancing Transparency and Trust in Supply Chains

One of the significant barriers to achieving a circular economy is the lack of transparency in supply chains. Blockchain addresses this issue by providing a secure and immutable record of all transactions, enabling stakeholders to trace the origin, movement, and handling of products and materials (Shojaei et al., 2021). This transparency ensures that all actors in the supply chain adhere to sustainability standards and ethical practices.

For example, blockchain can track the sourcing of raw materials to verify that they come from sustainable and ethical sources. In industries like fashion or electronics, where supply chains are often complex and global, blockchain ensures accountability at every stage, from raw material extraction to product disposal or recycling.

13.3.1.2 Facilitating Product Life Cycle Tracking and Material Traceability

Effective circular economy practices rely on the ability to track products and materials throughout their life cycle. Blockchain enables end-to-end traceability by recording every transaction and transformation a product undergoes (Zhu et al., 2021). This capability is essential for managing resources, reducing waste, and optimizing recycling efforts.

For instance, in the manufacturing sector, blockchain can record data about a product's composition, usage, and maintenance history. When the product reaches the end of its life, this information helps determine the most efficient recycling or repurposing methods. Similarly, in the food industry, blockchain can track perishable goods from farm to table, reducing food waste and ensuring safety (Haji et al., 2020).

13.3.1.3 Ensuring Product Authentication and Origin Verification

Counterfeit and substandard products undermine the goals of the circular economy by creating inefficiencies and waste. Blockchain ensures product authentication by providing a verifiable record of a product's origin and journey (Montecchi et al., 2019). Consumers can access this information to confirm the authenticity and sustainability of the products they purchase.

This feature is particularly valuable in industries like pharmaceuticals and luxury goods, where product authenticity is critical. For example, blockchain can authenticate a recycled product's compliance with safety standards or verify that a luxury item has been sourced and manufactured sustainably (Nygaard, 2024).

13.3.1.4 Streamlining Circular Economy Processes

Blockchain also simplifies circular economy processes through the use of smart contracts and tokenization. Smart contracts automate complex tasks, such as triggering payments for recycled materials or enforcing compliance with environmental regulations. Tokenization, on the other hand, creates economic incentives for sustainable practices, such as rewarding consumers for recycling or adopting eco-friendly behaviors.

For instance, waste management companies can use blockchain to track the collection, sorting, and recycling of waste materials, ensuring that all processes are documented and compliant. Consumers who recycle correctly can be rewarded with tokens that can be exchanged for goods, services, or discounts, fostering greater participation in circular economy initiatives.

Blockchain's decentralized, immutable, and transparent nature makes it a powerful enabler of the circular economy. By addressing critical challenges such as supply chain opacity, lack of traceability, and product authentication, blockchain enhances the efficiency, accountability, and sustainability of circular economy systems (Upadhyay et al., 2021). The integration of smart contracts and tokenization further streamlines processes and incentivizes sustainable behaviors, creating a dynamic and collaborative ecosystem for stakeholders.

As industries and governments increasingly adopt blockchain technology, its potential to transform the circular economy becomes evident. By combining the power of blockchain with circular economy principles, stakeholders can create more resilient, resource-efficient, and sustainable systems that benefit both the society and environment.

13.4 Real-World Use Cases of Blockchain in Smart Circular Economy

13.4.1 Use Case 1: Waste Management and Recycling

Efficient waste management and recycling are critical pillars of the smart circular economy. Blockchain technology provides innovative solutions for tracking waste streams, optimizing recycling processes, and incentivizing participation in waste recovery (Ahmad et al., 2021).

- *Blockchain for Tracking Waste Streams and Recycling Processes*
- In traditional waste management systems, tracking the journey of waste from generation to recycling or disposal is often fragmented, leading to inefficiencies and noncompliance. Blockchain addresses this issue by offering a transparent and immutable record of waste flows (Ahmad et al., 2021). Using IoT-enabled sensors and smart contracts, waste streams can be tracked in real time, providing

stakeholders with actionable insights into waste generation patterns and recycling efficiencies (Shafik, 2024). As an example, systems utilizing blockchain technology can record information about the nature, amount, and whereabouts of collected waste, thereby ensuring accountability throughout the entire process (Ahmad et al., 2021). This level of transparency helps mitigate the risk of unlawful waste disposal and improves adherence to regulatory standards.

- ***Case Study***: *Waste Disposal Tracking (Helsinki Finland)*
- Cities and municipalities worldwide are turning to blockchain technology to address inefficiencies in waste management and ensure proper disposal practices. One notable example is Helsinki, Finland. In 2022, the city launched a forward-thinking initiative to explore blockchain-based platforms as part of its efforts to tackle circular economy challenges, including waste management (Lepore et al., 2023). This initiative focuses on improving supply chain traceability and transparency by integrating blockchain into waste management systems. By using blockchain, Helsinki can monitor waste streams in real time, from collection to processing, ensuring that waste is handled responsibly and processed at certified facilities. The transparent and immutable nature of blockchain helps to verify the performance of waste contractors, ensuring timely collections and adherence to environmental standards.

13.4.2 Use Case 2: Product Life Cycle and Extended Producer Responsibility (EPR)

Manufacturers are held responsible for their products' entire lifespan, from creation to disposal, under extended producer responsibility (EPR) frameworks. The implementation of blockchain technology is vital in ensuring adherence to EPR systems by facilitating comprehensive monitoring of a product's life cycle.

- *Blockchain to Ensure Compliance and Transparency in EPR Systems*
- Blockchain technology offers a comprehensive, see-through ledger that documents and authenticates information spanning a product's entire lifespan, from the procurement of raw materials to its ultimate disposal. This level of openness ensures that producers adhere to extended producer responsibility (EPR) guidelines, such as meeting specified recycling targets or material reclamation standards (Gazeau et al., 2024). According to Omar et al. (2020), the implementation of smart contracts can streamline the process of compliance monitoring, automatically enacting consequences like penalties or incentives based on observed performance.
- ***Case Study***: *Tracking Materials from Production to End-of-Life (Dell)*
- Blockchain technology is increasingly being adopted by electronics manufacturers to oversee their products' life cycles, promoting sustainability and adherence

to regulations. A prime example is Dell, which utilizes blockchain to track the materials in its laptops throughout their entire lifespan, from manufacture to disposal (Aggarwal et al., 2023). This blockchain-based system generates a comprehensive and unalterable digital log for each device, recording information about the sourcing, construction, utilization, and ultimate recycling or disposal of its components. Through this blockchain application, Dell can guarantee the responsible recycling or reuse of crucial elements like batteries and metals, thereby reducing environmental harm. This strategy not only aids in meeting extended producer responsibility (EPR) legislation requirements but also enhances Dell's standing as a sustainability pioneer, offering transparency and fostering trust among stakeholders and consumers.

13.4.3 Use Case 3: Supply Chain Transparency and Circular Supply Chains

The integration of blockchain into supply chains transforms them into circular supply chains, where materials flow continuously through loops of use, recovery, and reuse. Blockchain enhances transparency and efficiency in managing these material flows.

- *Blockchain for Circular Supply Chain Integration and Material Flow Tracking*

- In a circular supply chain, materials are tracked across multiple cycles of use and recovery. Blockchain ensures the traceability of these materials, enabling stakeholders to identify bottlenecks, optimize resource flows, and verify sustainability practices (Centobelli et al., 2022). The technology also fosters collaboration among supply chain partners by providing a shared, tamper-proof record of transactions.
- ***Case Study***: *Companies Utilizing Blockchain for Circular Supply Chain Management (Walmart)*

- Walmart has adopted blockchain technology to improve the transparency of its supply chain, particularly in food distribution (Astill et al., 2019). The retail giant utilizes this technology to monitor the source and transportation of food items, documenting each stage from production to retail. This method ensures adherence to sustainability criteria and enables Walmart to swiftly identify and rectify inefficiencies or disruptions within its supply network. For example, blockchain assists in locating contamination sources during food recalls, thereby reducing waste and safeguarding public health. Moreover, by offering consumers verifiable data about product origins, Walmart builds trust and showcases its dedication to sustainability. This cyclical supply chain approach not only diminishes waste but also supports the company's broader environmental objectives and bolsters its image as a socially conscious retailer.

13.4.4 Use Case 4: Tokenization and Circular Economy Incentives

The circular economy is being revolutionized by the introduction of tokenization, which provides financial incentives for sustainable practices. Through the use of blockchain technology, a concrete reward system is established, linking environmentally friendly actions to digital tokens. This approach motivates both individuals and organizations to embrace circular economy principles and adopt more sustainable behaviors.

- *Blockchain for Incentivizing Circular Behaviors Through Cryptocurrency and Tokens*
- Through tokenization, organizations can incentivize stakeholders for engaging in circular economy practices, such as recycling, minimizing waste, or utilizing eco-friendly products (Mhaskey, 2024). These digital tokens can be redeemed for various benefits, including merchandise, services, or price reductions, thereby encouraging broader involvement in sustainable economic initiatives.
- **Case Study 1** Platforms Encouraging Circular Economy Participation via Tokens (Plastic Bank)
- Platforms such as Plastic Bank are revolutionizing plastic waste management by utilizing blockchain technology to encourage recycling (Bhubalan et al., 2022). Plastic Bank functions in multiple nations, allowing people to gather and submit plastic waste at authorized collection points. In exchange, they receive tokens that are documented on a blockchain, guaranteeing transparency and deterring fraud. These tokens may be exchanged for necessary products, services, or even monetary assistance, enabling low-income communities and tackling environmental issues. The system fosters trust among participants and stakeholders by connecting rewards to validated recycling actions. It additionally promotes responsibility in the waste recovery procedure, establishing a sustainable framework that minimizes plastic destined for the ocean. This method shows how blockchain can serve as an effective instrument in promoting global involvement in the circular economy.

13.5 Challenges and Barriers to Blockchain Adoption in the Circular Economy

Blockchain technology could revolutionize the circular economy, but its broad adoption faces significant challenges. These obstacles, covering technical, regulatory, financial, and societal areas, must be tackled to completely utilize its transformative potential.

13.5.1 Technical Challenges

- *Scalability*
- Blockchain networks, particularly public ones using consensus methods such as Proof of Work (PoW), frequently face scalability challenges (Khan et al., 2023; Rejeb et al., 2022). The significant transaction volumes characteristic of circular economy systems, which include supply chains and recycling methods, may likely overwhelm these networks. The restricted throughput capacity of platforms like Bitcoin or Ethereum poses major challenges for monitoring millions of product life cycles in real time, thus resulting in operational inefficiencies and delays.
- *Interoperability*
- The execution of circular economy strategies frequently requires the incorporation of blockchain technologies with current digital frameworks, including enterprise resource planning (ERP) systems, Internet of Things (IoT) devices, and various digital platforms. Nevertheless, the lack of consistent standards and interoperability between blockchain networks and traditional systems hinders seamless integration (Shaikh et al., 2024). The absence of effective communication among various technologies can result in fragmented and less efficient circular economy efforts.
- *Energy Consumption*
- The sustainable principles of the circular economy clash with the high energy consumption of blockchain networks such as Bitcoin and Ethereum, which rely on Proof of Work (PoW) systems (Wendl et al., 2023). Although more energy-efficient alternatives like Proof of Stake (PoS) exist, their widespread implementation in blockchain applications remains limited.

13.5.2 Economic and Organizational Barriers

- *High Initial Costs*
- The adoption of blockchain technology necessitates substantial financial investments in hardware, software development, and employee training (Prager et al., 2021). These costs pose significant difficulties for small- and medium-sized enterprises (SMEs), which frequently possess limited financial assets yet play an essential role in circular economy initiatives. For example, small- and medium-sized enterprises involved in recycling or materials recovery might find it challenging to set aside resources for implementing blockchain while overseeing their daily operations (Singh et al., 2024). Moreover, the ongoing maintenance and improvements add to the financial strain, potentially obstructing widespread adoption. It is crucial for government entities and major corporations to

collaborate in providing subsidies, financial support, or public infrastructure to ease the financial strain on SMEs and guarantee widespread technology adoption.

- *Stakeholder Collaboration*

- The effectiveness of blockchain technology in circular economy projects depends on the cooperative actions of multiple stakeholders, such as manufacturers, suppliers, recycling organizations, and regulatory agencies (Shojaei et al., 2021). However, achieving consensus on data sharing, management systems, and technical standards is a significant challenge. For instance, competing firms may hesitate to share sensitive supply chain details on a blockchain network, worried about the risks of losing confidential information. Additionally, differing priorities among stakeholders may lead to extended decision-making timelines and postponed implementation. To overcome these challenges, it is essential to create fair governance systems, trust-building frameworks, and standardized protocols. These actions are essential for promoting collaboration and guaranteeing seamless integration within the system.

13.5.3 Public Perception and Trust Issues

- *Lack of Understanding*

- The complexities of blockchain technology frequently lead to confusion and hesitation among the public and key decision-makers. Numerous people see blockchain only as the basis for volatile cryptocurrencies, overlooking its capabilities in domains like supply chain transparency or waste management. For instance, corporate leaders lacking knowledge in blockchain may question its relevance or consider it a risky endeavor. This lack of comprehension hinders adoption, particularly in sectors where technological skills are scarce (Isiaku & Adalier, 2024). To tackle this knowledge gap and promote broader acceptance, it is essential to employ clear communication methods, educational programs, and simple demonstrations of blockchain's benefits in circular economy contexts.
- *Trust in Technology*
- Although blockchain is designed to promote trust via transparency and immutability, its association with the volatile nature of cryptocurrencies and significant fraud cases has led to concerns regarding its dependability (Isiaku et al., 2024). Participants in the circular economy, such as manufacturers and consumers, may be hesitant to adopt blockchain solutions because of concerns regarding data security and possible abuse. For instance, fears regarding disclosing sensitive supply chain details or relying on an untested technology may deter participation. Building trust requires continual efforts to highlight blockchain's secure and verifiable features through pilot projects, transparency in implementation, and partnerships with respected entities.
- *Resistance to Change*

- Many organizations and individuals often hold onto established systems and processes, even if they may be inefficient. This reluctance to adopt new technologies frequently stems from concerns about disruption, cost effects, and doubts about long-term benefits. A notable example is the hesitation of traditional recycling or waste management companies to adopt blockchain technology, due to its perceived complexity or the possible need to overhaul current operational practices (Isiaku et al., 2024; Rejeb et al., 2022). To address this resistance, it is essential to present clear proof of blockchain's tangible advantages, like lower costs, better regulatory compliance, or increased environmental sustainability. The execution of pilot programs and gradual integration can allow stakeholders to directly realize the benefits and encourage a more seamless shift to blockchain-supported circular economy systems.

13.6 Future Directions and Opportunities

As the blockchain technology matures, its influence on the circular economy is set to grow through technological progress, worldwide adoption patterns, and new commercial prospects. By combining with complementary technologies and encouraging innovative business approaches, blockchain can tackle existing obstacles and reveal fresh opportunities in the pursuit of sustainability.

13.6.1 Technological Innovations

- *Integration of AI and IoT*
- The combination of blockchain, artificial intelligence (AI), and the Internet of Things (IoT) signals revolutionary opportunities for the circular economy. IoT sensors collect real-time data on resource use, waste generation, and material transfers, whereas blockchain technology ensures the secure documentation and sharing of this data among stakeholders (Ahmad et al., 2021). AI systems subsequently analyze these datasets to enhance processes like waste sorting, recycling logistics, and supply chain management. For example, a blockchain-enabled IoT network might track the condition and location of resources in a circular supply chain, while AI algorithms predict the best reuse or recycling strategies based on usage trends. This merging of technologies not only improves operational efficiency but also strengthens decision-making abilities and accountability, promoting a more adaptable and robust circular economic framework.
- *Blockchain-Based Digital Twins*
- Digital twins integrated with blockchain signify a revolutionary advancement in managing resources. These digital replicas of tangible goods or systems gather real-time life cycle information, providing unmatched insights. Utilizing

blockchain technology, these digital replicas can safely document and share information regarding a product's origins, parts, and usage trends, thus guaranteeing transparency and traceability throughout its entire life cycle. The electronics sector, for example, might gain from blockchain-backed digital twins to track device lifespans, promoting effective recycling methods and compliance with sustainability guidelines. These technological innovations encourage collaboration among stakeholders and aid in creating closed-loop systems, which ultimately lead to more sustainable practices throughout various industries.

13.6.2 Global Adoption Trends

- *Regional Variations in Adoption*
- The application of blockchain technology in circular economy practices varies among different geographical areas and industrial sectors. European countries, boasting advanced economies, lead in employing blockchain for managing circular supply chains and ensuring regulatory compliance, driven by stringent environmental policies. Conversely, developing countries are exploring blockchain's capability to address waste management challenges and enhance informal recycling systems. For instance, specific regions in Asia and Africa are using blockchain technology to address issues like e-waste and plastic pollution, providing token rewards to motivate involvement (Ajala, 2023; Dua et al., 2020). These trends highlight blockchain's ability to tackle various challenges tailored to local needs, thus fostering global progress in circular economy initiatives.
- *Predictions for Blockchain's Future Role*
- As blockchain technology advances, its implementation in the circular economy is anticipated to grow more prevalent and cohesive. Improvements in scalability and energy efficiency, along with regulatory backing, will enhance blockchain accessibility for companies of all sizes. Over the next 10 years, blockchain is expected to be crucial in facilitating completely transparent and interconnected international supply chains, guaranteeing accountability across various sectors. Additionally, the rise of blockchain-powered circular business models, including product-as-a-service and tokenized recycling initiatives, will speed up the shift toward sustainable practices. Governments and global entities are anticipated to work together more vigorously in creating structures that promote blockchain use while tackling ethical and regulatory issues.

13.6.3 Opportunities for New Business Models

- *Circular Business Innovations*

- Blockchain technology offers a fertile ground for developing new business strategies that align with the principles of a circular economy. Both new enterprises and established firms are exploring blockchain-based systems to promote resource distribution, product recycling, and waste reduction (Khan et al., 2024). An example of this is businesses adopting token-based systems to encourage consumers to return used products for recycling or refurbishment. For example, a company could provide blockchain-based tokens as incentives to customers who bring back old electronic devices, which can then be utilized for discounts on new items. This approach not only boosts customer loyalty but also encourages eco-friendly practices.
- *Product-as-a-Service Models*
- Blockchain technology enables service-oriented product models, permitting customers to pay for access to products instead of full ownership. Organizations can improve product life cycle management by leveraging blockchain to securely record usage data and contractual agreements, ensuring that maintenance, reuse, or recycling happens at the right intervals. For instance, a furniture rental business might track item utilization and optimize refurbishment or redistribution processes, thus minimizing waste.
- *Peer-to-Peer Marketplaces*
- Peer-to-peer marketplaces are evolving due to decentralized blockchain technology, enabling individuals to trade excess goods or upcycled products directly without middlemen. These blockchain-driven platforms utilize the technology's fundamental transparency and immutable characteristics to foster trust among users, reducing the need for third-party participation. For instance, individuals can reliably sell refurbished electronics or distribute surplus materials, certain that their transactions are secure and easily traceable. This approach not only simplifies the exchange process but also promotes collaboration among local communities. These marketplaces promote circular economy principles by improving resource use and reducing waste, while allowing users to engage actively in sustainability initiatives.

13.7 Conclusion

The integration of blockchain technology into the smart circular economy offers an innovative approach to address global sustainability challenges. This chapter has shown how the fundamental characteristics of blockchain such as transparency, immutability, and decentralized governance align seamlessly with the principles of a circular economy. Blockchain enhances the monitoring of products across supply chains, encourages accountability in waste management, and supports creative business approaches, allowing stakeholders to develop more efficient and sustainable systems. Real-world examples have demonstrated its ability to improve resource use, inspire recycling initiatives, and ensure compliance with environmental regulations.

Although the potential of blockchain technology is clear, its extensive adoption encounters significant challenges. To completely leverage the benefits of this innovation, it is crucial to address technical challenges like scalability and power usage, navigate complicated regulatory frameworks, and surmount economic barriers. Moreover, the achievement of a genuinely cohesive and efficient blockchain-based circular economy depends on promoting collaboration among diverse stakeholders, such as manufacturers, consumers, government representatives, and researchers.

To accelerate this transition, companies need to lead the adoption of blockchain technologies that improve sustainability, showcasing their effectiveness and advantages. On the other hand, lawmakers ought to create favorable regulatory frameworks and promote adoption via financial incentives and partnerships between public and private entities. Scholars have an essential role in discovering new uses, enhancing blockchain's technical capabilities, and addressing ethical concerns to guarantee equitable application. By working together, these stakeholders can overcome current obstacles and create a more inclusive path for blockchain implementation in the circular economy.

In the coming years, blockchain technology can revolutionize circular economy practices, bringing about a new era of transparency, trustworthiness, and efficiency. As blockchain evolves, its integration with artificial intelligence, the Internet of Things, and other digital technologies will enhance its capabilities, making circular systems more efficient and effective. However, achieving this vision requires continuous commitment, innovation, and involvement from every part of the society.

References

Agarwal, U., Rishiwal, V., Tanwar, S., Chaudhary, R., Sharma, G., Bokoro, P. N., & Sharma, R. (2022). Blockchain technology for secure supply chain management: A comprehensive review. *IEEE Access, 10*, 85493–85517. https://doi.org/10.1109/ACCESS.2022.3194319

Aggarwal, R., Singh, H., & Garg, P. (2023). Dell Technologies Inc.: Innovating toward circularity. Journal of Information Technology Teaching Cases, 20438869231215089. https://doi.org/10.1177/20438869231215089

Ahmad, R. W., Salah, K., Jayaraman, R., Yaqoob, I., & Omar, M. (2021). Blockchain for waste management in smart cities: A survey. *IEEE Access, 9*, 131520–131541. https://doi.org/10.1109/ACCESS.2021.3113380

Ajala, O. (2023). Assessing plastic circular economy policies and the use of digital technology in Africa. In *Digital innovations for a circular plastic economy in Africa* (p. 222). Routledge.

Alcayaga, A., Wiener, M., & Hansen, E. G. (2019). Towards a framework of smart-circular systems: An integrative literature review. *Journal of Cleaner Production, 221*, 622–634. https://doi.org/10.1016/j.jclepro.2019.02.085

Alnuaimi, E., Alsafi, M., Alshehhi, M., Debe, M., Salah, K., Yaqoob, I., Zemerly, M. J., & Jayaraman, R. (2023). Blockchain-based system for tracking and rewarding recyclable plastic waste. *Peer-to-Peer Networking and Applications, 16*(1), 328–346. https://doi.org/10.1007/s12083-022-01413-5

Anastasiadis, F., Manikas, I., Apostolidou, I., & Wahbeh, S. (2022). The role of traceability in end-to-end circular agri-food supply chains. *Industrial Marketing Management, 104*, 196–211. https://doi.org/10.1016/j.indmarman.2022.04.021

Andoni, M., Robu, V., Flynn, D., Abram, S., Geach, D., Jenkins, D., McCallum, P., & Peacock, A. (2019). Blockchain technology in the energy sector: A systematic review of challenges and opportunities. *Renewable and Sustainable Energy Reviews, 100*, 143–174. https://doi.org/10.1016/j.rser.2018.10.014

Astill, J., Dara, R. A., Campbell, M., Farber, J. M., Fraser, E. D., Sharif, S., & Yada, R. Y. (2019). Transparency in food supply chains: A review of enabling technology solutions. *Trends in Food Science & Technology, 91*, 240–247. https://doi.org/10.1016/j.tifs.2019.07.024

Bhubalan, K., Tamothran, A. M., Kee, S. H., Foong, S. Y., Lam, S. S., Ganeson, K., Vigneswari, S., Amirul, A.-A., & Ramakrishna, S. (2022). Leveraging blockchain concepts as watermarkers of plastics for sustainable waste management in progressing circular economy. *Environmental Research, 213*, 113631. https://doi.org/10.1016/j.envres.2022.113631

Bodkhe, U., Tanwar, S., Parekh, K., Khanpara, P., Tyagi, S., Kumar, N., & Alazab, M. (2020). Blockchain for industry 4.0: A comprehensive review. *IEEE Access, 8*, 79764–79800. https://doi.org/10.1109/ACCESS.2020.2988579

Centobelli, P., Cerchione, R., Del Vecchio, P., Oropallo, E., & Secundo, G. (2022). Blockchain technology for bridging trust, traceability and transparency in circular supply chain. *Information & Management, 59*(7), 103508. https://doi.org/10.1016/j.im.2021.103508

Dua, A., Dutta, A., Zaman, N., & Kumar, N. (2020). Blockchain-based E-waste management in 5G smart communities. In *IEEE INFOCOM 2020-IEEE Conference on Computer Communications Workshops (INFOCOM WKSHPS)* (pp. 195–200). https://doi.org/10.1109/INFOCOMWKSHPS50562.2020.9162845

Gazeau, B., Zaman, A., Minunno, R., & Shaikh, F. (2024). Developing traceability systems for effective circular economy of plastic: A systematic review and meta-analysis. *Sustainability, 16*(22), 9973. https://doi.org/10.3390/su16229973

Haji, M., Kerbache, L., Muhammad, M., & Al-Ansari, T. (2020). Roles of technology in improving perishable food supply chains. *Logistics, 4*(4), 33. https://doi.org/10.3390/logistics4040033

Huynh, P. H. (2022). Enabling circular business models in the fashion industry: The role of digital innovation. *International Journal of Productivity and Performance Management, 71*(3), 870–895. https://doi.org/10.1108/IJPPM-12-2020-0683

Isiaku, L., & Adalier, A. (2024). Analyzing the barriers to blockchain adoption in educational sectors: A comprehensive literature review using the technology-organizational-environment (TOE) framework. *On the Horizon: The International Journal of Learning Futures, 33*(1), 32–60. https://doi.org/10.1108/OTH-05-2024-0029

Isiaku, L., Muhammad, A. S., Kefas, H. I., & Isiaku, H. H. (2024). Documenting trends in blockchain technology adoption: A systematic review. *Information Discovery and Delivery*. https://doi.org/10.1108/IDD-05-2024-0064

Junejo, A. Z., Hashmani, M. A., & Alabdulatif, A. A. (2020). A survey on privacy vulnerabilities in permissionless blockchains. *International Journal of Advanced Computer Science and Applications (IJACSA), 11*(9), 130–139.

Kahya, A., Avyukt, A., Ramachandran, G. S., & Krishnamachari, B. (2021). Blockchain-enabled personalized incentives for sustainable behavior in smart cities. In *2021 International Conference on Computer Communications and Networks (ICCCN)* (pp. 1–6). https://doi.org/10.1109/ICCCN52240.2021.9522340

Khan, S., Ali, S. S., & Singh, R. (2022). Determinants of remanufacturing adoption for circular economy: A causal relationship evaluation framework. *Applied System Innovation, 5*(4), 62.

Khan, S., Khan, M. I., & Singh, R. (2023). Modeling the barriers of blockchain technology implementation in supply chain. *Journal of Industrial Integration and Management, 8*(3), 367–385.

Khan, S., Singh, R., Alnahas, J., Abbate, S., & Centobelli, P. (2024). Navigating the smart circular economy: A framework for manufacturing firms. *Journal of Cleaner Production, 480*, 144007.

Komalavalli, C., Saxena, D., & Laroiya, C. (2020). Overview of blockchain technology concepts. In *Handbook of research on blockchain technology* (pp. 349–371). Elsevier. https://doi.org/10.1016/B978-0-12-819816-2.00014-9

Kouhizadeh, M., Zhu, Q., & Sarkis, J. (2020). Blockchain and the circular economy: Potential tensions and critical reflections from practice. *Production Planning & Control, 31*(11–12), 950–966. https://doi.org/10.1080/09537287.2019.1695925

Lepore, D., Testi, N., & Pasher, E. (2023). Building inclusive smart cities through innovation intermediaries. *Sustainability, 15*(5), 4024. https://doi.org/10.3390/su15054024

Mageto, J. (2021). Big data analytics in sustainable supply chain management: A focus on manufacturing supply chains. *Sustainability, 13*(13), 7101. https://doi.org/10.3390/su13137101

Mhaskey, S. V. (2024). Leveraging tokenization in blockchain-based circular economy: A paradigm shift towards sustainable resource management. *International Journal of Science and Research Archive, 13*(1), 293–300. https://doi.org/10.30574/ijsra.2024.13.1.1628

Montecchi, M., Plangger, K., & Etter, M. (2019). It's real, trust me! Establishing supply chain provenance using blockchain. *Business Horizons, 62*(3), 283–293. https://doi.org/10.1016/j.bushor.2019.01.008

Nama, P., Pattanayak, S., & Meka, H. S. (2023). AI-driven innovations in cloud computing: Transforming scalability, resource management, and predictive analytics in distributed systems. *International Research Journal of Modernization in Engineering Technology and Science, 5*(12), 4165.

Nawari, N. O., & Ravindran, S. (2019). Blockchain technology and BIM process: Review and potential applications. *Journal of Information Technology in Construction, 24*, 209–238.

Nižetić, S., Šolić, P., Gonzalez-De, D. L.-I., & Patrono, L. (2020). Internet of things (IoT): Opportunities, issues and challenges towards a smart and sustainable future. *Journal of Cleaner Production, 274*, 122877. https://doi.org/10.1016/j.jclepro.2020.122877

Nygaard, A. (2024). Authenticity, blockchain technology and green marketing. In *Green marketing and entrepreneurship* (pp. 155–165). Springer. https://doi.org/10.1007/978-3-031-50333-7_9

Nzuva, S. (2019). Smart contracts implementation, applications, benefits, and limitations. *Journal of Information Engineering and Applications, 9*(5), 63–75. https://doi.org/10.7176/JIEA/9-5-07

Omar, I. A., Jayaraman, R., Salah, K., Simsekler, M. C. E., Yaqoob, I., & Ellahham, S. (2020). Ensuring protocol compliance and data transparency in clinical trials using blockchain smart contracts. *BMC Medical Research Methodology, 20*, 1–17. https://doi.org/10.1186/s12874-020-01109-5

Parida, P. R., Ratnala, A. K., & Kondaveeti, D. (2024). Integrating IoT with AI-driven real-time analytics for enhanced supply chain management in manufacturing. *Journal of Artificial Intelligence Research and Applications, 4*(2), 40–84.

Peng, L., Feng, W., Yan, Z., Li, Y., Zhou, X., & Shimizu, S. (2021). Privacy preservation in permissionless blockchain: A survey. *Digital Communications and Networks, 7*(3), 295–307. https://doi.org/10.1016/j.dcan.2020.05.008

Politou, E., Casino, F., Alepis, E., & Patsakis, C. (2019). Blockchain mutability: Challenges and proposed solutions. *IEEE Transactions on Emerging Topics in Computing, 9*(4), 1972–1986. https://doi.org/10.1109/TETC.2019.2949510

Prager, F., Martinez, J., & Cagle, C. (2021). Blockchain and regional workforce development: Identifying opportunities and training needs. In *Blockchain and the public sector: Theories, reforms, and case studies* (pp. 47–72). Springer. https://doi.org/10.1007/978-3-030-55746-1_3

Rejeb, A., Rejeb, K., Keogh, J. G., & Zailani, S. (2022). Barriers to blockchain adoption in the circular economy: A fuzzy Delphi and best-worst approach. *Sustainability, 14*(6), 3611. https://doi.org/10.3390/su14063611

Rodríguez-Espíndola, O., Cuevas-Romo, A., Chowdhury, S., Díaz-Acevedo, N., Albores, P., Despoudi, S., Malesios, C., & Dey, P. (2022). The role of circular economy principles and sustainable-oriented innovation to enhance social, economic and environmental performance: Evidence from Mexican SMEs. *International Journal of Production Economics, 248*, 108495. https://doi.org/10.1016/j.ijpe.2022.108495

Sánchez-García, E., Martínez-Falcó, J., Marco-Lajara, B., & Manresa-Marhuenda, E. (2024). Revolutionizing the circular economy through new technologies: A new era of sustainable progress. *Environmental Technology & Innovation, 33*, 103509. https://doi.org/10.1016/j.eti.2023.103509

Schroeder, P., Anggraeni, K., & Weber, U. (2019). The relevance of circular economy practices to the sustainable development goals. *Journal of Industrial Ecology, 23*(1), 77–95. https://doi.org/10.1111/jiec.12732

Sedlmeir, J., Lautenschlager, J., Fridgen, G., & Urbach, N. (2022). The transparency challenge of blockchain in organizations. *Electronic Markets, 32*(3), 1779–1794. https://doi.org/10.1007/s12525-022-00536-0

Shafik, W. (2024). IoT-enabled model and waste management technologies for sustainable agriculture. In *IoT-based models for sustainable environmental management: Sustainable environmental management* (pp. 137–163). Springer. https://doi.org/10.1007/978-3-031-74374-0_8

Shaikh, Z. A., Dahri, K., Memon, A. A., Tahseen, S., & Abbasi, F. A. (2024). Interoperability: Blockchain's solution to SOA & traditional service-based integration challenges. In *2024 IEEE 1st Karachi Section Humanitarian Technology Conference (KHI-HTC)* (pp. 1–5). https://doi.org/10.1109/KHI-HTC60760.2024.10482337

Shojaei, A., Ketabi, R., Razkenari, M., Hakim, H., & Wang, J. (2021). Enabling a circular economy in the built environment sector through blockchain technology. *Journal of Cleaner Production, 294*, 126352. https://doi.org/10.1016/j.jclepro.2021.126352

Singh, S., Hosen, A. S., & Yoon, B. (2021). Blockchain security attacks, challenges, and solutions for the future distributed iot network. *IEEE Access, 9*, 13938–13959. https://doi.org/10.1109/ACCESS.2021.3051602

Singh, R., Khan, S., & Centobelli, P. (2022). Investigating the interplay between social performance and organisational factors supporting circular economy practices. *Sustainability, 14*(24), 16781.

Singh, R., Khan, S., & Dsilva, J. (2023). A framework for assessment of critical factor for circular economy practice implementation. *Journal of Modelling in Management, 18*(5), 1476–1497.

Singh, R., Gupta, A., Bajpai, A., & Kandpal, V. (2024). Exploring challenges of circular economy initiatives for smart cities. In *Smart cities and circular economy: The future of sustainable urban development* (pp. 59–69). Emerald Publishing Limited.

Souza, E. B., Carlos, R. L., de Mattos, C. A., & Scur, G. (2024). The role of blockchain platform in enabling circular economy practices. *Corporate Social Responsibility and Environmental Management, 31*(6), 5730–5743. https://doi.org/10.1002/csr.2885

Sunyaev, A., Kannengießer, N., Beck, R., Treiblmaier, H., Lacity, M., Kranz, J., Fridgen, G., Spankowski, U., & Luckow, A. (2021). Token economy. *Business & Information Systems Engineering, 63*(4), 457–478. https://doi.org/10.1007/s12599-021-00684-1

Tyagi, A. K., Kukreja, S., Richa, & Sivakumar, P. (2024). Role of blockchain technology in smart era: A review on possible smart applications. *Journal of Information & Knowledge Management, 23*, 2450032. https://doi.org/10.1142/S0219649224500321

Upadhyay, A., Mukhuty, S., Kumar, V., & Kazancoglu, Y. (2021). Blockchain technology and the circular economy: Implications for sustainability and social responsibility. *Journal of Cleaner Production, 293*, 126130. https://doi.org/10.1016/j.jclepro.2021.126130

Wendl, M., Doan, M. H., & Sassen, R. (2023). The environmental impact of cryptocurrencies using proof of work and proof of stake consensus algorithms: A systematic review. *Journal of Environmental Management, 326*, 116530. https://doi.org/10.1016/j.jenvman.2022.116530

Zhu, P., Hu, J., Li, X., & Zhu, Q. (2021). Using blockchain technology to enhance the traceability of original achievements. *IEEE Transactions on Engineering Management, 70*(5), 1693–1707. https://doi.org/10.1109/TEM.2021.3066090

Part IV
Intersection of Smart Circular Economy and Blockchain Technology

Chapter 14
Leveraging Blockchain Technology and Smart Circular Economy: Synergies and Challenges for Developing Countries

Rubee Singh, Mukesh Babu Gupta, and Neha Nupoor

Abstract The circular economy and sustainable development goals have been extensively debated in recent literature, with some of the works emphasized in this overview. Despite the topic's popularity, the relationship between circular economy (CE) and sustainable development goals (SDG) remains restricted. While CE is discussed in most studies in the context of the SDGs, the various works do not seek to analyse the link between CE and the SDGs. This chapter explores the potential of the R programming framework in facilitating the development and implementation of a smart circular economy. The study investigates the uses of R's data analysis, visualization, and machine learning capabilities in domains such as resource optimization, waste management, life cycle assessment, and predictive modelling. The study shows how R's broad libraries and capabilities may be used to effectively monitor resource flows, evaluate sustainability measures, and simulate circular economy methods. Blockchain technology improves circular economy closed-loop data management by enabling transparency, traceability, and security. Effective resource management requires increased operational competence and stakeholder trust. Blockchain has several benefits for circular economy data management, but scalability and regulatory difficulties remain. To maximize blockchain's sustainability potential, these issues must be addressed.

Keywords Smart circular economy · Blockchain · R framework · Reuse · Refuse · Developing countries

R. Singh (✉)
Institute of Business Management, GLA University, Mathura, India

Department of Management Studies, Kumaun University, Nainital, India

M. B. Gupta · N. Nupoor
Faculty of Management, Sarala Birla University, Ranchi, India

R. Singh et al. (eds.), *Blockchain Technologies for Smart Circular Economy and Organisational Sustainability*, Environmental Footprints and Eco-design of Products and Processes, https://doi.org/10.1007/978-3-031-89183-0_14

14.1 Introduction

The emerging concept called circular economy (CE) has been proposed as an alternative to the current linear manufacturing process and based on the systems redesign and closed loops. It has a relatively high potential for developing a less damaging business approach that may rectify the environmental and social impacts of the mainstream business strategy and might revolutionize the organization, creation, use, and recycling of products (Singh et al., 2024a, b, c, d). The concept of circular economy can be defined more specifically in view of a set of activities or practices represented by a range of initiatives oriented to prolong the life cycle of products and recover material flows which helps in increasing social performance of firm (Singh et al., 2022). However, there are also implementation imperfections like the lack of information openness along the value chain and the predominantly transactional character of dominant sales operations which hinder a full-fledged implementation (Karunaratna et al., 2024).

The circular economy framework is essential for prolonging the lifespan of products, components, and materials (PCMs). Different stakeholders across the value chain have interpreted and comprehended it in diverse ways. Some perceive it as a notion, while others regard it as a framework designed to enhance the circularity of the PCM (Khan et al., 2022a, b; Singh et al., 2024a, b, c, d). However, there is a slow adoption of CE, even though there is more awareness raised by various industries and authorities, owing to some reasons.

14.2 Review of Literature

The topics of circular economy and sustainable development goals have been widely discussed in the recent literature with some of the studies highlighted in the literature review. Despite the ubiquity of the topic, the linkage between circular economy (CE) and sustainable development goals (SDG) is still limited. While CE is explored in most studies in the context of SDGs, the various works do not aim at analysing the relationship between the CE and the SDGs. Only a limited number of academics have addressed the intersection of circular economy (CE) and sustainable development goals (SDGs) in their work; for instance, Khan et al. (2024) conducted a study to define framework for manufacturing firm for the development of circular economy. Kumar et al. (2024) discussed the strategies for achieving net-zero economy. This research mainly focuses on EU compliance as a benchmark regarding the SDGs. Singh et al. (2023) presented a new circular economy model to discuss the assessment of critical factor for circular economy.

Numerous researches examine various applications of the circular economy in relation to net zero. Singh et al. (2024a, b, c, d) conducted a study to identify the organizational factors for implementing CSR for the development of smart economy. They analysed various applications for attaining net-zero emissions. Khan et al. (2023) try to find out the barriers of blockchain technology in implementation in supply chain to develop smart economy. This study associates the transition to electric vehicles (EVs) with a circular and low-carbon economy to attain a net-zero economy. Khan et al. (2022a, b) examined determinants of remanufacturing for circular economy. This study presented a bottom-up methodology to advance a circular carbon economy in addressing end-of-life global plastics, encompassing carbon collection, storage, and significant reduction.

Singh et al. (2024a, b, c, d) provide a compressive view of AI for the development of smart economy. In comparison, circular economy (CE) research related to sustainable development goals (SDGs) is more extensive than that concerning net-zero aims, indicating substantial opportunity for further investigation into the relevance of CE in the context of net-zero objectives. Framework for smart circular economy is as follows:

14.3 Smart Circular Economy (SCE)

The framework for a smart circular economy (SCE) integrates digital technologies with circular economy principles to enhance sustainability and efficiency across various sectors. The R framework for smart circular economy integrates various components that collectively enhance sustainable development and emphasizes collaboration among multiple stakeholders, e.g. businesses houses, governments, and society. The crucial mechanisms of this framework include resource management strategies and the integration of environmental, social, and governance (ESG) criteria. Further research is needed to explore its effectiveness in real-world scenarios (Table 14.1).

Table 14.1 R framework for smart circular economy

R1	Refuse	Avoid using nonessential products or materials that generate waste
R2	Rethink	Redesign processes, products, and business models to be more sustainable
R3	Reduce	Minimize resource consumption and waste production
R4	Reuse	Extend the life cycle of products through reuse and repurposing
R5	Repair	Fix broken products instead of discarding them
R6	Refurbish	Restore used products to good condition for resale or reuse
R7	Remanufacture	Rebuild products using recycled components
R8	Repurpose	Use discarded products or materials for different purposes
R9	Recycle	Process waste into new materials or products
R10	Recover	Extract useful materials or energy from waste

14.3.1 Refuse

Refuse is critical in circular economy since it intensifies trash to valuable commodities hence reducing environmental detrimental effects. Waste is integrated into various functions effectively eliminating garbage and moreover assisting energy and material production due to refuse (Panda et al., 2010). Refuse method adheres to the principles of circular economy, resource intensity, and wastage reduction. Refuse can similarly be used in the making of construction materials such as bricks, cement among others. Wrapping garbage in plastic and shaping the compact mass to bricks provides effective method of recycling waste for construction, thus decreasing the need for new resources and minimizing use of landfill (Hebel et al., 2014). The use of trash in construction materials reduces environmental problems such as usage of land and pollution since it provides an appropriate use for material that would otherwise worsen land degradation (Lehmann, 2011). Refuse being converted to energy; there are a number of technologies for use in refuse power generation and RDF power generation. Refuse is one of these types of methods that propose general availability of a renewable energy type and at the same time would cause reduction of the CO_2 content without the need to increase system prices. Bio-waste, especially kitchen refuse, forms a major part of municipal waste, and they can be biologically converted into biodiesel and methane, to support clean energy efforts and to offset pollution as well (Sarker et al., 2024). This approach bolsters the recycling economy by optimizing resource utilization and reducing environmental pollution. Technologies engineered to process trash by thermal cracking and pyrolysis can transform garbage into heat, gas, and power (Chen et al., 2014).

14.3.2 Rethink

Rethink is crucial for advancing the circular economy by promoting innovative business models and sustainable practices. It emphasizes the imperative of reevaluating existing systems and integrating circular principles throughout many businesses, hence facilitating a transition from linear to circular economies (Zorpas, 2024). Restoration requires the use of circular models—reuse, repair, and remanufacturing as seen in the oil and gas industry (Adebayo et al., 2024). In a Circular Hub, which works as a Shared Resource Hub, we find that organizational contacts play a significant role in the implementation of circular initiatives (Mogos et al., 2021). Circularity in the built environment can be explained in a systematic way by such techniques as sharing, protecting, adapting, and rethinking existing resources to help real estate managers apply sustainable approaches (Kyrö, 2020). The received ethical principles of the circular economy in the light of human incentives suggest the importance of social responsibility for creating more sustainable conditions (Paletta, 2019; Bhushan & Singh, 2024). Learning factories, components of circular centres, incorporate education and local communities to bring sustainable development and concepts of circularity into the society to influence generations to come (Mogos et al., 2021).

14.3.3 Reduce

Reduce technical approach emphasizes the requirement of developing systems and processes that enhance managerial and sustainable efficiency instead of resources, thus promoting a wiser use of materials (Nižetić et al., 2019). Due to the large material consumption, the construction sector can benefit from the methods that focus on better resource use; an example is the design tool developed for pedestrian bridges that also incorporate the concept of morphological indicators for efficient use of materials (Anastasiades et al., 2022). Such instruments diminish the necessity for fresh materials, thus conserving prices and minimizing the energy consumption linked to production. The circular economy framework seeks to diminish plastic usage, mitigating the environmental hazards associated with microplastics (Syberg et al., 2022). To build upon sustainable practices for the plasticians, it is thus possible to go even further by extending the roles assumed by producers thus guaranteeing proper use of the material required. This concern about reducing the generation of wastes is in accord with the other objectives of circular economy and, hence, encourages the advancement of new paradigm of waste management business models (Madaan et al., 2024).

14.3.4 Reuse

Reuse supports the circular economy improvement, makes ideal use of resources, reduces waste, and decreases emissions of greenhouse gases (Gallego-Schmid et al., 2020). This methodology significantly improves efforts in sustainability because of the ability to reapply them to other industries and processes (Nidumolu & Knotts, 1998). The subsequent sections describe the elements of how reuse supports the circular economy at large. Reuse platforms facilitate the diversion of discarded items from landfills, thereby diminishing premature disposal and the necessity for new product acquisitions (Sarkhoshkalat et al., 2024). The implementation of incorporating materials from demolished structure in construction has impact raw material use or emission to some extent hence minimizing greatly the effects construction has to the environment (Bellini et al., 2024). Cultural heritage within the circular economy element proves that repurposing and bringing new productive use of the existing cultural heritage buildings improves the local economy by generating new employment opportunities and stimulating the tourism sector (Rudan, 2023). The focus on water reuse projects is beneficial in water conservation and also in return on costs that are used in water treatment processes and in supply. However, some challenges, for example, concern data management, and these regulatory barriers affect the extent of reuse strategies implemented. To explain these difficulties to address them, it is crucial for achieving the potential of reuse towards setting a circular economy model. On the other hand, there are some who say that too much focus on reuse may omit recycling and, in general, the reduction of demand, both of which are critical pillars of circular economy (Corona et al., 2019).

14.3.5 Repair

A concept of repair contributes to building a circular economy by extending durability of the products, reducing the levels of waste generation, and encouraging customers to adopt various sustainable consumption patterns (Shevchenko et al., 2023). However, it serves as a better replacement to disposal and recycling depending on the European legislation efforts towards promoting repair strategies (Balkenende et al., 2024). Several factors influence the integration of repair into circular economy including consumer trends, socioeconomic conditions, and legal requirements that define the effectiveness of repair activities (Parajuly et al., 2023). A concerning issue that reveals a clearer picture of promoting repair is that it fosters a closer association between consumers and products, supporting a culture of repair, and sustainability (Hernandez et al., 2024). The emergence of rapid fashion and a disposable culture has eroded the skills and values related to repair (Hernandez et al., 2024). Restricted access to repair services and tools may impede consumer engagement in repair activities (Parajuly et al., 2023).

14.3.6 Refurbish

Refurbishment is essential for promoting the circular economy by prolonging product lifespans and minimizing waste (Cooper, 2020). This technique conserves resources and reduces environmental impact through many strategies and technology. In the construction industry, refurbishing measures like the management of waste like the following and selective disassembly are effective for reducing raw material extraction and greenhouse gas emissions (Fernandes & Ferrão, 2023). New circular refurbishment model increases efficacy of the policy because it considers life cycle aspects and incorporated local stakeholders (Fernandes & Ferrão, 2023). It is expected that the use of blockchain technology to manage certificates of refurbishment will help make reused products safe for use and will scare away concerns about the safety and performance of reused products (Regueiro et al., 2023). However, the barriers still exist such as consumer scepticism and inherent difficulty in achieving end-of-life for products where legal support is needed to realize the full benefits of refurbishment in circular economy.

14.3.7 Remanufacturing

Remanufacturing is a method of returning used products to a near original state through element of disassembly, cleaning, and reassembly to minimize landfill waste and support sustainable manufacturing (Chiang et al., 2024). As a result of remanufacturing, consumption of raw materials is dramatically curtailed since remanufactured products are created by reclaiming components; thus Chile retains the country's natural resources (Palhares, 2024). This is evidenced by having shown that energy utilization as compared to conventional production is lower, making it more sustainable (Ding et al., 2021). The use of remanufacturing can then bring benefits in cost for enterprises as some remanufactured components are even cheaper than new ones maintaining quality characteristics (Palhares, 2024). Remanufacturing benefits economic development in India by improving the employment rate while also raising the efficiency of used products. The adoption of such instruments as the Remanufacturing Potential Index (RemPI) enhances the assessment of the feasibility of remanufacturing at the component level and thus optimizing the efficiency. Centralized remanufacturing hubs have been proposed to extend and improve advantages as well as increase efficiency through innovative business structures. However, there are still challenges in working on these methods at a wider industrial level even when remanufacturing has other benefits. Transition to circular economy means to solve challenges that are associated with it like consumer acceptance and regulatory hurdles that may slow down the process.

14.3.8 Repurpose

Repurposing is essential for promoting the circular economy by prolonging the lifespan of products and materials, therefore minimizing waste and preserving resources (Reddy et al., 2023). This method not only reduces environmental effects but also promotes new design approaches that correspond with sustainability objectives. The following sections trace how repurposing mediates other elements of the circular economy. The process of repurposing reduces emissions of green-house gases much more than the recycling process. Remanufacturing defers material demand from new resources and hence is an essential aspect of combating environmental pollution (Shekhar et al., 2022). Larger scale reuse requires a shift in design attitude that centres on producing items that can be transitioned to different uses. Studying (see Thompson et al., 2020) has shown that characteristic features of critical design for items with different product life cycles are important for recycling. Architecture of reuse raises the importance of architectural modularity for using back materials and can lead to new creative solutions in the construction works. Repurposing can result in reductions in manufacturing cost and waste management. Automobile industry's strategy of reusing damaged batteries for secondary applications exemplifies how corporations can minimize expenses while advancing sustainability. However, it also faces challenges, like the variability of recycled materials and the necessity for innovative design techniques (Viola & Diano, 2019). Overcoming these obstacles is crucial for fully actualizing the potential of repurposing in the circular economy.

14.3.9 Recycle

Mamudu et al. (2024) found that recycling is essential in CE as it converts waste materials into valuable resources, reducing environmental consequences and fostering sustainable practices. This technique decreases the quantity of waste directed to landfills while reintegrating materials into the manufacturing cycle, hence promoting economic growth and resource efficiency (Awogbemi et al., 2022). Recycling diminishes the ecological impact by redirecting garbage from landfills, so alleviating pollution and preserving natural resources. The recycling of plastics, including high-density polyethylene (HDPE) and high impact polystyrene (HIPS), has demonstrated the capability to generate new products while reducing the reliance on virgin materials. The use of Extended Producer Responsibility (EPR) incentivizes producers to create goods that facilitate recycling, hence improving the economic feasibility of recycling programs (Kunz et al., 2018). The commercial adoption of recycled products, including textiles and automotive components, illustrates the economic potential of recycling in a circular economy. It is expanding through legal tools that establish objectives and requirements for recycling and a circular economy within the legal landscape among the EU member states. While recycling is important for a circular economy, reuse and recovery must manage rebound effects since they could in fact decrease the gains in recycling if not well controlled (Allwood, 2014).

14.3.10 Recover

Under the recovery term, those material streams that transform trash into valuable resources can optimize the circular economy while reducing environmental harm and promoting sustainability (Singh & Ordoñez, 2016). It means that recovery procedures are considered to be critically important for several branches, such as textile, biomass, and wastewater treats differently contributing to the circular economy elements. Business produces substantial waste, and recovery methods such as repurposing, recycling, and upcycling prolong the lifespan of textiles (Stanescu, 2021). These techniques diminish the necessity for raw materials such as cotton and wool, hence preserving resources and mitigating environmental effect. Material recovery in businesses entails transforming waste into secondary raw materials that may be reintroduced into the manufacturing cycle, hence advancing sustainable development goals (Kurniawan et al., 2022). Biomass waste recovery by aerobic and anaerobic digestion transforms waste into bioenergy and bio-based products, including biogas and biohydrogen. This method offers a substitute for fossil fuels while also improving energy security and environmental sustainability through the use of renewable resources. Biorefineries are essential in this process, providing a framework for the effective valorization of biomass waste. The extraction of bio-based products from industrial effluents by biorefinery methods diminishes environmental impacts and enhances resource efficiency (Aryanfar et al., 2024). The amalgamation of circular economy principles with biorefinery technologies promotes sustainable industrial practices and facilitates the transition to a resource-efficient industrial sector.

14.4 Challenges in Circular Economy Categorized into Five Barriers

14.4.1 Technological Challenges

Challenges such as increased cost of capital, restricted funding, low returns, and shift from long-term vision both create and intensify bottlenecks such as inadequate investment and mismanagement and, to a great extent, stifle the progress of a circular economy. High implementation costs and limited capital prevent the growth of circular economy strategies, while low profits reduce sustainable investment in circular solutions. Additionally, the said methods signify the absence of a strategic vision and proper management techniques; cumulative uncoordinated approaches lower the effectiveness of circular economy activities. This requires policy approval for such endeavours, creative financial models for financing and development, and strengthened leadership focused on sustainability.

14.4.2 Financial Challenges

Challenges such as increased cost of capital, restricted funding, low returns, and shift from long-term vision both create and intensify bottlenecks such as inadequate investment, mismanagement, to a great extent stifle the progress of a circular economy. High implementation costs and limited capital prevent the growth of circular economy strategies, while low profits reduce sustainable investment in circular solutions. This requires policy approval for such endeavours, creative financial models for financing and development, and strengthened leadership focused on sustainability.

14.4.3 Infrastructural Challenges

Barriers including lack of sound business models, weak information management systems, limited innovation accessibility, and insufficient technical support greatly hinder the realization of circular economy. Consequently, organizations get challenges when transitioning to circular structures due to unresolved profitability issues and sustainability paths, which does not allow business models to provide positive guidance. Weakness in handling information affects identification of resources, tracking, tracing, and optimality of resource use. The lack of technical help hinders enterprises from obtaining the necessary skills to properly implement circular strategies, hence impeding overall progress toward a sustainable and resource-efficient economy.

14.4.4 Institutional Challenges

Obstacles such as being 'ill-suited for global harmony', 'regulatory constraints', 'restricted circular procurement', and 'insufficient recycling regulations' considerably impede the implementation of a circular economy. The lack of global alignment results in disjointed efforts, diminishing the efficacy of sustainability measures. Regulatory constraints generate jurisdictional complexity, impeding cross-border circular processes. The constraints of limited circular procurement impede the market for recycled products, hence diminishing economic incentives for enterprises to embrace circular models. Finally, the lack of adequate recycling requirements compromises the disposal system resulting in low material collection and increasing the ecology's pollution level.

14.4.5 Societal Challenges

Innovative solutions, timid organizational culture, insufficient leadership and changer attitudes, and low customer understanding are strong obstacles to the implementation of the circular economy. An organization with a risk-averse culture impedes innovation, hindering the shift towards sustainable practices. Insufficient leadership leads to ambiguous strategic direction, hindering effective policy execution. The consumers' ignorance hinders the demand for eco-friendly products, hence diminishing market incentives for enterprises to use circular models. In addition, stakeholder anxiety resulting from uncertainty of economic profitability and technology sustainability discourages investment and partnership required for a circular economy.

14.5 Techniques for Implication of Smart Circular Cities

Advancing a smart circular economy requires an integrated approach encompassing policy, technology, design, resource management, consumer behaviour, financial mechanisms, data analytics, and collaboration. Establishing regulatory frameworks is essential to encourage recycling, reuse, and reduced consumption, while public-private partnerships (PPPs) foster collaboration, shared investments, and capacity-building. Enforcing compliance and sustainability standards ensures accountability and responsible production. Technology integration plays a critical role, with IoT enabling real-time monitoring of resources, blockchain enhancing supply chain transparency, and AI optimizing processes and predicting maintenance needs. Sustainable design focuses on eco-design, 3D printing to reduce waste, and modular designs for upgrades and repairs. Resource management prioritizes closed-loop supply chains, material substitution with biodegradable or recycled options, and industrial symbiosis to minimize waste. Consumer engagement strategies like product-as-a-service models, gamification, and awareness campaigns encourage sustainable behaviours. Financial mechanisms such as green financing, carbon credits, and investment in R&D drive innovation and fund circular projects. Data and analytics, including life cycle assessment (LCA), big data analytics, and digital twins, support monitoring, impact evaluation, and process simulation for efficiency. Collaboration through industry alliances, research consortia, and open innovation platforms fosters knowledge sharing and collective problem-solving. Together, these strategies build a robust ecosystem that transitions industries, communities, and consumers toward a sustainable, smart circular economy.

14.6 Role of Blockchain

The smart circular economy has to do with the use of technology such as the blockchain in areas of transparency, accountability, and managing limited resources. Due to its highly decentralized nature, it enables safe tracking of all material throughout their life cycle, helping in reduction, reuse, recycling, and recovery. The following are the main ways through which blockchain has supported this change.

14.6.1 Improved Transparency

Blockchain enhances product traceability because it is nearly impossible to alter the records that are written into the ledger thereby warranting sustainable sourcing and reuse of the products. Fostering circular economy engagements requires building stakeholder trust, which is done through transparency.

14.6.2 Smart Contracts and Automation

Automation, smart contracts automation, and enforcement of smart contracts make the performance of resource exchange less costly. Supply chains are known to be dynamic in that they ensure that innovative circular economy business models are promoted to meet new demands.

14.6.3 Better Data Management

Blockchain allows closed-loop systems to track resources and waste in real time. The combination of blockchain, IoT, and AI improves resource efficiency and waste reduction.

14.6.4 Better Incentivization Mechanisms

Blockchain smart contracts allow incentive structures for recycling and reuse. Incentives can boost collector engagement and improve closed-loop system operations.

14.6.5 Better Recycling

Token-based payment schemes encourage recycling. A mobile app may track tokens earned for recycling, encouraging households to participate more.

14.6.6 Improve Resource Management

Smart contracts automate and streamline recycling, making stakeholder engagement easier without complicated procedures. Efficiency improves recycling and resource management.

14.7 Conclusion

This chapter explained the use of R framework in the development of smart circular economy. R framework is a versatile and impactful tool for advancing the principles of a smart circular economy. This framework has the capacity to manage large datasets, do advanced analytics, and create actionable insights that make it a useful resource for governments, corporations, and researchers. R's open-source nature and flexibility make it ideal for collaborative projects and large-scale deployment of circular techniques. By incorporating R into decision-making processes, organizations may improve resource efficiency, minimize waste, and contribute to sustainability goals. The chapter emphasizes the need of more intensive exploration of R's framework usefulness in circular economy applications.

References

Adebayo, Y. A., Ikevuje, A. H., Kwakye, J. M., & Esiri, A. E. (2024). Circular economy practices in the oil and gas industry: A business perspective on sustainable resource management. *GSC Advanced Research and Reviews, 20*(3), 267–285.

Allwood, J. M. (2014). Squaring the circular economy: The role of recycling within a hierarchy of material management strategies. In *Handbook of recycling* (pp. 445–477). Elsevier.

Anastasiades, K., Lambrechts, T., Mennes, J., Audenaert, A., & Blom, J. (2022). Formalising the R of reduce in a circular economy oriented design methodology for pedestrian and cycling bridges. *J, 5*(1), 35–50. https://doi.org/10.3390/j5010003

Aryanfar, Y., Castellanos, H. G., Hammoodi, K. A., Farifteh, M., Keçebaş, A., Ghasemlou, S. M., et al. (2024). Cleaner and sustainable circular economy approaches for bio-based product recovery from industrial effluents in a biorefinery. *Environmental Progress & Sustainable Energy, 43*(6), e14483.

Awogbemi, O., Kallon, D. V. V., & Bello, K. A. (2022). Resource recycling with the aim of achieving zero-waste manufacturing. *Sustainability, 14*(8), 4503.

Balkenende, R., Bakker, C., Blondel, É., & Henneberry, Y.. (2024). Repair in the circular economy. https://doi.org/10.59490/mg.93

Bellini, A., Andersen, B., Klungseth, N. J., & Tadayon, A. (2024). Achieving a circular economy through the effective reuse of construction products: A case study of a residential building. *Journal of Cleaner Production, 450*, 141753. https://doi.org/10.1016/j.jclepro.2024.141753

Bhushan, M., & Singh, R. (2024). Talent management facilitates net-zero transition through employee green behavior. In R. Singh & D. Crowther (Eds.), *Transition towards a sustainable future: Net zero policies and environmental sustainability* (pp. 117–129). Springer.

Chen, D., Yin, L., Wang, H., & He, P. (2014). Pyrolysis technologies for municipal solid waste: A review. *Waste Management, 34*(12), 2466–2486.

Chiang, P. C., Ma, H. W., Wen, L., & Lin, C. H. (2024). Reduction, reuse, recycling, recovery, redesign. In *Introduction to green science and technology for green economy: Principles and applications* (pp. 255–285). Springer Nature Singapore.

Cooper, T. (2020). Slower cycles: An essential characteristic of the circular economy. In *The circular economy in the European Union: An interim review* (pp. 99–116). Springer.

Corona, B., Shen, L., Reike, D., Carreón, J. R., & Worrell, E. (2019). Towards sustainable development through the circular economy—A review and critical assessment on current circularity metrics. *Resources, Conservation and Recycling, 151*, 104498.

Ding, Q., Khattak, S. I., & Ahmad, M. (2021). Towards sustainable production and consumption: Assessing the impact of energy productivity and eco-innovation on consumption-based carbon dioxide emissions (CCO2) in G-7 nations. *Sustainable Production and Consumption, 27*, 254–268.

Fernandes, J., & Ferrão, P. (2023). A new framework for circular refurbishment of buildings to operationalize circular economy policies. *Environments, 10*(3), 51. https://doi.org/10.3390/environments10030051

Gallego-Schmid, A., Chen, H. M., Sharmina, M., & Mendoza, J. M. F. (2020). Links between circular economy and climate change mitigation in the built environment. *Journal of Cleaner Production, 260*, 121115.

Hebel, D. E., Wisniewska, M. H., & Heisel, F. (2014). *Building from waste: Recovered materials in architecture and construction*. Birkhäuser.

Hernandez, L., Petreca, B., Baurley, S., Bianchi-Berthouze, N., & Cho, Y. (2024). Examining clothing repair practices, core competences, techniques, tools and community structures involved in extending the life of garments. *International Journal of Sustainable Fashion & Textiles, 3*(1), 99–121. https://doi.org/10.1386/sft_00041_1

Karunaratna, K. S. S. N., Dewasiri, N. J., Singh, R., & Rathnasiri, M. S. H. (2024). What does artificial intelligence-powered ChatGPT bring to academia? A review. In R. Singh, S. Khan, A. Kumar, & V. Kumar (Eds.), *Artificial intelligence enabled management* (pp. 1–13). De Gruyter. https://doi.org/10.1515/9783111172408-001

Khan, S., Ali, S. S., & Singh, R. (2022a). Determinants of remanufacturing adoption for circular economy: A causal relationship evaluation framework. *Applied System Innovation, 5*(4), 62. https://doi.org/10.3390/asi5040062

Khan, S., Singh, R., Sá, J. C., Santos, G., & Ferreira, L. P. (2022b). Modelling of determinants of logistics 4.0 adoption: Insights from developing countries. *Machines, 10*(12), 1242. https://doi.org/10.3390/machines10121242

Khan, S., Khan, M. I., & Singh, R. (2023). Modeling the barriers of blockchain technology implementation in supply chain. *Journal of Industrial Integration and Management, 8*(3), 367. https://doi.org/10.1142/S2424862223500070

Khan, S., Singh, R., Alnahas, J., Abbate, S., & Centobelli, P. (2024). Navigating the smart circular economy: A framework for manufacturing firms. *Journal of Cleaner Production, 480*, 144007. https://doi.org/10.1016/j.jclepro.2024.144007

Kumar, K., Singh, R., Sharma, A., & Lehri, B. P. S. (2024). Climate change adaptation strategies for achieving net-zero economy. In R. Singh & D. Crowther (Eds.), *Transition towards a sustainable future: Net zero policies and environmental sustainability* (pp. 163–183). Springer.

Kunz, N., Mayers, K., & Van Wassenhove, L. N. (2018). Stakeholder views on extended producer responsibility and the circular economy. *California Management Review, 60*(3), 45–70.

Kurniawan, T. A., Maiurova, A., Kustikova, M., Bykovskaia, E., Othman, M. H. D., & Goh, H. H. (2022). Accelerating sustainability transition in St. Petersburg (Russia) through digitalization-based circular economy in waste recycling industry: A strategy to promote carbon neutrality in era of industry 4.0. *Journal of Cleaner Production, 363*, 132452.

Kyrö, R. (2020). Share, preserve, adapt, rethink—A focused framework for circular economy. *IOP Conference Series Earth and Environmental Science, 588*(4), 042034. https://doi.org/10.1088/1755-1315/588/4/042034

Lehmann, S. (2011). Optimizing urban material flows and waste streams in urban development through principles of zero waste and sustainable consumption. *Sustainability, 3*(1), 155–183.

Madaan, G., Singh, A., Mittal, A., & Shahare, P. (2024). Reduce, reuse, recycle: Circular economic principles, sustainability and entrepreneurship in developing ecosystems. *Journal of Small Business and Enterprise Development, 31*, 1041. https://doi.org/10.1108/jsbed-01-2023-0009

Mamudu, U. U., Obasi, C. D., Awuye, S. K., Danso, H., Ayodele, P., & Akinyemi, P. (2024). Circular economy in the manufacturing sector: Digital transformation and sustainable practices. *International Journal of Science and Research Archive, 12*(2), 129–141.

Mogos, M. F., Vildåsen, S., Sørumsbrenden, J., & Powell, D. (2021). Rethinking circular business models: The role of the learning factory. In *Advances in production management systems. Artificial intelligence for sustainable and resilient production systems* (pp. 402–410). Springer. https://doi.org/10.1007/978-3-030-85906-0_45

Nidumolu, S. R., & Knotts, G. W. (1998). The effects of customizability and reusability on perceived process and competitive performance of software firms. *MIS Quarterly, 22*, 105–137.

Nižetić, S., Djilali, N., Papadopoulos, A., & Rodrigues, J. J. (2019). Smart technologies for promotion of energy efficiency, utilization of sustainable resources and waste management. *Journal of Cleaner Production, 231*, 565–591.

Paletta, A. (2019). Rethinking economics in a circular way in the light of encyclical "Laudato Sì". In *Sustainability and the humanities* (pp. 339–357). Springer. https://doi.org/10.1007/978-3-319-95336-6_19

Palhares, T. (2024). Operational excellence: Re-think beyond reuse, reduce, recycle in the circular economy. https://doi.org/10.2118/223004-ms.

Panda, A. K., Singh, R. K., & Mishra, D. K. (2010). Thermolysis of waste plastics to liquid fuel: A suitable method for plastic waste management and manufacture of value added products—A world prospective. *Renewable and Sustainable Energy Reviews, 14*(1), 233–248.

Parajuly, K., Green, J., Richter, J., Johnson, M., Rückschloss, J., Peeters, J., Kuehr, R., & Fitzpatrick, C. (2023). Product repair in a circular economy: Exploring public repair behavior from a systems perspective. *Journal of Industrial Ecology, 28*, 74. https://doi.org/10.1111/jiec.13451

Reddy, V. V., Kumar, P., Rao, A. L. N., Kumar, R., Singh, S., Asha, V., & Kareem, S. H. (2023). Waste to wealth generation: Innovative methodologies in resource utilization and minimization in circular economy. In *E3S Web of Conferences* (Vol. 453, p. 01035). EDP Sciences.

Regueiro, C., Gómez-Goiri, A., Pedrosa, N., Semertzidis, C., Iturbe, E., & Mansell, J. (2023). Blockchain-based refurbishment certification system for enhancing the circular economy. *Blockchain: Research and Applications, 5*(1), 100172. https://doi.org/10.1016/j.bcra.2023.100172

Rudan, E. (2023). Circular economy of cultural heritage—Possibility to create a new tourism product through adaptive reuse. *Journal of Risk and Financial Management, 16*(3), 196. https://doi.org/10.3390/jrfm16030196

Sarker, T. R., Ethen, D. Z., Asha, H. H., Islam, S., & Ali, M. R. (2024). Transformation of municipal solid waste to biofuel and bio-chemicals—A review. *International Journal of Environmental Science and Technology, 22*, 3811–3832.

Sarkhoshkalat, M. M., Afkham, A., Bonyadi Manesh, M., & Sarkhosh, M. (2024). Circular economy and the recycling of E-waste. In *New technologies for energy transition based on sustainable development goals: Factors contributing to global warming* (pp. 319–354). Springer Nature Singapore.

Shekhar, A. R., Parekh, M. H., & Pol, V. G. (2022). Worldwide ubiquitous utilization of lithium-ion batteries: What we have done, are doing, and could do safely once they are dead? *Journal of Power Sources, 523*, 231015.

Shevchenko, T., Saidani, M., Ranjbari, M., Kronenberg, J., Danko, Y., & Laitala, K. (2023). Consumer behavior in the circular economy: Developing a product-centric framework. *Journal of Cleaner Production, 384*, 135568.

Singh, J., & Ordoñez, I. (2016). Resource recovery from post-consumer waste: Important lessons for the upcoming circular economy. *Journal of Cleaner Production, 134*, 342–353.

Singh, R., Khan, S., & Centobelli, P. (2022). Investigating the interplay between social performance and organisational factors supporting circular economy practices. *Sustainability, 14*(24), 16781. https://doi.org/10.3390/su142416781

Singh, R., Khan, S., & Dsilva, J. (2023). A framework for assessment of critical factor for circular economy practice implementation. *Journal of Modelling in Management, 18*(5), 1476–1497. https://doi.org/10.1108/JM2-06-2021-0145

Singh, R., Crowther, D., & Ravesangar, K. (2024a). Harnessing human capital for a sustainable future: HR's pivotal role toward net zero policies. In R. Singh & D. Crowther (Eds.), *Transition towards a sustainable future: Net zero policies and environmental sustainability* (pp. 27–47). Springer.

Singh, R., Khan, S., Dsilva, J., et al. (2024b). Modelling the organisational factors for implementation of corporate social responsibility: A modified TISM approach. *Global Journal of Flexible Systems Management, 25*, 283–301. https://doi.org/10.1007/s40171-024-00388-x

Singh, R., Kumar, K., & Khan, S. (2024c). A comprehensive view of artificial intelligence (AO)-based technologies for sustainable development goals (SDGs). In R. Singh, S. Khan, A. Kumar, & V. Kumar (Eds.), *Artificial intelligence enabled management* (pp. 183–193). De Gruyter.

Singh, R., Gupta, A., Bajpai, A., & Kandpal, V. (2024d). Exploring challenges of circular economy initiatives for smart cities. In V. Kandpal, E. D. Santibanez-Gonzalez, P. Chatterjee, & M. K. Nallapaneni (Eds.), *Smart cities and circular economy* (pp. 59–69). Emerald Publishing Limited. https://doi.org/10.1108/978-1-83797-957-820241005

Stanescu, M. D. (2021). State of the art of post-consumer textile waste upcycling to reach the zero waste milestone. *Environmental Science and Pollution Research, 28*(12), 14253–14270.

Syberg, K., Nielsen, M. B., Oturai, N. B., Clausen, L. P. W., Ramos, T., & Hansen, S. F. (2022). Circular economy and reduction of micro(nano)plastics contamination. *Journal of Hazardous Materials Advances, 5*, 100044. https://doi.org/10.1016/j.hazadv.2022.100044

Thompson, D. L., Hartley, J. M., Lambert, S. M., Shiref, M., Harper, G. D., Kendrick, E., et al. (2020). The importance of design in lithium ion battery recycling—A critical review. *Green Chemistry, 22*(22), 7585–7603.

Viola, S., & Diano, D. (2019). Repurposing the built environment: Emerging challenges and key entry points for future research. *Sustainability, 11*(17), 4669.

Zorpas, A. A. (2024). The hidden concept and the beauty of multiple "R" in the framework of waste strategies development reflecting to circular economy principles. *Science of the Total Environment, 952*, 175508.

Chapter 15
Sustainable Product Life Cycle Management for Smart Circular Economy and Organizational Sustainability

Michael Oyedele Oyenuga and Solomon Jeresa

Abstract Sustainable product life cycle management (SPLCM) is an essential approach through which sustainability principles may be integrated into the design, production, use, and disposal of products. This paper presents the importance of SPLCM in promoting environmental stewardship, economic viability, and social responsibility throughout the product life cycle. Employing life cycle thinking on the part of organizations offers the opportunity to evaluate any environmental and social impacts that emanate from raw materials extraction to end-of-life disposal. The study offers key principles of the SPLCM, such as adopting a circular economy approach, followed by eco-design and green manufacturing, which allow them to evaluate their sustainability performance and opportunities for improvement. It further mentions the barriers that are linked with the implementation of SPLCM, which include the requirement of interdisciplinary collaboration, data availability, and stakeholder engagement. In addition, the paper presents best practices and case studies across various countries and how countries can incorporate SPLM into their operational strategies to meet the goal of net zero.

Keywords SDGs · Sustainable product life cycle management · Sustainability

M. O. Oyenuga (✉)
School of Business, Woxsen University, Hyderabad, Telangana, India
e-mail: oyenuga.michael@woxsen.edu.in

S. Jeresa
University of Huddersfield, Huddersfield, UK
e-mail: solomon.jeresa@hud.ac.uk

R. Singh et al. (eds.), *Blockchain Technologies for Smart Circular Economy and Organisational Sustainability*, Environmental Footprints and Eco-design of Products and Processes, https://doi.org/10.1007/978-3-031-89183-0_15

15.1 Introduction to Sustainable Product Life Cycle Management (SPLCM)

SPLCM aims to streamline the flow of product and related process data across a product's full life cycle so that relevant information may be accessed when it is needed (Stark, 2022). According to Tanpure et al. (2023), product life cycle management aims to improve the administration of a product from its inception to its final disposal. To achieve climate leadership, reduced emissions, and improved compliance, product life cycle decisions should prioritize sustainability.

Incorporating sustainability principles into the product life cycle at every stage is crucial for achieving business and environmental goals (Singh et al., 2024a, b, c, d). Incorporating stakeholders at every step of a product's lifecycle from introduction to growth, maturity, and decline presents opportunities for firms to cut resource consumption, prevent pollution, and produce things with a circular design (Stark, 2022). A company's growth and integration of all levels and elements need a grasp of SPLCM's strategic aspects beyond those of an IT system to conduct effective PLM deployment planning (Singh et al., 2024a, b, c, d).

15.1.1 Core Principles of SPLCM

15.1.1.1 The Circular Economy Approach

The circular economy approach to SPLCM is based on the following core principles:

Principles	Meaning	Authors
Design for circularity	This involves adopting sustainable design principles like design for disassembly, design for recycling, and design for remanufacturing within the product development life cycle process to recover, reuse, and recycle products and parts	Singh et al. (2024a, b, c, d)
Sustainable resource sourcing	To source raw materials sustainably and ethically, to prevent resource depletion, and to counter social and environmental problems in the supply chain	Singh et al. (2024a, b, c, d)
Closed-loop manufacturing	Energy and water consumption, waste, emissions, worker safety, and wellbeing during manufacturing/production processes—making the manufacturing system more circular	Wallbaum and Ebrahimi (2021)
Optimized distribution and logistics	Improving the efficiency of transport routes, reducing fuel consumption, and minimizing the environmental impacts of packaging and distribution within the circular flow of materials	Knauer and Möslang (2018)
Longer product life	Encouraging sustainable consumer behaviours, practising product lifetime, and using phase environmental impact reduction to prolong the useful lifetime of products	Beng and Omar (2013)

15.1.2 Benefits of the Circular Economy Approach

The circular economy approach towards SPLCM provides four major benefits. First, this circular economy model helps save natural materials and reduces adverse effects of product lifecycles on the natural environment due to reduced waste, which also promotes reuse and recycling of materials. In addition, becoming more circular can help save costs through lower new raw material demand and fewer energy-intensive production processes that go into remanufacturing and refurbishment. The circular economy serves as a roadmap for the organization to differentiate its brand, elevate market presence, and respond to the increased demand for sustainable products and services. Furthermore, by adopting the principles of the circular economy, business will better position themselves in response to changes in environmental regulations and the need to provide attention from customers and investors. Finally, the circular approach will enhance new jobs, local economy, social conditions, and environmental conditions along the product life cycle.

15.2 Core Principles of Sustainable Product Life Cycle Management: Eco-Design and Green Manufacturing

Eco-design and green manufacturing the pillars of SPLCM are essential to enable more sustainable products and production processes.

Eco-design, sometimes referred to as design for the environment (DfE), includes considering the environment when designing and developing products. The main components of eco-design are as follows:

Components	Meaning	Authors
Design for disassembly	Designing products that enable easy disassembly and repair which allow recycling at the end of its useful life	Everard (2023)
Design for recycling	Choosing materials and components that can be recycled effectively, reducing the use of non-recyclable or hazardous materials	Pimenta and Ball (2015)
Design for reuse and refurbishment	Allowing for the initial products and components to be reused for as long as possible	Bag and Pretorius (2020)
Life cycle assessment	Utilize LCA methodologies to assess the environmental implications of design choices over its life cycle	Chitaka et al. (2022)

Implementing eco-design concepts will help manufacturers to develop products with lower than average resource consumption, and lower than average environmental impact, and is easier to adopt for circular economy methods.

Green manufacturing (or sustainable manufacturing) refers to a style of production that seeks to reduce the environmental impact of manufacturing operations. Key aspects of green manufacturing are as follows:

Components	Meaning	Authors
Resource efficiency	Usage reduction [decreasing the use of energy, water, and raw materials, and reducing waste and emissions]	Nilsson-Lindén et al. (2019)
Cleaner production	Implementation of clean technologies, processes, and waste management to improve the environmental performance of manufacturing operations	Hapuwatte and Jawahir (2021)
Closed-loop systems	Using principles of the circular economy to ensure the continued value of materials and reduce waste through reuse, remanufacturing, and recycling	Akter et al. (2020)
Supplier collaborations	Working with suppliers to promote sustainability across the entire supply chain, from the sourcing of raw materials to product distribution	Guo et al. (2017)

In green manufacturing, organizations focus on minimizing waste generated during production by optimizing processes and using cleaner technologies.

15.2.1 Benefits of Eco-Design and Green Manufacturing

SPLCM leads to several significant advantages by incorporating eco-design and green manufacturing principles into SPLCM. Natural resources (minimization and reuse and recycling) of products reduce the environmental impact of the life cycle of the product. Further, resource efficiency and circularity often reduce waste and so long-term costs, which can increase the sustainability of businesses for end-users as well. Additionally, by adopting eco-design and sustainable manufacturing practices that align with international standards, organizations can proactively respond to changing environmental regulations, as well as standards of ecological improvement and sustainability in products, which can offer a competitive edge, improve brand perception and address meeting customer needs for sustainable products that contribute to green jobs, and develop local economies and social/environmental conditions over the product lifecycle.

15.3 Core Principles of Sustainable Product Life Cycle Management: Life Cycle Assessment (LCA)

Life cycle assessment (LCA) is one of the bases of sustainable product life cycle management (SPLCM); it provides a systematic framework for the assessment of the environmental performance of products and services throughout their life cycle. Implementing LCA as a key element of SPLCM enables organizations to make informed decisions, optimize their processes, and contribute to the larger objective of sustainability.

15.3.1 Understanding Life Cycle Assessment

Life cycle assessment (LCA) is a standardized method for evaluating the environmental impacts over the life cycle of a product or service, from raw material extraction to disposal or recycling (Everard, 2023; Lukman et al., 2021). Therefore, this comprehensive strategy enables the detection of possible compromises and the enhancement of environmental efficiency throughout the entire life cycle.

15.3.2 Key Principles of LCA

Principles	Meaning	Author
Cradle-to-Grave	LCA encompasses the environmental impacts, benefits, and contributions of a product or service throughout its entire life cycle	Saidani et al. (2022)
Holistic impact assessment	LCA studies multiple categories of environmental impact (climate change, pollution, resource depletion, land use, etc.) and, therefore, promotes a thorough understanding of multiple interventions	Li et al. (2018), Nikolić et al. (2019)
Continuous development	An iterative process, LCA enables organizations to identify hotspots, assess alternative scenarios, and ultimately improve products and processes' environmental performance	Saidani et al. (2022)
Transparency and standardization	An iterative process, LCA enables organizations to identify hotspots, assess alternative scenarios, and ultimately improve products and processes' environmental performance	Sherry and Tivona (2022), Hertwich (2005)

15.3.3 Benefits of Integrating LCA in SPLCM

LCA provides organizations with valuable and data-informed insights underlying the product design and processes (Hamdar et al., 2016), the generation of which is supported by the comprehensive evaluation of different alternatives, or hotspot identification (Biró & Csete, 2023), which also aids in the reduction of product and service environmental footprints of organizations and help them adhere to the governmental or regulatory guidelines as well as reporting LEED targets (Brand et al., 2022), EPD reporting (López et al., 2021), and overall building trust and transparency (Ren et al., 2022). Moreover, the implementation of LCA in the product development process will also foster innovation and give a competitive edge to the business (Finnveden et al., 2009).

15.4 Strategies for Implementing Sustainable Product Life Cycle Management

15.4.1 Cross-functional Collaboration

Cross-functional collaboration is a key strategy for the successful implementation of SPLCM, as it enables the coordination of efforts, the sharing of knowledge, and the alignment of goals across different departments.

15.4.1.1 Strategies for Implementing Cross-functional Collaboration

The implementations of the SPLCM initiatives should be managed by a cross-functional team (Mollenkopf et al., 2010; Vezzetti et al., 2013). Such a team should be comprising representatives from different departments, such as design, manufacturing, procurement, marketing, and sustainability. This encompasses investments in digital platform and tools that provide real-time collaboration and cooperation with data sharing for teams and cross-functional teams' collective and cooperative decision-making processes; cross-functional teams (Singh et al., 2022a, b, c; Alam et al., 2024); a sense of culture within an organization, which strives towards the pursuit of sustainability as motivation to encourage employees participation toward SPLCM initiatives. This can be through training, incentives, or recognition programs (Vadoudi et al., 2014; Duigou & Bosch-Mauchand, 2013). Another round of collaboration between suppliers and customers as external stakeholders will help in aligning the sustainability goals and including the entire product life cycle (Fielding et al., 2014; Oliveira et al., 2021) while reviewing the refining process shall be done regularly and the feedback and learning process to improve upon SPLCM implementation process (Wiesner et al., 2015; Liu et al., 2021).

15.4.1.2 Benefits of Cross-functional Collaboration in SPLCM

Cross-functional collaboration provides for broader assessments of the potential environmental, social, and economic impacts products can have in their lifetime (Saorín et al., 2019; Koomen et al., 2019), as well as integrate other functions of the organization towards the effective use of available resources (Menon et al., 2018; Bianchi et al., 2021) and communicate its strategy for sustainability to customers, suppliers, as well as other stakeholders in its value network (Kärkkäinen et al., 2012; Vezzetti et al., 2015) which, in the end, aid the organization in its differentiation strategy at the market level and create awareness among its markets as a pioneering sector toward innovation (Oyenuga et al., 2019) and sustainability (Kalverkamp et al., 2018). Focus on cross-functional collaboration as an important element for successful SPLCM implementation enables companies to exploit the rich knowledge and different perspectives of their employees to foster sustainable innovation, improve resource management, and yield long-lasting returns for the business, environment, and society.

15.4.2 Regulatory Compliance and Standards

Organizations have to adhere to several regulations, standards, and guidelines to implement SPLCM and ensure the environmental, social, and economic sustainability of products and processes. Therefore, in SPLCM deployment, compliance with regulatory frameworks and with industry standards is a fundamental element. Environmental organizations are subject to environmental laws, including those on emissions, waste management, and materials. Examples include the European Union's REACH regulation (Registration, Evaluation, Authorization and Restriction of Chemicals) or the RoHS directive of the European Union (Sonnemann et al., 2015). Active EPR policies, similar to the European Union's Waste Electrical and Electronic Equipment (WEEE) directive, obligate such producers to take responsibility for managing products at their end of life. These regulations demand that companies design with disassembly, reuse, and recyclability in mind (Hapuwatte & Jawahir, 2021). There is also mandatory sustainability reporting under the auspices of many jurisdictions, for instance, the European Union's Non-Financial Reporting Directive. This comprises of law forcing companies to report about the environmental, social, and governance performance (Koroneos et al., 2013). More policies and regulations are being rolled out by governments to encourage the shift to a circular economy, such as plastic bag bans and take-back and recycling requirements for products (Wallbaum & Ebrahimi, 2021).

Internationally, ISO 14040 and ISO 14044 set out the principles and framework for the conduct of life cycle assessment studies to assess the potential environmental impacts associated with products and services (Popescu, 2023). ISO 14006 and IEC 62430 standards give direction on incorporating environmental aspects into the product design and development process (Yue et al., 2013).

Many organizations can be certified under different sustainability standards like the Cradle to Cradle Certified™ program, the LEED (Leadership in Energy and Environmental Design) rating system, or GRI (the Global Reporting Initiative) standards (Mwaijande, 2024) and so does the Sustainable Apparel Coalition's Higg Index, a set of industry guidelines that can help designers and manufacturers assess and improve the sustainability performance of their respective supply chains (Fish, 2015).

15.4.2.1 Benefits of Regulatory Compliance and Standards Adoption

Compliance with regulations and adherence to industry standards help organizations avoid legal penalties and maintain a positive reputation with stakeholders (Gemechu et al., 2015). Also, aligning with sustainability standards can lead to process improvements, waste reduction, and more efficient resource utilization (Khan et al., 2022) while demonstrating compliance with regulations and certifications can provide a competitive advantage (Oyedele et al., 2024), particularly in markets where sustainability is a key purchasing criterion (Zhang et al., 2020). Shared standards and guidelines enable organizations to collaborate more effectively with suppliers, customers, and other stakeholders throughout the product life cycle (Tišma & Škrtić, 2023). The need to comply with evolving regulations and meet industry standards can spur organizations to develop innovative solutions and technologies that enhance the sustainability of their products and processes (Türkoğlu et al., 2015).

15.4.3 Stakeholder Engagement

Several strategies have been proposed to achieve SPLCM which demands constant cooperation among stakeholders during the product life cycle. Since SPLCM is a new concept and its successful application relies heavily on effective engagement among stakeholders to balance various interests, integrate sustainability initiatives, and allow for shared value creation, practitioner perspectives can provide valuable insights to further understanding of the challenges and benefits this approach offers.

15.4.3.1 Strategies for Effective Stakeholder Engagement

- **Stakeholder Identification and Analysis**: The first step in stakeholder management is to identify the stakeholders involved in the process along with their interests and the influence and relevance they possess (Bal et al., 2013).
- **Co-creation of Goals**: Involve stakeholders in the definition of sustainability objectives and targets, so that their concerns are considered in the SPLCM strategy (Shaukat et al., 2021).
- **Continuous Communication and Feedback**: Set open lines of communication with stakeholders, regularly providing an update on SPLCM initiatives along with solicitations to receive feedback to promote continual improvement (Ardiana, 2022).
- **Collaborative Product Design and Development**: Involve stakeholders (i.e. suppliers, customers, sustainability experts, etc.) in the product design and development process to embed sustainability considerations in the early stages of development (Yokokawa et al., 2020).
- **Share Effort and Accountability**: Create a culture of shared effort and accountability for SPLCM implementation and monitoring among the stakeholders (Eberhardt et al., 2020).
- **Capacity Building and Knowledge Sharing**: Further, numerous stakeholders (Theyel & Hofmann, 2012) should be trained, developed, and guided to have SPLCM understanding and contribute more towards sustainability.

15.4.3.2 Benefits of Stakeholder Involvement in SPLCM

It can elucidate a more holistic understanding of the environmental, social, and economic impacts of a product over its life cycle by engaging a diverse array of stakeholders—customers, suppliers, employees, community representatives, etc. (Rycroft et al., 2019). It classifies a participatory decision-making process, which enables the incorporation of different viewpoints and the search for agreement in the decision-making processes (Kumar et al., 2024). Stakeholders' engagement in the design, production, and end-of-life management of products improves sustainability performance along the value chain, creating value for all involved (Singh et al., 2024a, b, c, d) and enhancing the organization's accountability, thus building the organization's reputation and credibility (Koroneos et al., 2013).

Stakeholder engagement will bring identification of innovative solutions and optimization of resource utilization as well as environmental and social impacts across the life cycle of a product (Singh et al., 2022a, b, c). The recognition of the needs of diverse stakeholders and action taken to address them helps in strengthening relationships with various stakeholders and builds trust that can facilitate long-term collaboration and support (Ong et al., 2012) leading to increased competitiveness (Nduji et al., 2023).

15.5 The Role of SPLCM in Achieving the United Nations Sustainable Development Goals (SDGs)

Making sustainability a core component of product management and development is a way SPLCM contributes to the SDGs. Environmental, social, and economic considerations are all part of SPLCM, which begins with product conceptualization and continues through product disposal or recycling. One way that SPLCM contributes to the attainment of SDG 13, a goal of sustainable development, is by reducing emissions of greenhouse gases. This is by using less energy during production and consumption and using low-impact materials to make eco-friendly design choices and reducing the effect on the environment (Oyetunde et al., 2023; Oyedele et al., 2023; Stark, 2022). Supporting the development of eco-friendly products and technologies is an avenue where SPLCM is working to advance SDG 9. Also, supporting SDG 17 is SPLCM's advocacy for more collaboration between locals, companies, and governments to increase the prevalence of sustainable practices (Bhushan & Singh, 2024; John, 2024; Khan et al., 2024).

15.6 Barriers to Implementing Sustainable Product Life Cycle Management (SPLCM)

The implementation of SPLCM suffers from several barriers which prevent the successful integration of sustainability principles throughout a product life cycle. Such barriers, when known, could help the organizations to be acquainted with the challenges and implement SPLCM strategies effectively. The absence of holistic and integrated approaches to life cycle sustainability assessment (LCSA) hinders the overall sustainability performance assessment for products and processes (Alwaeli & Mannheim, 2022; Everard, 2023). Furthermore, sourcing credible and detailed data for life cycle assessment (LCA) and other sustainability analyses can often prove difficult and costly, especially for complex products/supply chains (Singh et al., 2024a, b, c, d). From these stages in the product life cycle, the availability and quality of data vary, making the assessment of sustainability difficult and unreliable (Wang & Su, 2022). In addition, some established organizations may be reluctant to embrace new SPLCM practices since they may challenge existing business models, processes, and organizational structures (Chew & Lee, 2021). Acquiring the buy-in and commitment of employees for SPLCM implementation may prove difficult, as organizational change can be met with resistance (Kravchenko et al., 2019), to successfully engage and diverse stakeholders, e.g. ensuring alignment from suppliers, customers, and regulatory bodies across the product life cycle can be a highly intricate and resource-intensive exercise (Wallbaum & Ebrahimi, 2019) while balancing the competing interests and priorities of various stakeholders can be a significant challenge in SPLCM implementation (Traverso et al., 2012). Weak or less stringent regulatory frameworks and policies regarding sustainability can lead to uncertainty and do not contribute to the future scalability of SPLCM practices (Costello et al., 2015). This is a challenge because firms having to comply with a range of legal obligations are faced with the difficulty of analysis and reporting of the same data in various ways (Kiss et al., 2021).

Other challenges include the availability and integration of advanced tools (Lenzo et al., 2018), the lack of sustainability competency (Ramani et al., 2010), poor infrastructure (Traverso et al., 2015), and the absence of training and capacity-building programs (Oyenuga & Omale, 2024).

15.7 Overcoming the Barriers

Through these considerations, the following SPLCM practices are expected to be facilitated: the multifaceted approach is integrated via sustainability assessment frameworks and tools and data management systems investing in and with stakeholders data availability and quality; implementing a culture change and the training of building sustainability competencies; reaching out with policymakers and the industry for harmonized regulatory frameworks; and utilizing technological innovations combined with infrastructure investments as support factors for executing SPLCM practice.

15.8 Conclusion

Sustainable product life cycle management is about a holistic approach towards managing the social, economic impacts of products and environmental issues over their entire life cycle. It starts from designing and developing the product up to the reverse logistics phase and end-of-life while allowing organizations to improve their competitiveness and increase long-term value and contribution toward sustainable development. Adopting successful SPLCM demands a multidimensional approach and includes critical components, namely eco-design, green manufacturing, life cycle assessment, stakeholder engagement, compliance, and interdepartmental collaboration. Further, it is also worth stating that to overcome the above SPLCM challenges and limitations like fragmented sustainability assessment, unavailability of data, poor quality of data, organizational inertia, and technological limitation, companies should look upon the four core attributes.

References

Akter, S., Ji, X., Sarker, M., Cai, L., Shao, Y., Hasan, M., et al. (2020). Clean manufacturing and green practices in the apparel supply chain. *Open Journal of Business and Management, 8*(1), 104–113. https://doi.org/10.4236/ojbm.2020.81007

Alam, M. N., Turi, J. A., Bhuiyan, A. B., Kharusi, S. A., Oyenuga, M., Zulkifli, N., & Iqbal, J. (2024). Factors influencing intention for reusing virtual reality (VR) at theme parks: The mediating role of visitors satisfaction. *Cogent Social Sciences, 10*(1), 2298898. https://doi.org/10.1080/23311886.2023.2298898

Alwaeli, M., & Mannheim, V. (2022). Investigation into the current state of nuclear energy and nuclear waste management—A state-of-the-art review. *Energies, 15*(12), 4275. https://doi.org/10.3390/en15124275

Ardiana, P. (2022). Conceptualising stakeholder engagement in sustainability reporting. *Jurnal Ilmiah Akuntansi Dan Bisnis, 17*(1), 1. https://doi.org/10.24843/jiab.2022.v17.i01.p01

Bag, S., & Pretorius, J. (2020). Relationships between industry 4.0, sustainable manufacturing and circular economy: Proposal of a research framework. *International Journal of Organizational Analysis, 30*(4), 864–898. https://doi.org/10.1108/ijoa-04-2020-2120

Bal, M., Bryde, D., Fearon, D., & Ochieng, E. (2013). Stakeholder engagement: Achieving sustainability in the construction sector. *Sustainability, 5*(2), 695–710. https://doi.org/10.3390/su5020695

Beng, L., & Omar, B. (2013). An axiomatic approach to design for sustainable end-of-life: A review of the literature. *Applied Mechanics and Materials, 315*, 705–709. https://doi.org/10.4028/www.scientific.net/amm.315.705

Bhushan, M., & Singh, R. (2024). Talent management facilitates net-zero transition through employee green behavior. In *Net zero economy, corporate social responsibility and sustainable value creation: Exploring strategies, drivers, and challenges* (pp. 117–129). Springer Nature Switzerland. https://doi.org/10.1007/978-3-031-55779-8_7

Bianchi, G., Testa, F., Boiral, O., & Iraldo, F. (2021). Organizational learning for environmental sustainability: Internalizing lifecycle management. *Organization & Environment, 35*(1), 103–129. https://doi.org/10.1177/1086026621998744

Biró, K., & Csete, M. (2023). Revealing the concept of sustainability in life cycle assessment. *Ecocycles, 9*(2), 49–58. https://doi.org/10.19040/ecocycles.v9i2.277

Brand, U., Papantoni, V., Trillos, J., Tippe, M., Lütkehaus, H., Oswald, M., et al. (2022). Sustainability and future trajectories: How is prospectivity integrated into life cycle sustainability assessment? *E3S Web of Conferences, 349*, 02006. https://doi.org/10.1051/e3sconf/202234902006

Chew, K., & Lee, W. P. (2021). A proposed RAMI 4.0 product life cycle framework using the manufacturing chain management platform (pp. 24–35). https://doi.org/10.56453/icdxa.2021.1002

Chitaka, T., Kock, L., & Blottnitz, H. (2022). Evolution of value chain and governance actor responses to the plastic leakage problem in South Africa. *Frontiers in Sustainability, 3*. https://doi.org/10.3389/frsus.2022.993011

Costello, C., Xue, X., & Howarth, R. (2015). Comparison of production-phase environmental impact metrics derived at the farm- and national-scale for United States agricultural commodities. *Environmental Research Letters, 10*(11), 114004. https://doi.org/10.1088/1748-9326/10/11/114004

Duigou, J., & Bosch-Mauchand, M. (2013). Implementing sustainable supply chain in PLM. In *Advances in production management systems. Competitive manufacturing for innovative products and services* (pp. 168–175). Springer. https://doi.org/10.1007/978-3-642-40361-3_22

Eberhardt, L., Stijn, A., Rasmussen, F., Birkved, M., & Birgisdóttir, H. (2020). Development of a life cycle assessment allocation approach for circular economy in the built environment. *Sustainability, 12*(22), 9579. https://doi.org/10.3390/su12229579

Everard, M. (2023). Assessment of the sustainable use of chemicals on a level playing field. *Integrated Environmental Assessment and Management, 19*(4), 1131–1146. https://doi.org/10.1002/ieam.4723

Fielding, E., McCardle, J., Eynard, B., Hartman, N., & Fraser, A. (2014). Product lifecycle management in design and engineering education: International perspectives. *Concurrent Engineering, 22*(2), 123–134. https://doi.org/10.1177/1063293x13520316

Finnveden, G., Hauschild, M., Ekvall, T., Guinée, J., Heijungs, R., Hellweg, S., et al. (2009). Recent developments in life cycle assessment. *Journal of Environmental Management, 91*(1), 1–21. https://doi.org/10.1016/j.jenvman.2009.06.018

Fish, L. (2015). *Managerial best practices to promote sustainable supply chain management & new product development*. IntechOpen. https://doi.org/10.5772/59581

Gemechu, E., Sonnemann, G., Remmen, A., Frydendal, J., & Jensen, A. (2015). How to implement life cycle management in business? In *Life cycle management* (pp. 35–50). Springer. https://doi.org/10.1007/978-94-017-7221-1_4

Guo, Z., Liu, H., Zhang, D., & Yang, J. (2017). Green supplier evaluation and selection in apparel manufacturing using a fuzzy multi-criteria decision-making approach. *Sustainability, 9*(4), 650. https://doi.org/10.3390/su9040650

Hamdar, Y., Chehab, G., & Srour, I. (2016). Life-cycle evaluation of pavements: A critical review. *Journal of Engineering Science and Technology Review, 9*(6), 12–26. https://doi.org/10.25103/jestr.096.02

Hapuwatte, B., & Jawahir, I. (2021). Closed-loop sustainable product design for circular economy. *Journal of Industrial Ecology, 25*(6), 1430–1446. https://doi.org/10.1111/jiec.13154

Hertwich, E. (2005). Life cycle approaches to sustainable consumption: A critical review. *Environmental Science & Technology, 39*(13), 4673–4684. https://doi.org/10.1021/es0497375

John, M. (2024). Climate change, food insecurity, peace and sustainable development in East Africa: Case study of South Sudan, Sudan, Ethiopia and Kenya. In *Climate change and socio-political violence in Sub-Saharan Africa in the Anthropocene: Perspectives from peace ecology and sustainable development* (pp. 141–165). Springer Nature Switzerland. https://doi.org/10.1007/978-3-031-48375-2_6

Kalverkamp, M., Pehlken, A., Wuest, T., & Young, S. (2018). Sustainability of cascading product lifecycles. In *Product lifecycle management. Towards knowledge-rich enterprises. PLM 2012* (pp. 159–168). Springer. https://doi.org/10.1007/978-3-030-01614-2_15

Kärkkäinen, H., Pels, H., & Silventoinen, A. (2012). Defining the customer dimension of PLM maturity. In *Product lifecycle management. Towards knowledge-rich enterprises* (pp. 623–634). Springer. https://doi.org/10.1007/978-3-642-35758-9_56

Khan, S., Singh, R., Sá, J. C., Santos, G., & Ferreira, L. P. (2022). Modelling of determinants of logistics 4.0 adoption: Insights from developing countries. *Machines, 10*(12), 1242. https://doi.org/10.3390/machines10121242

Khan, S., Singh, R., Alnahas, J., Abbate, S., & Centobelli, P. (2024). Navigating the smart circular economy: A framework for manufacturing firms. *Journal of Cleaner Production, 480*, 144007. https://doi.org/10.1016/j.jclepro.2024.144007

Kiss, N., Tamás, J., Szőllősi, N., Gorliczay, E., & Nagy, A. (2021). Assessment of composted pelletized poultry litter as an alternative to chemical fertilizers based on the environmental impact of their production. *Agriculture, 11*(11), 1130. https://doi.org/10.3390/agriculture11111130

Knauer, T., & Möslang, K. (2018). The adoption and benefits of life cycle costing. *Journal of Accounting and Organizational Change, 14*(2), 188–215. https://doi.org/10.1108/jaoc-04-2016-0027

Koomen, B., Hoogeboom, M., & Schellens, V. (2019). PLM implementation success rate in SME. An empirical study of implementation projects, preliminary findings. In *Product lifecycle management in the digital twin era* (pp. 47–57). Springer. https://doi.org/10.1007/978-3-030-42250-9_5

Koroneos, C., Achillas, C., Moussiopoulos, N., & Nanaki, E. (2013). Life cycle thinking in the use of natural resources. *Open Environmental Sciences, 7*(1), 1–6. https://doi.org/10.2174/1876325101307010001

Kravchenko, M., Pigosso, D., & McAloone, T. (2019). Towards the ex-ante sustainability screening of circular economy initiatives in manufacturing companies: Consolidation of leading sustainability-related performance indicators. *Journal of Cleaner Production, 241*, 118318. https://doi.org/10.1016/j.jclepro.2019.118318

Kumar, K., Singh, R., Sharma, A., & Lehri, B. P. S. (2024). Climate change adaptation strategies for achieving net-zero economy. In *Approaches to global sustainability, markets, and governance* (pp. 163–183). https://doi.org/10.1007/978-981-97-5756-5_8

Lenzo, P., Traverso, M., Mondello, G., Salomone, R., & Ioppolo, G. (2018). Sustainability performance of an Italian textile product. *Economies, 6*(1), 17. https://doi.org/10.3390/economies6010017

Li, T., Roskilly, A., & Wang, Y. (2018). Life cycle sustainability assessment of grid-connected photovoltaic power generation: A case study of Northeast England. *Applied Energy, 227*, 465–479. https://doi.org/10.1016/j.apenergy.2017.07.021

Liu, G., Man, R., & Wang, Y. (2021). A data management approach based on product morphology in product lifecycle management. *Processes, 9*(7), 1235. https://doi.org/10.3390/pr9071235

López, N., Sáenz, J., & Biedermann, A. (2021). Methodology of product sustainable redesign. Case study: Furniture of a clothing retail store. In *Advances on mechanics, design engineering and manufacturing III* (pp. 175–181). Springer. https://doi.org/10.1007/978-3-030-70566-4_28

Lukman, R., Omahne, V., & Krajnc, D. (2021). Sustainability assessment with integrated circular economy principles: A toy case study. *Sustainability, 13*(7), 3856. https://doi.org/10.3390/su13073856

Menon, K., Kärkkäinen, H., Wuest, T., & Gupta, J. (2018). Industrial internet platforms: A conceptual evaluation from a product lifecycle management perspective. *Proceedings of the Institution of Mechanical Engineers Part B Journal of Engineering Manufacture, 233*(5), 1390–1401. https://doi.org/10.1177/0954405418760651

Mollenkopf, D., Stolze, H., Tate, W., & Ueltschy, M. (2010). Green, lean, and global supply chains. *International Journal of Physical Distribution & Logistics Management, 40*(1/2), 14–41. https://doi.org/10.1108/09600031011018028

Mwaijande, F. (2024). Social life cycle assessment of solar dryer house for postharvest loss management technology in Tanzania. *African Journal of Empirical Research, 5*(2), 1–9. https://doi.org/10.51867/ajernet.5.2.1

Nduji, R., Orji, M., Oyenuga, M., & Oriaku, C. (2023). Assessing e-business and organisational performance in Nigeria today: Evidence from Jumia Ltd, Lagos. *Britain International of Humanities and Social Sciences Journal, 5*(2), 81–92. https://doi.org/10.33258/biohs.v5i2.897

Nikolić, D., Jovanović, S., Skerlić, J., Šušteršič, J., & Radulović, J. (2019). Methodology of life cycle sustainability assessment. *Proceedings on Engineering Sciences, 1*(2), 793–800. https://doi.org/10.24874/pes01.02.084

Nilsson-Lindén, H., Rosén, M., & Baumann, H. (2019). Product chain collaboration for sustainability: A business case for life cycle management. *Business Strategy and the Environment, 28*(8), 1619–1631. https://doi.org/10.1002/bse.2388

Oliveira, M., Andreatta, L., Stjepandić, J., & Canciglieri, O. (2021). Product lifecycle management and sustainable development in the context of industry 4.0: A systematic literature review. https://doi.org/10.3233/atde210100

Ong, H., Mahlia, T., Masjuki, H., & Honnery, D. (2012). Life cycle cost and sensitivity analysis of palm biodiesel production. *Fuel, 98*, 131–139. https://doi.org/10.1016/j.fuel.2012.03.031

Oyedele, M. O., Marcus, G. O., & Ahungwa, A. I. (2023). Do consumers care about green marketing practices? Insight from A developing nation. *Budapest International Research and Critics Institute Journal (BIRCI Journal), 6*(3), 1424–1436. https://doi.org/10.33258/birci.v6i3.7665

Oyedele, M. O., Sunday, A. O., & Abuh, A. I. (2024). Fostering technological-enhanced training and development for business survival and performance in the new normal. *Journal of Propulsion Technology, 45*(3), 1858–1869. https://propulsiontechjournal.com/index.php/journal/article/view/7479

Oyenuga, M. O., & Omale, S. A. (2024). Is Africa jinxed? Exploring the challenges of technology access and adoption in Africa. *African Journal of Economics and Sustainable Development, 7*(4), 142–161. https://doi.org/10.52589/AJESDULN1LRNF

Oyenuga, M. O., Andah, R. A., Marcus, G. O., & Agabi, A. U. (2019). Effects of customer relationship management on product innovation in Nigeria. A case study of Xerox HS Nigeria Limited. *American Journal of Theoretical and Applied Business, 5*(4), 113–126. https://doi.org/10.11648/j.ajtab.20190504.15

Oyetunde, T. O., Oyenuga, M. O., & Adoga, G. J. (2023). Green management practices and organisational efficiency in higher institution. *Budapest International Research and Critics in Linguistics and Education (BirLE) Journal, 6*(3), 276–290. https://doi.org/10.33258/birle.v6i3.7716276

Pimenta, H., & Ball, P. (2015). Analysis of environmental sustainability practices across upstream supply chain management. *Procedia CIRP, 26*, 677–682. https://doi.org/10.1016/j.procir.2014.07.036

Popescu, I. (2023). Investment funds are responsible for substantial environmental and social impacts with trade-offs. *Communications Earth & Environment, 5*, 355. https://doi.org/10.21203/rs.3.rs-3345219/v1

Ramani, K., Ramanujan, D., Bernstein, W., Zhao, F., Sutherland, J., Handwerker, C., et al. (2010). Integrated sustainable life cycle design: A review. *Journal of Mechanical Design, 132*(9). https://doi.org/10.1115/1.4002308

Ren, Z., Zhang, D., & Zhu, G. (2022). The sustainable design strategy of cosmetic packaging in China based on life cycle assessment. *Sustainability, 14*(13), 8155. https://doi.org/10.3390/su14138155

Rycroft, T., Wood, M., Zemba, V., Kennedy, A., Weiss, C., Desmet, D., et al. (2019). Assessing the sustainability of advanced materials using multicriteria decision analysis and the triple bottom line. *Integrated Environmental Assessment and Management, 15*(6), 1021–1028. https://doi.org/10.1002/ieam.4205

Saidani, M., Cluzel, F., Leroy, Y., Pigosso, D., Kravchenko, M., & Kim, H. (2022). Nexus between life cycle assessment, circularity and sustainability indicators—Part II: Experimentations. *Circular Economy and Sustainability, 2*(4), 1399–1424. https://doi.org/10.1007/s43615-022-00160-2

Saorín, J., Cantero, J., Díaz, D., & López-Chao, V. (2019). Cloud-based collaborative 3d modeling to train engineers for the industry 4.0. *Applied Sciences, 9*(21), 4559. https://doi.org/10.3390/app9214559

Shaukat, M., Latif, K., Sajjad, A., & Eweje, G. (2021). Revisiting the relationship between sustainable project management and project success: The moderating role of stakeholder engagement and team building. *Sustainable Development, 30*(1), 58–75. https://doi.org/10.1002/sd.2228

Sherry, J., & Tivona, S. (2022). Reducing the environmental impact of food service in universities using life cycle assessment. *International Journal of Sustainability in Higher Education, 23*(7), 1469–1481. https://doi.org/10.1108/ijshe-06-2021-0224

Singh, R., Khan, S., & Centobelli, P. (2022a). Investigating the interplay between social performance and organisational factors supporting circular economy practices. *Sustainability, 14*(24), 16781. https://doi.org/10.3390/su142416781

Singh, R., Khan, S., & Dsilva, J. (2022b). A framework for assessment of critical factor for circular economy practice implementation. *Journal of Modelling in Management, 18*(5), 1476–1497. https://doi.org/10.1108/jm2-06-2021-0145

Singh, R., Khan, S., & Dsilva, J. (2024a). Modelling the organisational factors for implementation of corporate social responsibility: A modified TISM approach. *Global Journal of Flexible Systems Management, 25*, 283–301. https://doi.org/10.1007/s40171-024-00388-x

Singh, R., Crowther, D., & Ravesangar, K. (2024b). Harnessing human capital for a sustainable future: HR's pivotal role toward net zero policies. In *Transition towards a sustainable future: Net zero policies and environmental sustainability* (pp. 27–47). Springer Nature. https://doi.org/10.1007/978-981-97-5756-5_2

Singh, R., Kumar, K., & Khan, S. (2024c). A comprehensive view of artificial intelligence (AI)–based technologies for sustainable development goals (SDGs). In *Artificial intelligence enabled management: An emerging economy perspective* (p. 183). De Gruyter. https://doi.org/10.1515/9783111172408-012

Singh, R., Gupta, A., Bajpai, A., & Kandpal, V. (2024d). Exploring challenges of circular economy initiatives for smart cities. In *Smart cities and circular economy* (pp. 59–69). Emerald Publishing Limited. https://doi.org/10.1108/978-1-83797-957-820241005

Sonnemann, G., Gemechu, E., Remmen, A., Frydendal, J., & Jensen, A. (2015). Life cycle management: Implementing sustainability in business practice. In *Life cycle management* (pp. 7–21). Springer. https://doi.org/10.1007/978-94-017-7221-1_2

Stark, J. (2022). Product lifecycle management (PLM). In *Product lifecycle management (volume 1) 21st century paradigm for product realisation* (pp. 1–32). Springer International Publishing. https://doi.org/10.1007/978-3-030-98578-3_1

Tanpure, G., Yadav, V., Jain, R., & Soni, G. (2023). Framework for cloud product lifecycle management system: A case study of an automotive industry. *Business Process Management Journal, 29*(6), 1920–1937. https://doi.org/10.1108/BPMJ-10-2022-0527

Theyel, G., & Hofmann, K. (2012). Stakeholder relations and sustainability practices of us small and medium-sized manufacturers. *Management Research Review, 35*(12), 1110–1133. https://doi.org/10.1108/01409171211281255

Tišma, S., & Škrtić, M. (2023). Blockchain technology in the environmental economics: A service for a holistic and integrated life cycle sustainability assessment. *Journal of Risk and Financial Management, 16*(3), 209. https://doi.org/10.3390/jrfm16030209

Traverso, M., Finkbeiner, M., Jørgensen, A., & Schneider, L. (2012). Life cycle sustainability dashboard. *Journal of Industrial Ecology, 16*(5), 680–688. https://doi.org/10.1111/j.1530-9290.2012.00497.x

Traverso, M., Kim, P., Brattig, S., & Wagner, V. (2015). Managing life cycle sustainability aspects in the automotive industry. In *Life cycle management* (pp. 331–339). Springer. https://doi.org/10.1007/978-94-017-7221-1_24

Türkoğlu, A., Peng, Q., & Gu, P. (2015). A benchmark-based method for sustainable product design. *Benchmarking an International Journal, 22*(4), 643–664. https://doi.org/10.1108/bij-09-2014-0092

Vadoudi, K., Allais, R., Reyes, T., & Troussier, N. (2014). Sustainable product lifecycle management and territoriality: New structure for PLM. In *Product lifecycle management for a global market* (pp. 475–484). Springer. https://doi.org/10.1007/978-3-662-45937-9_47

Vezzetti, E., Violante, M., & Marcolin, F. (2013). A benchmarking framework for product lifecycle management (PLM) maturity models. *The International Journal of Advanced Manufacturing Technology, 71*(5–8), 899–918. https://doi.org/10.1007/s00170-013-5529-1

Vezzetti, E., Alemanni, M., & Macheda, J. (2015). Supporting product development in the textile industry through the use of a product lifecycle management approach: A preliminary set of guidelines. *The International Journal of Advanced Manufacturing Technology, 79*(9–12), 1493–1504. https://doi.org/10.1007/s00170-015-6926-4

Wallbaum, H., & Ebrahimi, B. (2019). Life cycle management of infrastructures. In *Industry, innovation and infrastructure* (pp. 1–16). Springer. https://doi.org/10.1007/978-3-319-71059-4_22-1

Wallbaum, H., & Ebrahimi, B. (2021). Life cycle management of infrastructures. In *Industry, innovation and infrastructure* (pp. 678–693). Springer. https://doi.org/10.1007/978-3-319-95873-6_22

Wang, S., & Su, D. (2022). Sustainable product innovation and consumer communication. *Sustainability, 14*(14), 8395. https://doi.org/10.3390/su14148395

Wiesner, S., Freitag, M., Westphal, I., & Thoben, K. (2015). Interactions between service and product lifecycle management. *Procedia CIRP, 30*, 36–41. https://doi.org/10.1016/j.procir.2015.02.018

Yokokawa, N., Masuda, Y., Amasawa, E., Sugiyama, H., & Hirao, M. (2020). Systematic packaging design tools integrating functional and environmental consequences on product life cycle: Case studies on laundry detergent and milk. *Packaging Technology and Science, 33*(11), 445–459. https://doi.org/10.1002/pts.2526

Yue, D., Kim, M., & You, F. (2013). Design of sustainable product systems and supply chains with life cycle optimization based on functional unit: General modeling framework, mixed-integer nonlinear programming algorithms and case study on hydrocarbon biofuels. *ACS Sustainable Chemistry & Engineering, 1*(8), 1003–1014. https://doi.org/10.1021/sc400080x

Zhang, A., Zhong, R., Farooque, M., Kang, K., & Venkatesh, V. (2020). Blockchain-based life cycle assessment: An implementation framework and system architecture. *Resources Conservation and Recycling, 152*, 104512. https://doi.org/10.1016/j.resconrec.2019.104512

Chapter 16
Blockchain for Sustainable Finance and Green Investments in the Era of Sustainability

Donkor Nawaah, Joseph Tufuor Kwarteng, and Michael Ayikwei Quarshie

Abstract Blockchain technology has started to attract the attention of an array of industries due to the decentralized and immutable nature of it. It has, however, yet to live up to its potential to reshape green investments and sustainable finance. This chapter explores the relevance of blockchain in addressing critical challenges such as greenwashing, lack of transparency, and inefficiencies in financing green projects. Through the lens of real-world scenarios, we break down the potential for blockchain to become a disruptive force in carbon trading, green bonds, and renewable energy markets. The chapter concludes with actionable recommendations for developers, the financial services sector, and policymakers on how to maximize the positive impact of blockchains towards global sustainability goals.

Keywords Blockchain · Blockchain applications · Environmental sustainability · Green investments · Sustainable finance

---

D. Nawaah (✉)
Department of Finance and Accounting, Woxsen University, Hyderabad, Telangana, India
e-mail: donkor.nawaah@woxsen.edu.in

J. T. Kwarteng
Department of Business and Social Sciences Education, University of Cape Coast, Cape Coast, Ghana
e-mail: Jtkwarteng@ucc.edu.gh

M. A. Quarshie
Department of Entrepreneurship and Business Sciences, University of Energy and Natural Resources, Sunyani, Ghana
e-mail: michael.quarshie@uenr.edu.gh

R. Singh et al. (eds.), *Blockchain Technologies for Smart Circular Economy and Organisational Sustainability*, Environmental Footprints and Eco-design of Products and Processes, https://doi.org/10.1007/978-3-031-89183-0_16

16.1 Introduction

One of the most significant problems of the world today is the interconnected crises of resource depletion, climate change, and environmental degradation (Malakar et al., 2023; Khan et al., 2022a, b). The need to move towards sustainable economic practices is more important than ever as countries across the globe work to fulfil their obligations under the Paris Agreement and the Sustainable Development Goals (SDGs) of the UN (Sachs et al., 2019). To promote growth that is both lucrative and in line with global sustainability concerns, sustainable finance, a method that incorporates environmental, social, and governance (ESG) considerations into investment decisions, is essential to this shift. Nevertheless, inefficiencies, a lack of transparency, and widespread problems like greenwashing have made it difficult for traditional financial channels to effectively drive this change. With its decentralized, transparent, and impenetrable design, blockchain technology provides a revolutionary answer to a number of these problems.

Blockchain has the ability to completely transform green investments and sustainable finance by facilitating real-time data exchange, automating procedures using smart contracts, and producing unchangeable records. It offers strategies to counteract greenwashing, builds confidence among many stakeholders, and makes it easier to allocate funds effectively for environmentally beneficial initiatives (Kumar et al., 2022). These features establish blockchain as a tool for accomplishing global sustainability objectives as well as a technological advancement. This chapter examines the revolutionary nexus between blockchain technology and sustainable finance, exploring practical applications in sectors including renewable energy markets, carbon trading, and green bonds. By blending theoretical insights with practical cases, this chapter illustrates how blockchain can bring to bear on pressing challenges in finance and the natural world. This chapter aims to equip developers, financial institutions, and policymakers alike with actionable strategies on how to translate the promise of blockchain into practice, to help build a more sustainable and green future.

16.2 Blockchain: Fundamentals and Features

16.2.1 Core Principles of Blockchain

Decentralization, transparency and immutability, and smart contracts are some of the key principles of blockchain (Lin et al., 2022).

(a) *Decentralization*: Its decentralized nature is one of its most fundamental characteristics; no central entity means no interference and influence of central authorities and it makes the system more reliable and stronger.

(b) *Transparency and Immutability*: Transparency and immutability are two features that provide that transactions are accessible to limited participants only, and entered transactions are not alterable, which ensure accountability and trust.
(c) *Smart Contracts:* Smart contracts are self-executing contracts. They are written to enforce terms of a contract, and they automatically execute relevant actions when the conditions stated in the contract are met.

All these traits grant blockchain its unique efficiency, security, and trust, making it a disruptive tool across industries, particularly in finance and sustainability.

16.2.2 Blockchain's Alignment with Sustainability Goals

The decentralized nature and immutable, transparent nature of blockchain technology aid sustainability goals through providing enhanced accountability and efficiency in resource allocation. This reduces the risk of misrepresentation and greenwashing of sustainability claims by enabling real-time tracking of environmental, social, and governance (ESG) measures (Singh et al., 2024a, b, c, d; Gu et al., 2023). Also, blockchain empowers underfunded and marginalized communities by providing access to financing through decentralized crowdfunding, asset tokenization, and less reliance on traditional financial intermediaries. Innovative financing of green investments: smart contracts automate flows and secure and transparent flows of funds help serve long-term sustainable development and fairer financial systems.

16.3 Challenges in Sustainable Finance

16.3.1 Overview

Although sustainable finance is becoming more and more well-known for tackling global environmental and social issues, some obstacles hinder its implementation:

(a) Greenwashing: May be the largest issue of the three, where companies or financial products are masquerading as green (Pimonenko et al., 2020). This fake portrayal leads to a loss of trust from investors, regulators, and stakeholders and creates scepticism of the authenticity of ESG initiatives. This creates an issue around distinguishing genuine sustainability projects, from those merely taking advantage of the growing appetite for green investment.
(b) *Inefficiencies*: Inefficiency in the ecosystem of sustainable financing is another major obstacle (Rizos et al., 2016). Investor commitment to ESG-focused initiatives is deterred by high transaction costs and bureaucratic hold-ups. The assessment and comparison of investments are made more difficult by the

absence of standardized frameworks and measures, which also impedes scaling and introduces uncertainty.

(c) *Exclusion of marginalized groups*: Traditional financial systems often overlook the needs of underrepresented communities, limiting their access to capital and participation in sustainable initiatives (Phillips & Johnson, 2021). Thus, this exclusion does not only repeat the social inequities but also continues to dilute sustainable finance impacts on global development goals.

The sustainable finance space needs to be more transparent, efficient, and inclusive to address these challenges.

16.4 Blockchain Applications in Sustainable Finance

The blockchain technology is changing the face of sustainable finance as never before by solving the challenges across the value chain related to transparency, trust, and efficiency. Blockchain is enabling innovation in carbon credits, green bonds, renewable energy markets, and sustainable supply chain traceability by leveraging its unique properties including decentralization and immutability (Singh et al., 2023). In this section, we further explore these applications and highlight some real-world use cases to show how blockchain has the potential to transform a variety of sectors.

16.4.1 Carbon Credits and Emissions Trading

Carbon credit systems fall into double counting, transparency, and fraud issues (Grau Ruiz, 2024). This is where blockchain technology comes into the picture by offering a reliable, transparent record verifying the uniqueness, traceability, and inalterability of each carbon credit. The system also boosts carbon trading by fostering greater trust and accountability between stakeholders. Blockchain technology will allow tracking the carbon credits real-time at the same time making it compliant with international standards, hence making it efficient instead of losing efficiencies. It also facilitates automated verification, which lowers administrative costs and delays.

- *Real World Case Study: Climate Trade (2021)*
- Climate Trade, a blockchain-based platform, makes it easier for businesses looking to offset their emissions to buy direct carbon credit. Through the benefits of blockchain, Climate Trade can make carbon credits unique, traceable, and auditable. Since its launch just over a year ago, more than 100 companies have lever-

aged the platform to offset one million tons of CO_2. By exposing the carbon credits through a transparent marketplace, Climate Trade increases companies' confidence on the carbon offset channel.

16.4.2 Green Bonds

Green bonds are financial instruments designed to raise funds for projects that deliver environmental benefits. Yet, these bonds have faced criticism over their transparency, particularly around the distribution of bonds and how the proceeds are used (Bhutta et al., 2022). The solutions to these challenges facilitated the use of the blockchain, providing an open record with near real-time data on the distribution and effect of the green bond. Issuers can directly demonstrate their project information related to environmental impacts to the investors to prove the veracity of their information in a world backed by blockchain. This level of transparency not only contributes to its credibility but also gains the interest of more green investors. Moreover, a decentralized platform makes administrative functions cost-effective by streamlining the processes.

- *Case Study*: HSBC Green Bond (2022)
- In 2022, HSBC made green bonds available on a blockchain, which allowed investors immediate and verifiable access to information on the use and impact of the proceeds, according to HSBC. This phenomenon fixed the perpetual transparency issues faced by the green bond market. The integration that relied on blockchain was facilitated by HSBC, which thereby ensured greater efficiency and a decrease in administrative expenses. It represented a world-first project demonstrating a new way for financial institutions to incentivize green investment using blockchain.

16.4.3 Renewable Energy Markets

Blockchain technology is transforming renewable energy markets by enabling peer-to-peer (P2P) energy trading (Khan et al., 2022a, b). This distributed system allows an individual home or small business owner to directly sell excess renewable energy to a residential or commercial end-user, rather than going through traditional intermediaries. Blockchain technology grants transparency and fairness when conducting transactions, where all parties trust each other. It also allows for renewable energy integration with real-time coordination between generation and consumption (Khan et al., 2024). This motivates efficiency and creates balance between supply and demand.

- *Case: Sun Exchange (2023)*
- Sun Exchange is a solar panel marketplace running on the blockchain that enables anyone to purchase and lease solar panels to communities in need in return for affordable energy. The solar cells are paid for by donations from investors that are leased to schools, businesses, and communities. Payments are done over clear payment flows, and the benefit is that you can trace back this energy production as well thanks to blockchain technology. By 2023, Sun Exchange had financed over 50 different solar projects across Africa, demonstrating how blockchain can help broaden access to renewable energy.

16.4.4 Sustainable Supply Chain Traceability

Ensuring the sustainability of supply chains is a critical challenge, as fraudulent claims often undermine product certifications. Blockchain technology addresses this challenge by providing an immutable ledger of the product from each point in its life cycle. Blockchain keeps sustainable certifications unique by nature, maximizing the effectiveness of responsible sourcing from raw material, processing, and ensuring the final distribution. Blockchain integration also enables companies to offer their consumers transparency and confidence about the sustainability of their purchase. It enhances the brand image, but also increases the demand for eco-sensitive products.

- Case: Everledger (2022)
- Everledger uses blockchain to verify the provenance of sustainable diamonds, ensuring that conflict diamonds do not enter the market. Each diamond is assigned its own digital identity, which is recorded on the blockchain. It also allows stakeholders to follow the diamond the entire way from mine to market, confirming that ethical and sustainable practices are followed throughout the supply chain. In 2022, Everledger set a new standard for honesty of diamond sourcing, creating a foundation for more to follow across sectors.

16.5 Challenges and Risks

Although the potential for transformation from the technology is massive, several barriers exist which may prove to hinder its uptake in sustainable finance (Khan et al., 2023). For blockchain to realize its full potential for sustainable global impact, these uncertainties need to be addressed.

16.5.1 Technological Barriers

- *Energy Usage*: Blockchain systems that utilize proof-of-work (PoW), consensus mechanisms consume massive amounts of energy. Such high consumption of energy runs against the sustainable goals blockchain is intended for. To get in line with environmental goals, there is the need for energy-free alternatives like proof-of-stake (PoS) in its place.
- Scalability Problems: Most of the existing blockchain networks lack the ability to provide sustainable scalability for transaction processing. As green finance projects scale up, this could mean performance bottlenecks for blockchain systems, leading to a slowdown and increasing costs. To meet those constraints, scalable solutions such as layer-two protocols (i.e. systems constructed on top of existing systems to increase their speed and efficiency are known as layer-two protocols) or sharding must be built (i.e. the technique of dividing big data sets into smaller, easier-to-manage units to speed up processing).

16.5.2 Regulatory Uncertainty

- *Regulatory Ambiguity*: There is a critical issue for blockchain conduct in renewable finance to engage with impact; there has not been sufficient regulatory daylight. This leads to ambiguity on the part of the stakeholders and hampers innovation with disparate practices being followed for blockchain governance across jurisdictions.
- *Privacy and Data Security Issues*: The transparent nature of blockchain is a significant advantage, but it also causes privacy and data security issues. If left uncontrolled, it will leak sensitive information such as banking account or information or personal information. Transparency and safety concerns should be managed to develop trust in the blockchain.

16.5.3 Economic Limitations

- *High Initial Investment*: Setting up the infrastructure, software development, and knowledge that blockchain would need is expensive. These potential costs may lead to small and medium enterprises (SMEs) to be excluded from blockchain-based sustainable finance.
- *Accessibility Issues*: Most of the blockchain tech is convoluted and non-intuitive for the common person, particularly those in the developing world. This has to do with making blockchain platforms easy and affordable to interact with.

16.5.4 *Social and Ethical Considerations*

- *Digital Divide*: The implementation of blockchain in sustainable finance could exacerbate existing disparities due to the potential for capital to migrate away from vulnerable sectors, as disenfranchised communities may lack both resources and technical expertise to access this technology. Bridging the digital divide is essential to make blockchain accessible.
- *Governance Issues*: It can be difficult to find a good governance model in decentralized systems. Networking blocks or decision-making processes fighting among themselves, or the same player controlling multiple blockchains, could impair trust in a given blockchain and its capacity.

16.6 Future Directions and Innovations

The governance of sustainable finance (both public and private) will also be more integrated with the constant expedition of a borderless digital currency and blockchain hardware. These new methods and forward-looking solutions will be applied more often by the future bank. This section outlines just a few of the many areas in which blockchain could help support global sustainability initiatives.

16.6.1 *Transition to Energy-Efficient Models*

Among these multifaceted barriers, energy consumption, especially by networks using proof-of-work (PoW) consensus mechanisms, is one of the most pressing challenges confronted by blockchain. In addition, the deployment of blockchain needs to fit within sustainability objectives and should support the shift to energy-efficient models such as proof-of-stake (PoS) and other emerging protocols. Blockchain networks have a significantly smaller environmental footprint due to energy-efficient blockchain models that have been designed well (Singh et al., 2022). For instance, PoS eliminates electrode mining, making it far more appropriate for green finance applications. As a result, new technologies such as directed acyclic graphs (DAGs) and multilayered blockchains can achieve even higher energy efficiency without sacrificing security and scalability. Even this would further open up blockchain to a broader range of sustainable finance initiatives without conflicting with ecological priorities. It is for governments and industry leaders to sustain investments in R&D to speed up the transition to greener adoption in blockchain technologies.

16.6.2 Integration with AI and IoT

The blockchain combined with AI and IoT is the big opportunity to turn the sustainable finance upside down. Combined, these can enhance ESG (environmental, social, and governance) monitoring, analytics, and automated decision-making.

- *Integration with AI*: Within blockchain networks, big data can be stored that AI systems can analyze to find trends, assess risk, and use smaller resources where needed. AI algorithms can examine the consequences of green investments, thus helping more informed decision-making by taking advantage of historical data while performing predictive analysis.
- *IoT functionality*: When real-life data from Internet of Things devices are transmitted to our blockchain, it adds transparency and accountability. IoT sensors can measure carbon emissions, water consumption, or energy consumption, ensuring that ESG metrics are recorded accurately on a blockchain record.

Collectively, these technologies together form one integrated ecosystem and are all supporting a sustainable finance ecosystem. This in turn will help them automate ESG reporting, enhance supply chain traceability, and facilitate climate adaptive strategy.

16.6.3 Decentralized Autonomous Organizations (DAOs)

Decentralized Autonomous Organizations (DAOs): DAOs are a possible way to democratize green finance. DAOs utilize the decentralization of blockchain technology to enable stakeholders to have a direct say in the decision-making processes, fostering increased inclusiveness and transparency to a certain extent. DAOs are run through smart contracts, i.e. utilizing smart contracts that can execute certain well-known rules without the use of an intermediary. Such framework provides an avenue for transparent governance as well as fair resource allocation. Such framework provides an avenue for transparent governance as well as fair resource allocation which in the context of green finance, the current DAOs can:

(a) Community-based renewable energy projects that share resources and make collaborative decisions.
(b) Allocate conservation expenditure in line with stakeholder votes.
(c) Monitor the performance of sustainability projects in real time to contractually prevent underperformance.
(d) By supporting local communities and a diverse set stakeholder, DAOs can create the bridges over the gaps that exist in traditional finance systems and move collective behaviour towards sustainability.

16.6.4 Global Interoperability Standards

Blockchain technology may be essential to global green finance projects if sufficient efforts are made to create global interoperability standards. There are frequently major inefficiencies and regulatory complications with standardized blockchain protocols intended to support cross-border transactions. Bidirectional communication across various blockchain networks is made possible by interoperability standards, which provide the easy cross-border transfer of data and digital assets. This is especially important for programs like carbon trading, which require consistency and uniformity across jurisdictions in order to be successful. Although adoption of these solutions is still progressing, interoperability is still a top priority.

16.6.4.1 Key Steps for Achieving Global Interoperability

- Accord on data standards, transaction protocols, as well as governance frameworks.
- Aligning blockchain standards and initiatives with UN sustainable development goals in collaboration with international partners and organizations including the United Nations and World Bank alliance.
- Incentivizing innovation and standardization initiatives through private-public partnerships.

Blockchain technology has a bright future in sustainable finance. A shift to more energy-efficient models, integration with cutting-edge technologies like AI and IoT, enabling decentralized autonomous organizations (DAOs), and actively participating in the creation of international interoperability standards are all necessary to realize this promise. If attention is paid to these factors, blockchain can become the essential component of a more enduring, equitable world financial system. To realize this potential, there is the need to empower adopters to use the technology to accelerate the pace of innovation and collaboration, unlocking transformative change to realign financial systems with the climate and sustainable development challenges of our era.

16.7 Strategic Recommendations for Blockchain Integration in Sustainable Finance

The growing awareness of sustainability and the urgent need to combat climate change have spurred the demand for green finance initiatives globally. Blockchain technology has the potential to revolutionize numerous industries. It can be a consideration for a more streamlined green finance system that allows for that first intention behind every transaction to be laid bare along with providing a transparency, efficiency, and accountability unparalleled in any system today (Bhushan & Singh, 2024). However, realizing the potential of blockchain for sustainable finance requires responsible steering from regulators, finance, and the developer community. Each of these stakeholders has an important part to play in ensuring that in the first instance the blockchain can help and in the second instance the blockchain can deliver sustainable development goals, particularly in the areas of Environmental, Social, and Governance (ESG) and sustainable investment practice.

16.7.1 Strategic Recommendations for Policymakers

Policymakers have the crucial responsibility of creating an enabling environment for the adoption of emerging technologies, such as blockchain, in sustainable finance. To effectively harness blockchain's potential in this space, it is necessary for governments and regulatory bodies to develop appropriate frameworks, foster innovation, and incentivize the use of blockchain solutions. Below are key strategic recommendations for policymakers:

16.7.1.1 Develop Comprehensive Blockchain Regulations That Incentivize Its Use in Sustainable Finance

For blockchain technology to be broadly adopted in the finance, including the sustainable financial sector, there should be a clear and supportive regulatory context. Currently, the regulations seem to be either vague or nonexistent in many countries, and this inhibits the applicability of blockchain in innovative formats, such as green finance. Policymakers should address this problem by adopting regulations that would comprehensively cover the potential issues and opportunities associated with blockchain. These regulations should involve the following aspects:

- *Data Privacy and Security*: While blockchain is a secure method of storing and processing transactions, the standards of data privacy and security should be elaborated in more detail. Since, when it comes to sustainable finance, the data is often virtually sensitive referring to ESG criteria, the level of privacy of this information should be appropriately high.
- *Legal Recognition of Blockchain Transactions*: For a blockchain to be adopted in green finance, the transactions processed by it should be formally recognized by the law. This is why the governments need to develop standards to classify actions taken in blockchain systems as enforceable contracts or agreements.
- *Incentives for Adoption*: Governments may also use their soft power functions to promote the use of blockchain for sustainable finance. This can take the form of tax breaks or subsidies for financial institutions, firms, and startups in case they choose to use blockchain technology to reach environmental or other sustain-

ability goals. For instance, the government may provide financial rewards for companies integrating blockchain with sustainable development, implement, and promote emission reduction projects or renewable energy investments.

16.7.1.2 Promote Public-Private Partnerships to Pilot Blockchain-Based Green Finance Initiatives

To realize the full potential of blockchain in sustainable finance, it is necessary for policymakers to promote public-private partnerships (PPPs) where state actors and the private sector can collaborate. Such alliances can be helpful for the piloting and scaling of blockchain-based green finance projects.

- *Co-funding R&D*: Make funds available for joint projects between private firms and public projects, especially to research and develop blockchain applications for green finance. They could be about blockchain-powered carbon credit systems, tokenization of clean energy assets, or open and transparent green bonds utilizing the immutable and traceable nature of blockchain ledger as well.
- *Regulatory Sandbox Initiatives*: Policymakers may design regulatory sandboxes that allow blockchain-based green finance projects to be tested in a controlled setting while real-world market conditions are mimicked. That would enable financial institutions and startups to test out innovative solutions like blockchain-based tracking of ESG without being required to follow the full regulatory demands. Feedback from such initiatives could also be used to improve regulation ahead of full-scale implementation.
- *Shared Blockchain Infrastructure for Green Solutions*: Governments can partner with private sector stakeholders to create shared blockchain infrastructure, including public ledgers for the tracking of green investments, carbon credits, or other green assets. This infrastructure could result in lower blockchain implementation costs, increasing SME and startup access to the blockchain in the green finance space.

PPPs on the other hand can channel the power of policies to scale up the use of blockchain in sustainable finance by creating, promoting, and incentivizing blockchain solutions that reinforce rather than contradict the goals of sustainable development.

16.7.2 Strategic Recommendations for Financial Institutions

Financial institutions are significate players in the green finance ecosystem; the adoption of blockchain technology will reshuffle sector. Blockchain relates to the ESG by providing transparency, security, and efficiencies of transactions that can be instrumental in more effective ESG reporting and compliance and in the offering of resilient and sustainable financial product (Basarkod, 2024; Singh et al., 2024a, b, c, d). The following strategic recommendations provide direction for those financial institutions interested in capturing the potential added value of blockchain as it applies to sustainable finance.

16.7.2.1 Block-Chaining ESG Reporting and Compliance

There has been some focus on ESG criteria in financial decisions over the last couple of years. How can we trust the credibility of ESG investments when traditional ESG reporting and monitoring systems are highly opaque to manipulation? While the popularity of ESG investments grows by the day, the information about environmental and social affairs is often inaccurate and unreliable (Kotsantonis et al., 2016). Blockchain is the best possible solution for ESG reporting given its immutable and transparent nature:

(a) *Blockchain-Based Transparency of Climate Data:* Blockchain is capable of great transparency, but it is not possible to modify the records of ESG data on a blockchain. This can help and depend on verification of derivatives to give good climate transparency between assets. At the identical time, finance can monitor this through the blockchain to make sure all ESG claim is trackable, minimizing the greenwashing risk. For instance, blockchain can assist in the verification of a company's carbon emissions, the audited sourcing of materials, or the social impact of projects financed by green bonds.
(b) *Automated Compliance with ESG Regulations*: Blockchain can be used to automate the tracking and reporting process making ESG compliance automated. Once prices change based on certain ESG conditions, smart contracts will enable these financial intuitions to automatically execute transactions. It would reduce the manual effort of reporting and ensure the compliance of sustainability standards is met in real time.
(c) *Tokenization of ESG Assets*: Financial institutions can use tokenization to issue ESG assets like sustainable real estate, renewable energy projects, or green bonds via blockchain; tokenization renders these assets more tradable and investable and will attract a broader base of sustainable finance-focused investors. Such tokenized assets can use the blockchain for trading, which adds to their value.

To justify their sustainability claims, attract more investors, and limit the risk of fraud or misreporting, financial institutions can adopt blockchain solutions for ESG reporting and compliance.

16.7.2.2 Working with IT Companies to Create Blockchain-Based Green Finance Solutions

The collaboration between financial institutions and IT companies can help to unleash the full potential of blockchain to fundamentally change the paradigm of green finance. Blockchain developers have the technical expertise needed by financial organizations to develop solutions that cater to the needs of sustainable finance:

(a) *Collaborate to Create Blockchain-Based Green Finance Platforms*: Financial institutions should partner with tech companies to develop platforms designed specifically for the sustainable finance market. Such platforms may focus on areas such as sustainable investment monitoring, integrated ESG disclosures, or efficient carbon credit trading mechanisms. Such partnerships would tailor blockchain solutions to the particularities of the green finance ecosystem, enabling a targeted and therefore more effective application of the technology.
(b) *Shared Expertise in Blockchain and Financ*e: Financial organizations make use of the blockchain innovation to provide even more transparent and also economical monetary items that serve wider sustainability objectives while technology firms deal with competency in blockchain production and also execution.

These sectors can together generate more innovative solutions that tackle the barriers to sustainable finance, including high transaction costs, inefficient reporting mechanisms, and scepticism on ESG statements.

16.7.3 Strategic Recommendations for Innovators and Developers

The providers of the platforms, methods, and technologies that aid in blockchain adoption in sustainable finance are primarily developers and inventors. They enable blockchain solutions to be usable, effective, and sustainable.

16.7.3.1 Focus on Developing Blockchain Interfaces That Are Easy to Use to Boost Accessibility

Complexity of blockchain technology is one of the major hurdles in achieving mass adoption. The problem is that a lot of people, especially in the finance sector, do not have the technological excellence needed to understand and engage with blockchain in an efficient way (Abbas et al., 2020). By emphasizing the creation of intuitive blockchain interfaces, developers may assist in removing this obstacle:

(a) *User Experience:* User experience must be the top focus of developers to build the user interface that makes communication between blockchain systems and banks, businesses, and individuals easier. That means developing graphical user interfaces (GUIs) which ease the interaction with tokenized assets.
(b) *Resources for Education and Training*: In addition to designing intuitive user interfaces, developers may work with academic institutions and trade associations to offer resources and training that assist stakeholders in comprehending the advantages and applications of blockchain technology in green finance

(Kumar et al., 2024a, b; Singh et al., 2024a, b, c, d). Clear instructions, tutorials, and customer service would promote uptake and guarantee that the blockchain is available to a larger audience.

By broadening access to blockchain interfaces, developers will ultimately be closing the gates to sustainable finance on a transformative journey in which mass adoption of blockchain can indeed be promoted.

16.7.3.2 Prioritize the Development of Energy-Efficient Blockchain Systems

The energy consumption of blockchain networks, especially those using proof-of-work, is one of the most common negativities against blockchain technology (Schinckus, 2021). In the context of green finance, with the emphasis on environmentally sustainable development, this is becoming a particularly pressing problem. Hence, it is on the developers to make the energy consumption of the blockchain systems more sustainable and in line with the sustainability targets:

(a) *Switch to Energy-Efficient Consensus Methods*: Since proof-of-stake (PoS) uses less energy than proof-of-work (PoW), developers should give priority to using more energy-efficient consensus methods. Blockchain platforms may lessen their carbon footprint and make blockchain a more sustainable green finance solution by switching to these technologies.
(b) *Sustainable Infrastructure and Carbon Offset*: Blockchain networks may be powered by sustainable infrastructure and carbon offsetting strategies in addition to energy-efficient consensus processes (Singh et al., 2024a, b, c, d). To better link blockchain with the goals of sustainable finance, companies may, for example, make sure that the servers and data centres that underpin blockchain platforms are fuelled by renewable energy sources.

By emphasizing the development of blockchain in an energy-efficient way, we may end up with a future in which blockchain technology is both helpful and environmentally friendly.

16.8 Conclusion

Blockchains can change the landscape of sustainable finance and green investments by providing greater transparency, accountability, and efficiency between parties to a financial transaction. Using the blockchain allows participants to build an ecosystem closer to frictionless in terms of tracking and verification of sustainable investments, thus enabling the trustworthiness of ESG data and limiting/mitigating the risk of whitewashing. In addition, blockchain is advantageous in simplifying the procedures surrounding trade of carbon credits, management of green bonds, and certification of sustainable projects—processes that are crucial in scaling up global green finance. However, there are challenges to be solved to realize the full potential of blockchain for sustainable finance. They include technical limitations, in terms of scalability and energy use, and regulatory challenges linked to the legal framework of blockchain transactions and the safeguarding of personal data. Moving forward to tackle these issues will require government intervention, bank facilitation of blockchain systems through a clear line of regulation, and sector-wide contribution, along with energy-efficient types of blockchain crypto. This chapter suggests a strategic framework to stakeholders on how to exploit the presented advantages/disadvantages of blockchain. To achieve this goal, stakeholders must implement these recommendations to create a financial ecosystem that is able to sustainably underpin the global transition to a new, sustainable, and inclusive economy.

References

Abbas, Y., Martinetti, A., Moerman, J. J., Hamberg, T., & van Dongen, L. A. (2020). Do you have confidence in how your rolling stock has been maintained? A blockchain-led knowledge-sharing platform for building trust between stakeholders. *International Journal of Information Management, 55*, 102228. https://doi.org/10.1016/j.ijinfomgt.2020.102228

Basarkod, P. I. (2024). A survey on blockchain security for electronic health record. In *Multimedia tools and applications* (pp. 1–35). Springer. https://doi.org/10.1007/s11042-024-19883-5

Bhushan, M., & Singh, R. (2024). Talent management facilitates net-zero transition through employee green behavior. In *Net zero economy, corporate social responsibility and sustainable value creation: Exploring strategies, drivers, and challenges* (pp. 117–129). Springer Nature Switzerland. https://doi.org/10.1007/978-3-031-55779-8_7

Bhutta, U. S., Tariq, A., Farrukh, M., Raza, A., & Iqbal, M. K. (2022). Green bonds for sustainable development: Review of literature on development and impact of green bonds. *Technological Forecasting and Social Change, 175*, 121378. https://doi.org/10.1016/j.techfore.2021.121378

Grau Ruiz, M. A. (2024). International fiscal cooperation to better integrate public and private efforts on sustainability: The case of carbon offset credits. In *Sustainable finances and the law: Between public and private solutions* (pp. 281–294). Springer Nature Switzerland. https://doi.org/10.1007/978-3-031-49460-4_12

Gu, Y., Dai, J., & Vasarhelyi, M. A. (2023). Audit 4.0-based ESG assurance: An example of using satellite images on GHG emissions. *International Journal of Accounting Information Systems, 50*, 100625. https://doi.org/10.1016/j.accinf.2023.100625

Khan, S., Ali, S. S., & Singh, R. (2022a). Determinants of remanufacturing adoption for circular economy: A causal relationship evaluation framework. *Applied System Innovation, 5*(4), 62. https://doi.org/10.3390/asi5040062

Khan, S., Singh, R., Sá, J. C., Santos, G., & Ferreira, L. P. (2022b). Modelling of determinants of logistics 4.0 adoption: Insights from developing countries. *Machines, 10*(12), 1242. https://doi.org/10.3390/machines10121242

Khan, S., Khan, M. I., & Singh, R. (2023). Modeling the barriers of blockchain technology implementation in supply chain. *Journal of Industrial Integration and Management, 8*(3), 367–385. https://doi.org/10.1142/S2424862223500070

Khan, S., Singh, R., Alnahas, J., Abbate, S., & Centobelli, P. (2024). Navigating the smart circular economy: A framework for manufacturing firms. *Journal of Cleaner Production, 480*, 144007. https://doi.org/10.1016/j.jclepro.2024.144007

Kotsantonis, S., Pinney, C., & Serafeim, G. (2016). ESG integration in investment management: Myths and realities. *Journal of Applied Corporate Finance, 28*(2), 10–16. https://doi.org/10.1111/jacf.12169

Kumar, A., Srivastava, S. K., & Singh, S. (2022). How blockchain technology can be a sustainable infrastructure for the agrifood supply chain in developing countries. *Journal of Global Operations and Strategic Sourcing, 15*(3), 380–405. https://doi.org/10.1108/JGOSS-08-2021-0058

Kumar, A., Singh, R., Khan, S., & Kumar, V. (2024a). Artificial intelligence enabled management: An emerging economy perspective. De Gruyter. Retrieved December 23, 2024 from https://www.google.co.in/books/edition/Artificial_Intelligence_Enabled_Manageme/wgkFEQAAQBAJ?hl=en&gbpv=1&dq=info:JX62jkTBv_sJ:scholar.google.com&pg=PA3&printsec=frontcover .

Kumar, K., Singh, R., Sharma, A., & Lehri, B. P. S. (2024b). Climate change adaptation strategies for achieving net-zero economy. In *Transition towards a sustainable future: Net zero policies and environmental sustainability* (pp. 163–183). Springer Nature Singapore. https://doi.org/10.1007/978-981-97-5756-5_8

Lin, S. Y., Zhang, L., Li, J., Ji, L. L., & Sun, Y. (2022). A survey of application research based on blockchain smart contract. *Wireless Networks, 28*(2), 635–690. https://doi.org/10.1007/s11276-021-02874-x

Malakar, K. D., Kumar, M., Anand, S., & Kuzur, G. (2023). Climate crisis and socio-ecological sustainability. In *Climate change and socio-ecological transformation: Vulnerability and sustainability* (pp. 439–494). Springer Nature Singapore. https://doi.org/10.1007/978-981-99-4390-6_9

Phillips, S. D., & Johnson, B. (2021). Inching to impact: The demand side of social impact investing. *Journal of Business Ethics, 168*(3), 615–629. https://doi.org/10.1007/s10551-019-04241-5

Pimonenko, T., Bilan, Y., Horák, J., Starchenko, L., & Gajda, W. (2020). Green brand of companies and greenwashing under sustainable development goals. *Sustainability, 12*(4), 1679. https://doi.org/10.3390/su12041679

Rizos, V., Behrens, A., Van der Gaast, W., Hofman, E., Ioannou, A., Kafyeke, T., et al. (2016). Implementation of circular economy business models by small and medium-sized enterprises (SMEs): Barriers and enablers. *Sustainability, 8*(11), 1212. https://doi.org/10.3390/su8111212

Sachs, J. D., Schmidt-Traub, G., Mazzucato, M., Messner, D., Nakicenovic, N., & Rockström, J. (2019). Six transformations to achieve the sustainable development goals. *Nature Sustainability, 2*(9), 805–814. https://doi.org/10.1038/s41893-019-0352-9

Schinckus, C. (2021). Proof-of-work based blockchain technology and anthropocene: An undermined situation? *Renewable and Sustainable Energy Reviews, 152*, 111682. https://doi.org/10.1016/j.rser.2021.111682

Singh, R., Khan, S., & Centobelli, P. (2022). Investigating the interplay between social performance and organisational factors supporting circular economy practices. *Sustainability, 14*(24), 16781. https://doi.org/10.3390/su142416781

Singh, R., Khan, S., & Dsilva, J. (2023). A framework for assessment of critical factor for circular economy practice implementation. *Journal of Modelling in Management, 18*(5), 1476–1497. https://doi.org/10.1108/JM2-06-2021-0145

Singh, R., Crowther, D., & Ravesangar, K. (2024a). Harnessing human capital for a sustainable future: HR's pivotal role toward net zero policies. In *Transition towards a sustainable future: Net zero policies and environmental sustainability* (pp. 27–47). Springer. https://doi.org/10.1007/978-981-97-5756-5_2

Singh, R., Gupta, A., Bajpai, A., & Kandpal, V. (2024b). Exploring challenges of circular economy initiatives for smart cities. In V. Kandpal, E. D. Santibanez-Gonzalez, P. Chatterjee, & M. K. Nallapaneni (Eds.), *Smart cities and circular economy* (pp. 59–69). Emerald Publishing Limited. https://doi.org/10.1108/978-1-83797-957-820241005

Singh, R., Khan, S., Dsilva, J., Akram, U., & Haleem, A. (2024c). Modelling the organisational factors for implementation of corporate social responsibility: A modified TISM approach.

Global Journal of Flexible Systems Management, 25(2), 283–301. https://doi.org/10.1007/s40171-024-00388-x

Singh, R., Kumar, K., & Khan, S. (2024d). *A comprehensive view of artificial intelligence (AI)-based technologies for sustainable development goals (SDGs).* Walter de Gruyter GmbH & Co KG. Retrieved December 15, 2024 from https://www.google.co.in/books/edition/Artificial_Intelligence_Enabled_Manageme/wgkFEQAAQBAJ?hl=en&gbpv=1&dq=info:OikXNbmU7fEJ:scholar.google.com&pg=PA183&printsec=frontcover

Chapter 17
Empowering a Sustainable Future: Blockchain's Role in Economy, Organizations, and Finance Towards Smart Circular Economy

Rubee Singh, Kunwar Siddharth Tiwary, Kishore Kumar, and Amit Kumar

Abstract In the present era, blockchain is considered as driving force reshaping the global economy. It is redefining the organizational structure and the business activities and revolutionizing the financial market. Blocking-based technologies also have immense potential to address climate change, concerns, and resource constraints. This chapter provides insights into the multifaceted role of blockchain in driving sustainable development goals by fostering transparency, improving accountability, and integrating sustainability across organizational and financial domains. It analyses the blockchain potential to strengthen financial sustainability through decentralized finance, an equitable financial system. Further, it also highlights the instrumental role of blockchain by improving the supply chain, transparency, corporate governance, and incorporation of environmental, social, and governance (ESG) principles into the business model. Moreover, this chapter discusses the alignment of blockchain with the United Nations' sustainable development goals, highlighting its potential to address, pressing issues such as resource efficiency, climate, change, and inequality. It critically assesses the various challenges in the adoption of blockchain and underscores its importance in achieving long-term sustainability goals.

R. Singh
Institute of Business Management, GLA University, Mathura, India
e-mail: dr.rubeerajput.delhi@gmail.com

K. S. Tiwary
University of the Cumberlands, Williamsburg, KY, USA
e-mail: kunwarsiddharth92@gmail.com

K. Kumar (✉)
School of Management, Bennett University, Greater Noida, India
e-mail: akishore001@gmail.com

A. Kumar
Christ University, Bangalore, India
e-mail: amit.kumar@christuniversity.in

R. Singh et al. (eds.), *Blockchain Technologies for Smart Circular Economy and Organisational Sustainability*, Environmental Footprints and Eco-design of Products and Processes, https://doi.org/10.1007/978-3-031-89183-0_17

Keywords Blockchain · Circular economy · ESG · Triple Bottom Line (TBL) · Ethical investing · Sustainable finance · Sustainable development goals (SDGs)

17.1 Introduction

Sustainability development has assumed critical importance globally. Sustainability has shifted the focus from traditional businesses to those driven by environmental, social, and economic development norms. In recent years, several frameworks have evolved to accelerate the adoption of sustainable development. The Triple Bottom Line (TBL) framework talks about the shift from companies having a profit-earning focus earlier to wealth maximization, societal well-being, and ecological conservation. The long-term economic sustainability can be achieved by efficiently using the resources. The long-term social sustainability could be achieved by focusing on the well-being of the society (Sahoo & Upadhyay, 2024). These aspects go a long way in promoting value creation and resilience in a dynamically changing business world (Tarkar & Kumar, 2024). The challenges of the TBL framework can be addressed by blockchain technology, which would in turn benefit sustainable development in a big way. The economic, social, and environmental impacts of businesses can easily be transparently tracked by blockchain technology. It would help in increasing the trust of the shareholders in the system. Blockchain can go a long way in developing sustainable and ethical sourcing of raw materials, thereby ensuring a better supply chain process (Singh et al., 2024b; Upadhyay et al., 2021). It can help in carbon credit trading and help businesses reduce their carbon emissions and make them more accountable. The blockchain backed platforms could also help minorities by offering them better financial services, thereby reducing the social inequity in the market. The TBL approach coupled with blockchain can act as a key enabler in achieving sustainability (Sahoo & Upadhyay, 2024; Singh et al., 2024a). This integration not only promotes better accountability but also contributes to sustainable development globally. Blockchain technology can provide better innovative solutions to businesses, which would, in turn, drive economic growth and social well-being. These actions would make the world a better place to live for future generations (Nandi et al., 2021; Shojaei et al., 2021).

The remaining chapter is organized as follows: The next part outlines the unique role of blockchain across the sectors. Section 17.2 in this chapter presents the role of blockchain in driving sustainable economy. Section 17.4 discusses leveraging blockchain-based technologies in creating sustainable organization. Section 17.5 provides discussion related to the role of blockchain in sustainable finance and impact investing. Final section concludes the chapter.

17.1.1 Unique Role of Blockchain

Blockchain technology has greatly empowered cryptocurrencies and has helped in developing innovative financial systems, operational and financial efficiency of organizations, etc. Blockchain is a decentralized ledger that maintains transactions across computers in a decentralized manner. This very nature of blockchain ensures security and transparency, thereby making it an important tool in reducing inefficiencies and improving trust in economic processes (Basile et al., 2023; Khan et al., 2023). Blockchain applications in cryptocurrencies, supply chain management, identity verifications, cross-border payments, etc. make it a sought-after technological transformative tool. In a formal business setup, data integrity and automation of processes are bolstered using blockchain technology. One of the important features of blockchain is smart contracts which are fully automated and help in creating error-free workflows (Kouhizadeh et al., 2019). Smart contracts create a workflow for different processes, such as vendor payments, regulatory compliance, cost reduction, etc. Blockchain enables data sharing between the different departments of the organization through a secure data network. The application of blockchain in businesses can greatly help in recording the level of carbon footprints and ensuring that the raw materials are ethically sourced (Nandi et al., 2021). The financial services sector has also greatly appreciated the potential of blockchain as a technological platform. The look and feel of traditional banking have changed with the advances in technology such as blockchain and products such as digital asset trading and decentralized finance (DeFi) gaining wider importance. These financial products have greatly helped individuals and businesses to participate in global financial markets. Blockchain is also applied in Central Bank Digital Currencies (CBDCs), thereby signaling the relevance of the technology in the future (Alamsyah et al., 2024; Ozili, 2022). Auditing, reporting, and fraud reduction have been the key characteristics of the application of blockchain in business.

17.2 Leveraging Blockchain-Based Technologies for Sustainable Economy

Blockchain has significantly emerged as a driving force in making an economy more sustainable through enhanced accountability, efficiency, and transparency. The government and industries across the world are looking at achieving the sustainable development goals (SDGs) as prescribed by the United Nations (Singh et al., 2022; Khan et al., 2024). The blockchain-based technologies can act as a key building block in identifying the inefficiencies of the supply chain, misuse of resources, and even greenwashing. Blockchain provides records that cannot be changed, which ensures transparency of operations and provides a faith in the system to the shareholders to understand the legitimacy of claims. Supply chain systems that are backed by blockchain technology can identify the origin of materials and figure out whether it has been ethically sourced or not (Bhushan & Singh, 2024; Khan et al., 2022b). This helps to develop trust among customers as well as investors which internally encourages sustainable practices globally. A sustainable economy promotes the efficient utilization of resources and the reduction of waste (Kumar & Prakash, 2020). The combination of blockchain and sustainability can greatly benefit the circular economy (Kouhizadeh et al., 2023). Smart contracts help in various automated transactions that reduce indirect administrative expenses and promote better accountability. In developed energy markets, energy trading is also a common site, wherein households and businesses can exchange excess renewable energy smoothly (Yildizbasi, 2021). Blockchain shows a lot of promise in the sustainability domain, but the challenges of high energy consumption, economies of scale, and uncertainty around regulations remain an area of concern. Various energy-efficient methods and proof-of-stake (PoS) are meant to address these concerns, thereby making blockchain technology more relevant and scalable. These challenges can easily be overcome if private companies, government, and blockchain operators join hands and work towards achieving the full potential of the technology (Kumar & Chopra, 2022). The world is moving towards greener economies and the role of blockchain in achieving a sustainable future cannot be ignored.

17.2.1 Blockchain-Based Technologies for Waste Management, Recycling, Reuse and Resource Management

A circular economy is greatly transforming the landscape of sustainability by emphasizing waste elimination and continued usage of resources (Singh et al., 2023; Böckel et al., 2021). Blockchain is a great enabler in this process and provides a platform that increases transparency and security. It provides a decentralized system that helps in tracking and managing the resources. It uses advanced technology to track the different waste materials right from collection to its recycling (Singh et al., 2024b). This system provides better accountability and ensures that the environmental norms are better complied with. Blockchain offers immutable records which reduces fraud and illegal dumping to a large extent. The traceability feature of blockchain benefits the businesses involved in recycling. The life cycle of the raw materials from production to recycling is tracked efficiently by blockchain. Efficient tracking of raw materials also reduces the impurity levels in recycled products. Blockchain can also incentivize individuals and businesses by giving them rewards based on the level of their participation in recycling programs. This in turn greatly contributes to an efficient and sustainable supply chain (Khan et al., 2022a).

Blockchain is the key driver in the circular economy, and it advocates the efficient use of resources. The integration of Internet-of-Things (IoT) devices and smart contracts has contributed to sharing waste-related and resource data on a real-time basis. This would help companies to identify the areas in which these materials can be recycled. This centralized data can also prove useful for different companies that can make use of this information and could divert materials, which would create a symbiotic relationship between industries. Blockchain has the potential to help greatly in achieving long-term sustainability in a circular economy.

17.2.2 Enhancing Transparency in Trade Practices Through Blockchain

Fairtrade and ethical markets are important frameworks in ensuring transparency in trade practices globally. These systems try to balance power between different large-scale manufacturers in developing countries. There is a growing need for price determination that is transparent, and each manufacturer gets the right share from the proceeds of the sale of the goods. Ethical trade labels on a product could instill a sense of trust in the minds of the customers that these products have been sourced responsibly and have also met environmental and social norms (Kumar et al., 2024). These initiatives help build trust between the manufacturers and the customers and promote accountability between the different supply chain participants. Fair Trade is an important tool in ensuring that minorities can significantly increase their income levels. Artisans and small-scale farmers have been exploited in the markets for a long time. It provides a guarantee of minimum prices and a direct trading platform for the minority (Balzarova et al., 2022). Infrastructure, education, and healthcare sectors have been massive beneficiaries of such an initiative. Ethical markets encourage companies to follow sustainable practices right from raw material sourcing to reduce the current emissions. These efforts together contribute to achieving sustainable development by promoting social equity and conservation of the environment. Fairtrade and ethical markets are important frameworks in creating awareness for sustainability in the eyes of the customers as well as businesses (Kumar et al., 2021). Today, the consumers demand products that are in perfect alignment with their values and that also address the different environmental challenges and global inequalities. When customers choose buying goods that are ethically sourced, they indirectly choose to contribute towards a sustainable economy and it also forces companies to adopt fair trade practices (Balzarova et al., 2022; Sharma et al., 2021). These initiatives go a long way in bringing systemic change in fair trade and ethical markets, and it not only makes the system transparent and accountable but also safeguards the interests of all the stakeholders.

17.3 Blockchain and Sustainable Organizations

Blockchain technology has transformed the business landscape by promoting sustainable practices and ensuring accountability, efficiency, and transparency. It has greatly benefited supply chain management by tracing the materials from the origin till the point in time it reaches the end user. This transparency helps the company to identify the level of wastage, identify inefficiencies, and verify whether the products were ethically sourced (Sun et al., 2022). Blockchain technology provides decentralized ledgers which can act as evidence, and it can be presented to the customers and other stakeholders for verification as well as to develop trust (Castiglione et al., 2023). The characteristic of keeping immutable records helps to reduce the risk of fraud, noncompliance, etc., thereby creating a supply chain system that is sustainable and resilient. Smart contracts are an innovative tool and play an important role in achieving the sustainable development goals (SDGs) as prescribed by the UN (Singh et al., 2024c; Kumar & Tarkar, 2024). These contracts are self-executing and customized to follow the environmental and social compliance levels. The best part about such contracts is the fact that they involve less manual intervention. For example, smart contracts will allow the payment to be made to a supplier only if they meet the already laid down criteria around the use of renewable energy, carbon emissions, and sustainability. This level of customization reduces indirect administration expenses and promotes accountability in supply chain management. The transactions are streamlined and there is a reduction in delays, which further reduces the cost and saves resources for companies (Vacca et al., 2021; Schmidt et al., 2024).

Blockchain is a technology that not only affects the operational environment of a business but also develops the sustainability culture in a company. If the companies can integrate the different blockchain systems, they can get themselves aligned with the United Nations sustainable development goals (SDGs) framework (Singh et al., 2024d; Kouhizadeh et al., 2022). Blockchain feature of transparency in reporting sustainability data enhances the brand image and reputation of the company in the market. It also has a feature wherein the different stakeholders can exchange information on a platform that is transparent and fraud-proof. Search collaboration would promote innovation and would help in achieving the sustainability goals of the organization (Khan et al., 2021). Blockchain is a continuously evolving technology with a huge potential to drive innovation and change and could contribute a great deal to the sustainability of the organization.

17.3.1 Blockchain for ESG Reporting

With accountability and transparency gaining importance in the present era, the integration of environmental, social, and governance (ESG) reporting and blockchain is a landmark development. The reporting methods that were traditionally used had various disadvantages such as delays, inaccuracies, inefficiencies, etc. The

ESG framework plays an important role in estimating the ethical impact and sustainability of companies (Kumar et al., 2021; Wu et al., 2022). The ability of blockchain to collect data on a real-time basis addresses these challenges in a big way. The fact that it not only collects data on a real-time basis but also does data validation and data sharing fast as well adds to its importance. Companies can store the ESG data on the blockchain network, which makes the data fraud-proof and can also be used by the stakeholders whenever required (Wu et al., 2022). The different stakeholders can verify the legitimacy of the sustainability claims of the organization due to the technological capabilities of blockchain.

ESG reporting on a real-time basis backed by blockchain has the scope of making the compliance and decision-making process of an organization better. Companies face problems with unorganized data and a long process of verification which delays the disclosures required for sustainability (Kumar, 2022). The ability of blockchain to provide immutable records helps to tackle such challenges ensuring the integrity and efficiency of ESG data. The unique feature of blockchain smart contracts helps improve compliance by initiating reward actions if the company meets the carbon emission norms or undertakes social responsibility initiatives. Blockchain provided key insights into the performance of a company which helps them to actively manage the different opportunities and challenges thereby creating value for the organization as well as the society (Wu et al., 2022). The businesses of today are facing bigger challenges of social inequity, governance problems, and climate change, blockchain should be adopted for ESG reporting as it provides greater transparency and accountability (Mishra & Kumar, 2023). Stakeholders do expect that the companies would not only talk about sustainability but also try to achieve it. Companies feel empowered and they try to make a bigger impact with the data provided by blockchain for better ESG adoption (Almadadha, 2024). The combination of ESG reporting and blockchain will play an important role in the adoption of sustainable business practices and the development of a resilient and sustainable future.

17.4 Blockchain in Sustainable Finance

Blockchain is also affecting the financial sector by offering solutions that improve efficiency, accountability, and transparency in sustainable finance. The decentralized finance (DeFi) system is backed by blockchain technology, and it promotes financial inclusion. The DeFi promotes the use of a decentralized platform for distributing financial services to people across the country without intermediaries in between. Smart contracts facilitate better financial transactions and help in the safe transfer of assets so that capital access could be easier for people. The DeFi helps in funding projects that are eco-friendly and contributes to environmental sustainability through organizations in a significant manner. Crowdfunding and peer-to-peer lending are backed by blockchain and offer a decentralized and transparent mechanism which is an alternative to already existing financing techniques. This helps in driving different investments in different sustainability initiatives such as waste management, renewable energy, etc.

Blockchain application in green bonds has been a revelation owing to the role that it has played in the issue and management of green bonds. Green bonds are financial products that generate funding for environment-friendly projects. Blockchain helps in providing accountability and transparency in the sourcing and utilization of funds through green bonds (Malamas et al., 2024). The characteristic of an immutable ledger in blockchain develop a sense of trust in the minds of the funding parties, as they are in the loop about the projects that they are supporting. This aspect in turn helps in getting funding from more investors in the future. The process of issuing bonds traditionally was cost-intensive, so the issuance of green bonds backed by blockchain is significant in present era. This technology drives sustainable investment and addresses the challenges of greenwashing by providing a transparent and accountable platform to stakeholders for monitoring the impact of projects that are funded through green bonds (Giri et al., 2024; Malamas et al., 2024). Carbon credit is an area where the application of Blockchain can be huge. Carbon credits can be traded on the decentralized platform provided by blockchain where authenticity and traceability can be done for transactions that too fraud-free. The tokenization of carbon credits helps companies to get involved in programs of emission reduction. The verification of different processes is also done by blockchain, which helps in carbon offset and ensures that regulatory norms are complied with (Tian et al., 2020). This ensures accountability and helps to reward companies for adopting sustainable practices (Morrow & Zarrebini, 2019). Climate change has become a cause of concern for the world and carbon markets backed by blockchain could serve as a driving force to move toward a low carbon econom.

17.4.1 Blockchain and Ethical Investing

Blockchain has changed the world of ethical investing by addressing the issues of accountability, trust, and transparency. Ethical investing (also known as impact investing) deals with mapping the financial goals with ESG principles to benefit society. Traditionally, the investment tools that were used lacked transparency leading to wrong ESG claims. The fact that blockchain provides an immutable ledger and a decentralized platform helps in providing a solution by tracking investments in a transparent manner which ensures that funds are used for specified purposes and preventing false claims. The ability of blockchain to record data on a real-time basis helps in building trust among stakeholders thereby promoting ethical investments (Kim & Kim, 2021). Blockchain application in ethical investing improves data security and integrity which is important for the evaluation of ESG initiatives. Through smart contracts, blockchain makes sure that funds are distributed to people only when the eligibility criteria are met. The data provided by blockchain cannot be tampered. This provides investors with real-time accurate data through which they could also track their investments which helps them to make decisions faster.

In the area of renewable energy and energy production, carbon offset data could be easily tracked by blockchain which signifies its huge impact on the business. This level of application helps investors to help fund projects that are aligned with their SDG goals and sustainability values (Kumar & Prakash, 2020).

The collaboration and engagement of stakeholders is important in impact investing and blockchain helps with just that. It helps in effective communication between beneficiaries and investors through the decentralized platform for better reporting and monitoring. Through decentralized finance (DeFi), it helps in providing ethical investment options, and it also allows small investors to get involved in funding environmental projects that are beneficial for society (Ozili, 2022). Through the feature of tokenization in blockchain, one can take part ownership in assets which significantly ups the scalability aspect of impact investing. The innovations in blockchain are significant help in improving the effectiveness and accountability of ethical investing and bolstered the global transition towards a better sustainable financial system. Blockchain has redesigned the landscape of ethical investing by ensuring accountability, data integrity, and transparency (Kim & Kim, 2021). The ability to provide assessments that are verified and impactful helps in the collaboration of stakeholders; it becomes a major driving force in promoting sustainable investments (Alamsyah et al., 2024). There is a huge demand for ethical investing options, and the relevance of aligning the sustainability goals with financial systems has become increasingly important since it can have a massive societal and environmental impact. The ecosystem approach of sustainability focuses on maintaining relationships between the different stakeholders of the business so that the different challenges around the world can be addressed. To do this the different perspectives and resources have to be taken into account for addressing the needs of sustainable development. There is a need for greater collaboration between the stakeholders who can freely share information and make decisions based on the data available (Ozili, 2022). Blockchain could prove to be an essential component in such a collaboration as it focuses on decentralization, transparency, and accountability. Blockchain can go a long way in smoothening the communication between the stakeholders so that all these stakeholders have reliable and up-to-date data.

Blockchain can help in having good stakeholder interaction in a sustainable ecosystem which would allow the usage of resources, carbon footprint, and sustainability efforts to be tracked. It can help in establishing smart contracts and tokenization which incentivizes companies and individuals if they focus on sustainability, environmental, and social goals. It can verify the origin of raw materials and ensure that the supply chains are sustainable and ethical. The government can also use blockchain to oversee the compliance of regulatory norms and that the environmental norms are also adhered to. The blockchain helps in efficient data sharing, and the progress tracking towards the different goals of the organization also becomes easy. Blockchain and sustainability together can drive different business models that would promote environmental and social development. It helps the stakeholders to form a network where each stakeholder can contribute towards the goal of sustainability. Moreover, it provides a system that has integrity which reduces the number of frauds as well. This collaboration would lead to efficient and scalable solutions for different sustainability issues such as depletion, climate change, and societal inequality. Blockchain can greatly help in developing the ecosystem of sustainability by tackling issues and creating value for the entire planet.

17.5 Challenges and Barriers to Adoption of Blockchain

Blockchain faces many challenges in adoption by industries even though it offers several innovative solutions. The major challenge in the implementation of blockchain is that there are a lot of calculations and logarithms involved. There is a massive energy consumption in the implementation of blockchain which raises the issue of its environmental impact and the huge operational costs (Yildizbasi, 2021). The application of blockchain requires technical manpower who are skilled at the job, and the blockchain itself is a complex system which makes the adoption process slightly complicated (Balzarova et al., 2022). The governments around the world are still figuring out the regulations around blockchain applications, particularly in supply chain, healthcare, and finance. The legal systems are also unclear, and it looks ambiguous because of which the businesses might not adopt it. There is a lot of data that exchanges hands through blockchain, and there is a greater potential for the data getting misused which raises concerns around ethical issues and privacy (Tang et al., 2020). There are many technical and regulatory challenges in the adoption of blockchain, but the lack of awareness around the different issues about blockchain among the different stakeholders of the business is a major area of concern. Not many companies are comfortable with the implementation of a new technologies, as they concerned about volatility and threats that might arise because of the implementation of a new technologies. Blockchain implementation can only be successful if there is a collaboration between all the stakeholders (Böckel et al., 2021; Basile et al., 2023; Kouhizadeh et al., 2019). There might be trust issues between the stakeholders of an organization which can disrupt the process of blockchain implementation.

17.6 Conclusion

Blockchain has emerged as a potent instrument aimed at achieving a sustainable future impacting key sectors as well as the economy. The features of decentralization, traceability, and transparency have helped companies to follow sustainable practices. In the economic domain, blockchain influences efficiency and reduces waste through automation by eliminating intermediaries in due course. Moreover, it reduces the operational costs of the organization in a big way and fosters new business models to come up in the entire business ecosystem. The impact of blockchain on the company is huge and helps the organization to monitor the sustainability matrix more efficiently thereby promoting corporate governance norms coupled with accountability. The financial services sector has greatly benefited from the potential of blockchain in the areas of green investments, financial inclusion, and sustainable financing. The integration of blockchain in DeFi, carbon credits, and green bonds has huge potential for funding projects that are sustainable and reducing the effect of natural transactions on the environment as well. Tracking the supply chain in carbon credits on a real-time basis gives credibility to financial services products, and it also promotes more funding coming into projects that encourage sustainability goals. As this world is moving towards a future that is sustainable and the contribution of blockchain is extremely crucial. Financial institutions, policymakers, and organizations should join hands in blockchain implementation so that the challenges and barriers could be addressed effectively. The power of blockchain cannot be ignored, and it can create a lasting impact on companies, mitigate the risks of climate change, and ensure a resilient economic and environmental ecosystem that can make a very bright sustainable future.

References

Alamsyah, A., Kusuma, G. N. W., & Ramadhani, D. P. (2024). A review on decentralized finance ecosystems. *Future Internet, 16*(3), 76.

Almadadha, R. (2024). Blockchain technology in financial accounting: Enhancing transparency, security, and ESG reporting. *Blockchains, 2*(3), 312–333.

Balzarova, M., Dyer, C., & Falta, M. (2022). Perceptions of blockchain readiness for fairtrade programmes. *Technological Forecasting and Social Change, 185*, 122086.

Basile, D., D'Adamo, I., Goretti, V., & Rosa, P. (2023). Digitalizing circular economy through blockchains: The blockchain circular economy index. *Journal of Industrial and Production Engineering, 40*(4), 233–245.

Bhushan, M., & Singh, R. (2024). Talent management facilitates net-zero transition through employee green behavior. In *Net zero economy, corporate social responsibility and sustainable value creation: Exploring strategies, drivers, and challenges* (pp. 117–129). Springer Nature Switzerland.

Böckel, A., Nuzum, A. K., & Weissbrod, I. (2021). Blockchain for the circular economy: Analysis of the research-practice gap. *Sustainable Production and Consumption, 25*, 525–539.

Castiglione, A., Cimmino, L., Di Nardo, M., & Murino, T. (2023). A framework for achieving a circular economy using the blockchain technology in a sustainable waste management system. *Computers & Industrial Engineering, 180*, 109263.

Giri, G., Nivedhitha, K. S., & Manohar, H. L. (2024). Can blockchain enabled green bond issuance lead to intent to invest? A moderated mediation model. *Applied Economics, 56*(18), 2191–2208.

Khan, S. N., Loukil, F., Ghedira-Guegan, C., Benkhelifa, E., & Bani-Hani, A. (2021). Blockchain smart contracts: Applications, challenges, and future trends. *Peer-to-Peer Networking and Applications, 14*, 2901–2925.

Khan, S., Ali, S. S., & Singh, R. (2022a). Determinants of remanufacturing adoption for circular economy: A causal relationship evaluation framework. *Applied System Innovation, 5*(4), 62.

Khan, S., Singh, R., Sá, J. C., Santos, G., & Ferreira, L. P. (2022b). Modelling of determinants of logistics 4.0 adoption: Insights from developing countries. *Machines, 10*(12), 1242.

Khan, S., Khan, M. I., & Singh, R. (2023). Modeling the barriers of blockchain technology implementation in supply chain. *Journal of Industrial Integration and Management, 8*(3), 367–385.

Khan, S., Singh, R., Alnahas, J., Abbate, S., & Centobelli, P. (2024). Navigating the smart circular economy: A framework for manufacturing firms. *Journal of Cleaner Production, 480*, 144007.

Kim, J., & Kim, S. (2021). A survey of decentralized finance (DeFi) based on blockchain. *Journal of the Korea Society of Computer and Information, 26*(3), 59–67.

Kouhizadeh, M., Sarkis, J., & Zhu, Q. (2019). At the nexus of blockchain technology, the circular economy, and product deletion. *Applied Sciences, 9*(8), 1712.

Kouhizadeh, M., Zhu, Q., Alkhuzaim, L., & Sarkis, J. (2022). Blockchain technology and the circular economy: An exploration. In *Circular economy supply chains: From chains to systems* (pp. 189–213). Emerald Publishing Limited.

Kouhizadeh, M., Zhu, Q., & Sarkis, J. (2023). Circular economy performance measurements and blockchain technology: An examination of relationships. *The International Journal of Logistics Management, 34*(3), 720–743.

Kumar, K. (2022). Emerging phenomenon of corporate sustainability reporting: Evidence from top 100 NSE listed companies in India. *Journal of Public Affairs, 22*(1), e2368.

Kumar, N. M., & Chopra, S. S. (2022). Leveraging blockchain and smart contract technologies to overcome circular economy implementation challenges. *Sustainability, 14*(15), 9492.

Kumar, K., & Prakash, A. (2020). Managing sustainability in banking: extent of sustainable banking adaptations of banking sector in India. Environment, Development and Sustainability, 22, 5199–5217

Kumar, K., & Tarkar, P. (2024). Influence of sustainable human resource management on sustainability performance of the organization: Exploring the mediating role of organizational citizenship behaviour towards sustainability. *Journal of the Knowledge Economy*, 1–28.

Kumar, K., Kumari, R., & Kumar, R. (2021). The state of corporate sustainability reporting in India: Evidence from environmentally sensitive industries. *Business and Society Review, 126*(4), 513–538.

Kumar, K., Singh, R., Sharma, A., & Lehri, B. P. S. (2024). Climate change adaptation strategies for achieving net-zero economy. In *Transition towards a sustainable future: Net zero policies and environmental sustainability* (pp. 163–183). Springer Nature Singapore.

Malamas, V., Dasaklis, T. K., Arakelian, V., & Chondrokoukis, G. (2024). A blockchain framework for digitizing securities issuance: The case of green bonds. *Journal of Sustainable Finance & Investment, 14*(3), 569–595.

Mishra, P., & Kumar, K. (2023). Uncovering the sustainability reporting: Bibliometric analysis and future research directions. *International Journal of Disclosure and Governance, 21*(4), 520–542.

Morrow, M. J., & Zarrebini, M. (2019). Blockchain and the tokenization of the individual: Societal implications. *Future Internet, 11*(10), 220.

Nandi, S., Sarkis, J., Hervani, A. A., & Helms, M. M. (2021). Redesigning supply chains using blockchain-enabled circular economy and COVID-19 experiences. *Sustainable Production and Consumption, 27*, 10–22.

Ozili, P. K. (2022). Decentralized finance research and developments around the world. *Journal of Banking and Financial Technology, 6*(2), 117–133.

Sahoo, S., & Upadhyay, A. (2024). Improving triple bottom line (TBL) performance: Analyzing impacts of industry 4.0, lean six sigma and circular supply chain management. *Annals of Operations Research*, 1–32.

Schmidt, J. L., Sehnem, S., & Spuldaro, J. D. (2024). Blockchain and the transition to the circular economy: A literature review. *Corporate Social Responsibility and Environmental Management, 31*(3), 2010–2032.

Sharma, R., Kshetri, N., Wingreen, S., Shaikh, A. A., & Altamimi, F. (2021). UML artefacts for a blockchain-enabled platform for Fairtrade.

Shojaei, A., Ketabi, R., Razkenari, M., Hakim, H., & Wang, J. (2021). Enabling a circular economy in the built environment sector through blockchain technology. *Journal of Cleaner Production, 294*, 126352.

Singh, R., Khan, S., & Centobelli, P. (2022). Investigating the interplay between social performance and organisational factors supporting circular economy practices. *Sustainability, 14*(24), 16781.

Singh, R., Khan, S., & Dsilva, J. (2023). A framework for assessment of critical factor for circular economy practice implementation. *Journal of Modelling in Management, 18*(5), 1476–1497.
Singh, R., Crowther, D., & Ravesangar, K. (2024a). Harnessing human capital for a sustainable future: HR's pivotal role toward net zero policies. In *Transition towards a sustainable future: Net zero policies and environmental sustainability* (pp. 27–47). Springer.
Singh, R., Gupta, A., Bajpai, A., & Kandpal, V. (2024b). Exploring challenges of circular economy initiatives for smart cities. In *Smart cities and circular economy: The future of sustainable urban development* (pp. 59–69). Emerald Publishing Limited.
Singh, R., Khan, S., Dsilva, J., Akram, U., & Haleem, A. (2024c). Modelling the organisational factors for implementation of corporate social responsibility: A modified TISM approach. *Global Journal of Flexible Systems Management, 25*(2), 283–301.
Singh, R., Kumar, K., & Khan, S. (2024d). A comprehensive view of artificial intelligence (AI)–based technologies for sustainable development goals (SDGs). In *Artificial intelligence enabled management: An emerging economy perspective* (p. 183). De Gruyter.
Sun, Y., Shahzad, M., & Razzaq, A. (2022). Sustainable organizational performance through blockchain technology adoption and knowledge management in China. *Journal of Innovation & Knowledge, 7*(4), 100247.
Tang, Y., Xiong, J., Becerril-Arreola, R., & Iyer, L. (2020). Ethics of blockchain: A framework of technology, applications, impacts, and research directions. *Information Technology & People, 33*(2), 602–632.
Tarkar, P., & Kumar, K. (2024). Sustainable human resource management and sustainable performance: Mediating role of organization citizenship behavior. The bottom line.
Tian, Y., Lu, Z., Adriaens, P., Minchin, R. E., Caithness, A., & Woo, J. (2020). Finance infrastructure through blockchain-based tokenization. *Frontiers of Engineering Management, 7*(4), 485–499.
Upadhyay, A., Mukhuty, S., Kumar, V., & Kazancoglu, Y. (2021). Blockchain technology and the circular economy: Implications for sustainability and social responsibility. *Journal of Cleaner Production, 293*, 126130.
Vacca, A., Di Sorbo, A., Visaggio, C. A., & Canfora, G. (2021). A systematic literature review of blockchain and smart contract development: Techniques, tools, and open challenges. *Journal of Systems and Software, 174*, 110891.
Wu, W., Fu, Y., Wang, Z., Liu, X., Niu, Y., Li, B., & Huang, G. Q. (2022). Consortium blockchain-enabled smart ESG reporting platform with token-based incentives for corporate crowdsensing. *Computers & Industrial Engineering, 172*, 108456.
Yildizbasi, A. (2021). Blockchain and renewable energy: Integration challenges in circular economy era. *Renewable Energy, 176*, 183–197.

Chapter 18
Tokenization in Blockchain Technologies: Enabling Smart Circular Economy and Organizational Sustainability

Azamat Ali and Mikhail Tyurkin

Abstract This chapter explores the role of tokenization in blockchain technologies. Tokenization is a key enabler of the smart circular economy and organizational sustainability. Tokenization involves converting assets or value into digital tokens on a blockchain. This process offers new opportunities to enhance resource efficiency, transparency, and accountability within circular economy initiatives. By leveraging blockchain's immutable and decentralized nature, tokenization enables seamless tracking, trading, and sharing of resources. This creates incentives for sustainable practices. The chapter examines how tokenization intersects with sustainability. It highlights its potential in areas like waste management, energy sharing, and sustainable supply chains. The chapter also discusses how businesses can adopt tokenized systems to optimize operations. These systems help reduce waste and promote a regenerative model that minimizes environmental impact. Additionally, the chapter addresses the challenges of implementing tokenization at scale. It offers insights on how organizations can integrate blockchain solutions to drive long-term sustainability goals.

Keywords Tokenization · Blockchain · Smart circular economy · Sustainability · Resource efficiency

A. Ali (✉)
School of Liberal and Creative Arts (Journalism and Mass Communication), Lovely Professional University, Phagwara, Punjab, India

M. Tyurkin
School of Arts and Humanities, Department of Media, HSE University, St. Petersburg, Russia

R. Singh et al. (eds.), *Blockchain Technologies for Smart Circular Economy and Organisational Sustainability*, Environmental Footprints and Eco-design of Products and Processes, https://doi.org/10.1007/978-3-031-89183-0_18

18.1 Introduction

Blockchain technology has evolved beyond just cryptocurrencies, becoming a versatile tool for various applications. At its core, blockchain is a decentralized system that securely and transparently records data. Tokenization is an important blockchain aspect, involving the conversion of assets or values into digital tokens (Lamba, 2022). These tokens are stored and managed on the blockchain, making transactions faster, more secure, and transparent. Tokenization can be applied to physical assets like real estate or digital items such as copyrights or carbon credits with great potential to improve efficiency and reduce trade barriers (Eshghie et al., 2022). The smart circular economy is a sustainable model that focuses on reusing, recycling, and regenerating resources rather than discarding them. In this context, blockchain and tokenization are crucial. Blockchain technology can enable businesses to manage resources more effectively and align with circular economy principles by offering transparent, traceable, and decentralized systems (Khan et al., 2024). Tokenization plays a key role by tracking product and material flows, allowing businesses to monitor and engage in sustainable practices like recycling and reuse, helping reduce waste, improve resource management, and minimize environmental impact (Ranta & Saari, 2019).

Tokenization also helps improve organizational sustainability, encompassing social responsibility and efficient resource use beyond environmental goals (Richardson, 2022).

Tokenized assets allow companies to create new models for trading and sharing resources while ensuring sustainability principles. Businesses can use tokenization to trace product journeys from creation to disposal, ensuring sustainability standards, and manage supply chains more effectively, increasing transparency and reducing waste. Tokenized systems can incentivize sustainable practices, such as reducing energy consumption or using recycled materials (Yadav et al., 2023), crucial for reducing environmental impact and promoting sustainable business models (Khan et al., 2023). Furthermore, tokenization can create digital incentives for sustainable actions. For example, businesses can use blockchain-based tokens to trade renewable energy credits, making energy trading more efficient and transparent. Tokenization can also be used for carbon offset programs, making carbon trading more accountable and helping companies meet sustainability targets (Singh et al., 2024d; Kim & Huh, 2020). By linking the digital economy with real-world sustainability goals, tokenization supports environmentally responsible practices.

18.2 Understanding Tokenization

18.2.1 Definition and Process of Tokenization

Tokenization is a key blockchain concept that converts physical or digital assets into digital tokens, which are stored and managed on a decentralized, immutable blockchain (Guo & Zhou, 2023). This creates a digital representation of the asset, enabling secure transactions, ownership, and value exchange without centralized authorities. The process starts by identifying an asset, like real estate or intellectual property, and encoding its ownership and value into digital tokens. These tokens are then recorded and tracked on the blockchain, ensuring the asset's history and ownership are publicly visible, transparent, and immutable, providing enhanced trust and security compared to traditional systems (Sheldon, 2022). The main advantages of tokenization are increased liquidity, reduced transaction costs, and enhanced transparency. By representing assets digitally, tokenization allows for fractional ownership, enabling people to buy, sell, and trade smaller portions of high-value assets. This approach fosters accessibility to markets that may otherwise be difficult to enter, such as real estate and commodities, making tokenization particularly beneficial in a smart circular economy (Auer et al., 2023; Singh et al., 2022).

18.2.2 Types of Tokens

Blockchain technology enables the creation of different types of digital tokens, each serving a specific purpose within the ecosystem. The most common are utility tokens, asset-backed tokens, and security tokens.

- *Utility Tokens:* These tokens are used to access services or functionalities on a blockchain network. They are not tied to any underlying asset but facilitate transactions and access to network services. For example, Ethereum's ether is used to pay for transaction processing and execute smart contracts. Utility tokens drive blockchain adoption by creating incentives for participation and promoting sustainable practices in the circular economy (Auer et al., 2023).
- *Asset-Backed Tokens:* These tokens represent ownership of physical or digital assets, such as real estate, commodities, or intellectual property. The token value is directly linked to the underlying asset. Asset-backed tokens enable fractional ownership, increasing liquidity and accessibility to traditionally illiquid assets. For instance, tokenizing real estate allows dividing property into smaller, tradable units, lowering investment barriers. This makes asset-backed tokens essential for resource optimization in the circular economy, enabling efficient resource trading and sharing (Tian et al., 2022).
- *Security Tokens:* Security tokens are digital versions of traditional financial instruments, like stocks or bonds. They are regulated under existing laws and grant token holders rights similar to those of traditional securities, such as voting or dividend rights. Security tokens can be fractionalized, increasing their liquidity and reducing transaction costs compared to traditional markets. By bridging traditional finance and blockchain, security tokens facilitate the democratization of financial markets and support sustainability initiatives by improving access to capital (Manfrino, 2024; Auer et al., 2023).

18.2.3 How Tokenization Works on Blockchain Networks

Blockchain technology provides a decentralized and immutable record of transactions. The blockchain ledger ensures token ownership and transaction history cannot be altered, offering trust and transparency. Blockchain's decentralized nature eliminates the need for intermediaries, reducing transaction costs and increasing efficiency (Sharif & Ghodoosi, 2022). Smart contracts are self-executing contracts with predefined conditions encoded on the blockchain. They automatically execute token transfers when specific conditions are met, eliminating intermediaries and ensuring compliance with agreed terms. This automation streamlines the process, making transactions faster, more reliable, and transparent (Schär, 2021). Tokenization enhances resource sharing in the circular economy. For instance, in energy sharing, tokenization enables peer-to-peer exchanges of renewable energy credits, where each token represents a specific amount of energy produced or consumed. These tokens can be traded, creating a transparent and efficient system for promoting sustainable energy practices. Tokenization can be applied to other sectors, like waste management and sustainable supply chains, by enabling real-time tracking and trading of resources. Tokenized systems can help optimize supply chains, reduce waste, and encourage renewable resource use. Blockchain's transparency and traceability allow organizations to track resources from production to consumption, supporting the circular economy's regenerative model (Difrancesco et al., 2022; Khan et al., 2022b).

18.3 The Smart Circular Economy

18.3.1 Key Principles of the Circular Economy

The circular economy is a sustainable model that differs from the traditional linear economy, where products are made, used, and discarded. The key goals are to keep products, materials, and resources in use for as long as possible, reduce waste, and create value through reuse, refurbishment, remanufacturing, and recycling (Iacovidou et al., 2020). The circular economy operates on the idea that businesses can contribute to sustainability and drive economic growth by rethinking product design, production, and consumption.

The circular economy has three main principles:

- Design for longevity and repairability: Products are made to be durable, easy to maintain, and easily disassembled to enable material reuse.
- Resource efficiency: Minimize waste by reusing materials instead of discarding them, and reduce the environmental impacts of raw material extraction and manufacturing.

- Restorative and regenerative systems: Recycle and transform products or materials into new ones instead of disposing them in landfills (Kumar et al., 2024; Iacovidou et al., 2020).

18.3.2 Benefits of Transitioning to a Circular Economy Model

Transitioning to a circular economy model offers significant benefits for businesses and the environment. One key advantage is reduced resource consumption. By reusing materials and extending product lifespans, the circular economy lowers demand for raw materials, conserving natural resources and reducing environmental impacts from extraction and waste disposal (Singh et al., 2023; Ranta & Saari, 2019). This also decreases energy use, as recycling requires less energy than producing new materials. Furthermore, the circular economy provides financial benefits. By reusing materials and products, companies can lower operating costs, particularly in industries with expensive or scarce raw materials. Businesses adopting circular principles can also create new revenue streams through refurbishment, remanufacturing, and used product resale. This model encourages innovation in product design and production, often leading to more efficient, cost-effective, and sustainable solutions (Singh et al., 2024b; Cao et al., 2018). From an environmental standpoint, the circular economy helps reduce greenhouse gas emissions, waste, and pollution. Strategies that minimize raw material extraction and processing, and promote recycling, contribute to lowering carbon emissions associated with manufacturing. Additionally, the model reduces waste sent to landfills, easing pressure on waste management systems and decreasing environmental pollution (Bhushan & Singh, 2024; Nascimento et al., 2018).

18.3.3 How Blockchain and Tokenization Align with Circular Economy Goals

Blockchain and tokenization are vital for transitioning to a circular economy. Blockchain provides a transparent, secure, and decentralized platform to track assets, transactions, and supply chains, making it well-suited for managing circular economy systems. Tokenization, which converts physical or digital assets into digital tokens, plays a crucial role in ensuring efficient execution and monitoring of circular economy processes (Eshghie et al., 2022). The key advantage of blockchain is its ability to offer transparency and traceability across supply chains. Tracking materials is crucial in a circular economy to ensure proper reusing, remanufacturing, or recycling. Blockchain technology enables secure recording of resource-related transactions, ensuring accurate tracking from production to end-of-life. This transparency helps reduce fraud, enhance stakeholder trust, and verify sustainability claims. Tokenized systems can track product and material lifecycles, from production to recycling, ensuring business compliance with regulations and consumer verification of environmental impact (Kouhizadeh & Sarkis, 2018). Tokenized platforms enable decentralized markets for sharing and trading resources. These platforms facilitate businesses exchanging unused or excess resources, such as energy credits, waste materials, or recycled goods. Blockchain automation ensures efficient resource allocation without intermediaries, optimizing resource use and reducing waste. Tokenization also incentivizes participation in circular economy activities by creating financial rewards for sustainable practices. Tokens can represent value for renewable energy generation, waste reduction, or material recycling. Tokenization encourages responsible and sustainable behaviors, with token rewards driving change (Barclay et al., 2022).

18.4 Tokenization and Sustainability

18.4.1 Enhancing Resource Efficiency Through Tokenized Systems

Tokenization is crucial for enhancing resource efficiency, especially in industries with complex supply chains and resource management. By converting physical assets into digital tokens on a blockchain network, tokenization enables real-time tracking and management of resources, promoting greater efficiency in resource allocation (Tian et al., 2022). For example, tokenized systems can track the lifecycle of materials, ensuring optimal use and preventing waste. This is particularly beneficial in manufacturing, where raw materials can be reused or recycled, reducing the need for new resources and minimizing environmental impact (Aurisicchio et al., 2020). In the circular economy context, tokenization facilitates the flow of resources, such as raw materials, energy, or waste, between stakeholders, while ensuring transparency and accountability in their management. By assigning tokens to physical assets, businesses and individuals can more easily share and trade resources, incentivizing efficient material use. Tokenized systems help close the loop in the circular economy, ensuring resources continue to circulate rather than being discarded. This process of continuous value generation from existing resources significantly enhances resource efficiency, a key pillar of sustainability (Ranta & Saari, 2019).

18.4.2 Increasing Transparency and Accountability in Resource Management

Tokenization enhances transparency and accountability in resource management. Blockchain's features, such as immutability and decentralization, record all transactions involving tokenized assets on a public ledger. This provides greater visibility into how resources are handled, tracked, and traded throughout their lifecycle. In waste management, tokenized systems can track waste from origin to final disposal or recycling. By assigning a unique token to each waste unit, stakeholders can monitor its journey through the recycling process or disposal facility, ensuring proper handling. This transparency reduces fraud and assures consumers and regulators that waste practices are environmentally responsible (Singh et al., 2024a; Goldsmith et al., 2017). Additionally, blockchain-based tokenization can improve supply chain transparency. Tokenization can record and verify every supply chain step, from raw material sourcing to manufacturing and distribution. This allows businesses and consumers to access detailed information about the provenance of goods and services, ensuring they meet sustainability standards. This transparency can increase accountability, enabling companies to adhere to ethical sourcing, reduce environmental impact, and avoid greenwashing (Ali, 2024; McGrath et al., 2021).

18.5 Applications of Tokenization in the Circular Economy

18.5.1 Waste Management and Recycling Initiatives

Tokenization provides a transformative solution for waste management by creating systems that reward and incentivize recycling and waste reduction. Traditional waste management systems often face challenges such as inefficiency, lack of transparency, and difficulty tracking waste from origin to recycling or disposal. In contrast, tokenization allows for the digital tracking of waste materials through blockchain-based systems, where each waste unit is assigned a unique token. These tokens can be traded, redeemed, or used to claim rewards, making recycling and waste reduction more appealing for businesses and consumers (Taylor et al., 2020). A notable example is the initiative by Evergreen Recycling in Singapore. This platform assigns digital tokens to users for every recycling activity they perform. The tokens can be exchanged for rewards or used for further environmental initiatives. The blockchain-based system ensures transparent tracking of each piece of waste, guaranteeing proper management, sorting, and recycling. This incentivizes individuals and businesses to recycle and encourages closed-loop systems where waste is continually reused and repurposed, aligning with circular economy principles (Karunarathna, et al., 2024; Ranta & Saari, 2019).

Tokenization has also found significant applications in decentralized energy markets, enabling efficient energy sharing and consumption. With the rise of renewable energy sources like solar and wind, tokenization provides a way for individuals and businesses to trade their excess energy in a decentralized manner (Bañales, 2020). In a typical tokenized energy system, surplus energy generated from a home or business's solar panels can be converted into energy tokens. These tokens can then be sold or traded to other consumers within a local energy marketplace, reducing reliance on traditional energy providers and promoting more sustainable energy consumption. A notable example is the Power Ledger platform in Australia, which facilitates peer-to-peer energy trading through tokenization. Homeowners with solar panels can tokenize their excess energy and sell it directly to others in their community (Green & Newman, 2017). This empowers individuals to control their energy consumption and production while encouraging renewable energy sources. Tokenization ensures that every energy transaction is tracked, transparent, and verifiable on the blockchain, improving accountability and providing users with a sense of security about the provenance and sustainability of their energy. Furthermore, blockchain technology in decentralized energy markets allows for more resilient and efficient energy systems, making it a key enabler of sustainability.

18.5.2 Sustainable Supply Chain Management

Tokenization plays a crucial role in sustainable supply chain management. It helps track the flow of goods and resources from origin to end consumer. Traditional supply chains often lack transparency, making it difficult to verify the sustainability of materials, labor conditions, or carbon footprint (Nikolakis et al., 2018). Tokenization addresses this by assigning digital tokens to each supply chain stage, recording and verifying actions like sourcing, manufacturing, and transportation. For instance, Everledger uses blockchain to track luxury goods, such as diamonds, from source to sale. Tokenizing these goods ensures each step is transparent, verifiable, and aligned with sustainability goals. Tokenization allows businesses to demonstrate ethical sourcing, reduce environmental impact, and avoid greenwashing. This transparency enables consumers to make more informed, sustainable purchasing decisions (Rinaldi et al., 2022). Additionally, tokenized supply chain systems enable more efficient inventory management by facilitating real-time exchange and trade of goods. Businesses can track raw material and product availability, reducing excess production or waste. This creates a dynamic, responsive supply chain that can quickly adapt to changing demands while minimizing waste and inefficiency.

18.5.3 Tokenization of Physical Assets and Commodities

Tokenization expands beyond waste management, energy, and supply chains, reaching a broader range of physical assets and commodities. By converting physical assets like real estate, machinery, or agricultural products into digital tokens, new opportunities arise for managing, trading, and utilizing these assets more efficiently and sustainably. For instance, tokenizing real estate enables fractional ownership, making it easier for individuals to invest in and trade property. This provides liquidity to traditionally illiquid markets and ensures more efficient asset utilization (Markheim & Berentsen, 2021). Similarly, in agriculture, tokenization can track crop production and trade, ensuring sustainable farming practices. By tokenizing crops like wheat, coffee, or cotton, farmers can monitor their produce from seed to sale, verifying environmental responsibility and sustainability criteria. Tokenized systems can also improve farmers' financial access, allowing them to trade tokens representing future harvests or products, providing capital for sustainable investments (Singh et al., 2024c; Kamilaris et al., 2019). Tokenizing physical assets and commodities allows businesses and individuals to better manage asset life cycles, ensuring efficient use, transparent trade, and reuse whenever possible. This contributes to reducing waste and promoting a circular approach to resource management, maximizing the value, and minimizing the environmental impact of physical assets.

18.6 Business Adoption of Tokenized Systems

Businesses are increasingly adopting tokenized systems to improve efficiency, reduce waste, and promote sustainability. Tokenization converts physical assets, rights, or processes into digital tokens on a blockchain, providing transparency, security, and efficiency. These systems can help businesses lower costs, streamline operations, and encourage circular economy practices.

18.6.1 Implementing Tokenized Systems in Business

Implementing tokenized systems in businesses requires a strategic approach that combines blockchain technology with existing infrastructure. The first step is understanding the type of tokenization that aligns with the company's goals. Businesses can tokenize physical assets, intangible assets, or operational processes like waste management or energy use. The decision should align with the company's sustainability and efficiency objectives (Laarabi & Maach, 2020). One example is VeChain, a supply chain platform that uses blockchain to track products from production to sale. By assigning digital tokens to products, businesses gain transparency and can verify the origin, manufacturing processes, and environmental impact. Tokenization enables real-time tracking, reducing errors and fraud while increasing efficiency. Tokenized systems can help track raw material sources to ensure sustainability and monitor the carbon footprint of transportation. To implement tokenization, businesses must invest in blockchain infrastructure, either by developing their own network or partnering with a provider. They must also integrate the tokenized system with existing operations and educate stakeholders on how to interact with it. This involves training, user-friendly interfaces, and ensuring stakeholders understand the associated incentives. Incorporating tokenized systems into everyday processes can encourage participation in sustainable practices (Schletz et al., 2023).

18.6.2 Reducing Operational Costs and Waste Through Tokenization

One of the primary advantages of adopting tokenized systems is the potential to tokenization that helps businesses streamline operations and reduce waste. By creating digital tokens for physical or intangible resources, companies can track usage in real time. This allows them to identify inefficiencies, like overproduction, and take corrective action. Tokenized systems provide transparency, enabling data-driven decisions to minimize waste (Romero et al., 2018). For instance, businesses can tokenize energy consumption data to optimize usage and incentivize renewable energy. Tokenization can also automate processes and improve transparency, reducing costs associated with inventory management, quality verification, and audits. By eliminating these inefficiencies, companies can lower overall costs and waste.

18.6.3 Creating Incentives for Sustainability and Circular Economy Practices

Tokenized systems can encourage sustainable and circular economy practices by rewarding stakeholders for eco-friendly actions. For example, companies can offer customer's tokens for returning used items like packaging or electronics for recycling or reuse. These tokens can then be used for future purchases or exchanged for other rewards, fostering a culture of waste reduction (Kaviyarasu, 2024). Similarly, tokenized systems can reward individuals and organizations that generate renewable energy with digital tokens that can be traded in local energy markets, incentivizing the adoption of renewable energy. Moreover, tokenization can facilitate decentralized marketplaces for waste and recycled materials. Businesses can tokenize their waste and sell it to others who can reuse or repurpose it, closing the circular economy loop. This provides a financial incentive for businesses to adopt circular practices by monetizing waste and reducing disposal costs (Chowdhury et al., 2024). Tokenization also supports the transparency and traceability needed for businesses to demonstrate their sustainability efforts, helping them build trust with stakeholders and differentiate themselves in the market.

18.7 Challenges in Implementing Tokenization at Scale

Implementing tokenization on a larger scale poses various challenges for businesses. These challenges arise from technological, operational, regulatory, and adoption-related hurdles that organizations must overcome to successfully integrate tokenized systems into their operations. This section examines the main challenges businesses face when trying to scale up tokenization and discusses potential solutions.

18.7.1 Technological and Operational Barriers

One of the most significant barriers to scaling tokenization is the technological complexity involved in integrating blockchain systems with existing infrastructure. Blockchain technology is still relatively nascent, and many organizations lack the technical expertise to implement and maintain tokenized systems (Eshghie et al., 2022). For businesses to fully capitalize on the benefits of tokenization, they must invest in specialized blockchain technology that can handle the demands of a large-scale operation. This includes not only building or purchasing blockchain platforms but also ensuring the security and scalability of these systems. Blockchain networks, especially those based on proof-of-work consensus mechanisms, can struggle with scalability, leading to slow transaction processing times and high fees as the network grows. Moreover, integrating blockchain with legacy systems can be a time-consuming and costly process. Many businesses operate on traditional databases and have systems that are not designed to work with decentralized technologies. This lack of interoperability between existing systems and blockchain platforms can hinder the smooth adoption of tokenized solutions (Esmaeilian et al., 2023). Businesses may need to overhaul their technology infrastructure, which can be an expensive and resource-intensive endeavor. Additionally, maintaining these systems at scale requires significant investment in cybersecurity to safeguard sensitive data and prevent cyberattacks, as blockchain systems can become targets for hackers (Pandey, 2020).

18.7.2 Regulatory Challenges and Legal Considerations

The regulatory landscape poses a major challenge to widespread tokenization adoption. Blockchain and tokenization are new technologies, and governments are still learning how to regulate them. This creates uncertainty for businesses that want to implement tokenized systems. The decentralized and anonymous nature of blockchain transactions makes it difficult for regulators to track and control tokenized assets, leading to concerns about money laundering, fraud, and tax evasion (Schär, 2021). Existing laws often do not adequately address the unique aspects of tokenization. For example, tokenizing assets like real estate or intellectual property may not be explicitly covered by current legal frameworks. This uncertainty can make businesses hesitant to adopt tokenization, as they may face legal issues or need to consult lawyers to ensure compliance. Businesses operating across multiple jurisdictions must navigate a complex web of varying regulations. The lack of a unified global regulatory framework for tokenization also poses a challenge for businesses trying to scale their tokenized systems internationally. Companies that tokenize assets may have trouble ensuring compliance with different regulations on security tokens, privacy laws, and cross-border transactions (Ellul et al., 2020). The classification of tokens as securities or commodities can also vary by location, potentially leading to legal complications or additional compliance costs. To mitigate these risks, businesses may need to work closely with legal experts and regulators to ensure their tokenized systems comply with relevant laws and regulations.

18.7.3 Public and Business Adoption Hurdles

Widespread adoption of tokenization remains a major challenge, even with technological and regulatory progress. Many businesses and consumers lack familiarity with blockchain and tokenization, leading to hesitancy due to lack of understanding or trust. Businesses must invest in education and awareness campaigns to help stakeholders comprehend how these systems work and their benefits. Adopting tokenization involves a significant cultural shift for businesses, from traditional centralized models to decentralized systems. This transition often requires rethinking processes and workflows, which can face resistance from employees, stakeholders, and leadership. Integrating tokenization may also necessitate changes to business models and strategies, which can be met with reluctance if the potential return on investment is unclear (Yeung, 2023). Consumer adoption poses challenges too. While blockchain promises transparency and security, consumers may still struggle to trust and engage with tokenized systems. Many are unfamiliar with digital tokens and have concerns about cryptocurrency volatility or blockchain security. Consumers may be skeptical about using blockchain for everyday transactions or sustainability practices. Businesses need to develop user-friendly interfaces, create incentives, and build trust through initiatives like audits or transparency reports. Furthermore, the costs of adopting tokenization can be prohibitive for small- and medium-sized enterprises that lack resources to invest in blockchain infrastructure or navigate complex regulations (Eshghie et al., 2022). For these businesses, tokenization may seem out of reach, slowing widespread adoption. Government support, industry partnerships, and collaboration with service providers can help SMEs overcome these barriers and foster more inclusive tokenization adoption (Khan et al., 2022a).

18.8 Conclusion

This study has explored how tokenization in blockchain technologies can transform the circular economy and sustainability practices. Tokenization, which converts physical assets into digital tokens, enables better resource management, transparency, and accountability. Tokenized systems allow seamless tracking, trading, and sharing of resources, incentivizing sustainable practices, optimizing supply chains, and reducing waste and energy use. Businesses can adopt tokenization to improve efficiency, reduce waste, and support regenerative models that minimize environmental impact. Integrating tokenization with blockchain offers new opportunities for driving sustainability in industries like waste management, energy sharing, and sustainable supply chains. However, implementing these systems at scale faces technological, regulatory, and adoption challenges. Despite these obstacles, blockchain's ability to track sustainability performance and provide transparent, immutable records makes it a powerful tool for achieving long-term sustainability goals. Looking ahead, the continued evolution of blockchain and increased adoption of tokenized systems are likely to play a significant role in the future of sustainability. As blockchain becomes more scalable and accessible, its applications in the circular economy will grow, offering more innovative solutions for organizations to meet sustainability targets.

References

Ali, A. (2024). Navigating road for net zero policies to aligning policy, politics, and governance for environmental sustainability. In R. Singh & D. Crowther (Eds.), *Transition towards a sustainable future: Approaches to global sustainability, markets, and governance* (p. xx). Springer. https://doi.org/10.1007/978-981-97-5756-5_10

Auer, R., Haslhofer, B., Kitzler, S., Saggese, P., & Victor, F. (2023). The technology of decentralized finance (DeFi). *Digital Finance, 6*(1), 55. https://doi.org/10.1007/s42521-023-00088-8

Aurisicchio, M., van der Laan, A. Z., & Tennant, M. (2020). Material-service systems for sustainable resource management. In *Sustainable production, life cycle engineering and management* (p. 89). Springer Nature. https://doi.org/10.1007/978-981-15-6779-7_7

Bañales, S. (2020). The enabling impact of digital technologies on distributed energy resources integration. *Journal of Renewable and Sustainable Energy, 12*(4), 045301. https://doi.org/10.1063/5.0009282

Barclay, I., Cooper, M., Hackel, J., & Perrin, P. (2022). Tokenizing behavior change: A pathway for the sustainable development goals. *Frontiers in Blockchain, 4*. https://doi.org/10.3389/fbloc.2021.730101

Bhushan, M., & Singh, R. (2024). Talent management facilitates net-zero transition through employee green behavior. In *Net zero economy, corporate social responsibility and sustainable value creation: Exploring strategies, drivers, and challenges* (pp. 117–129). Springer Nature. https://doi.org/10.1007/978-3-031-55779-8_7

Cao, J., Chen, X., Zhang, X., Gao, Y., Zhang, X., Zhao, Y., Yang, X., Xu, J., Zhou, G., & Schnoor, J. L. (2018). Public awareness of remanufactured products in Yangtze River Delta of China: Present status, problems and recommendations. *International Journal of Environmental Research and Public Health, 15*(6), 1199. https://doi.org/10.3390/ijerph15061199

Chowdhury, M. J. M., Hassan, N. U., Tushar, W., Niyato, D., Saha, T. K., Poor, H. V., & Yuen, C. (2024). Blockchain-enabled circular economy—Collaborative responsibility in solar panel recycling. arXiv. https://doi.org/10.48550/arxiv.2403.09937

Difrancesco, R. M., Meena, P., & Kumar, G. (2022). How blockchain technology improves sustainable supply chain processes: A practical guide. *Operations Management Research, 16*(2), 620. https://doi.org/10.1007/s12063-022-00343-y

Ellul, J., Galea, J., Ganado, M., McCarthy, S., & Pace, G. J. (2020). Regulating blockchain, DLT and smart contracts: A technology regulator's perspective. *ERA Forum, 21*(2), 209. https://doi.org/10.1007/s12027-020-00617-7

Eshghie, M., Quan, L., Kasche, G. A., Jacobson, F., Bassi, C., & Artho, C. (2022). CircleChain: Tokenizing products with a role-based scheme for a circular economy. arXiv. https://doi.org/10.48550/arxiv.2205

Esmaeilian, B., Sarkis, J., Lewis, K., & Behdad, S. (2023). Blockchain for the future of sustainable supply chain management in Industry 4.0. *Resources, Conservation and Recycling, 163*, 105064. https://www.sciencedirect.com/science/article/pii/S0921344920303815

Goldsmith, W., Jeberg, S., Alex, J., Johnsen, W., Gurau, B., & Lindquist, E. (2017). Net zero waste: Issues, technologies, trends, and commercially viable solutions. In *NATO science for peace and security series. C, environmental security* (p. 219). Springer Nature (Netherlands). https://doi.org/10.1007/978-94-017-7600-4_11

Green, J., & Newman, P. (2017). Citizen utilities: The emerging power paradigm. *Energy Policy, 105*, 283–293. https://doi.org/10.1016/j.enpol.2017.02.004

Guo, D., & Zhou, P. (2023). The evolution of financial market infrastructure: From digitalization to tokenization. *International Journal of Innovation and Entrepreneurship, 2*(1), 1–12. https://doi.org/10.56502/ijie2010002

Iacovidou, E., Hahladakis, J. N., & Purnell, P. (2020). A systems thinking approach to understanding the challenges of achieving the circular economy. *Environmental Science and Pollution Research, 28*(19), 24785–24802. https://doi.org/10.1007/s11356-020-11725-9

Kamilaris, A., Fonts, A., & Prenafeta-Boldú, F. X. (2019). The rise of blockchain technology in agriculture and food supply chains. *Trends in Food Science & Technology, 91*, 640–652. https://doi.org/10.1016/j.tifs.2019.07.034

Karunarathna, K. S. S. N., Dewasiri, N. J., Singh, R., & Rathnasiri, M. S. H. (2024). What does artificial intelligence–powered ChatGPT bring to academia? A review. In *Artificial intelligence enabled management: An emerging economy perspective* (p. 3). Walter de Gruyter GmbH & Co KG.

Kaviyarasu, K. (2024). Performance of strategies implemented as part of a management approach from an ecological standpoint. *Discover Environment, 2*(1), 129. https://doi.org/10.1007/s44274-024-00161-y

Khan, S., Ali, S. S., & Singh, R. (2022a). Determinants of remanufacturing adoption for circular economy: A causal relationship evaluation framework. *Applied System Innovation, 5*(4), 62. https://doi.org/10.3390/asi5040062

Khan, S., Singh, R., Sá, J. C., Santos, G., & Ferreira, L. P. (2022b). Modelling of determinants of logistics 4.0 adoption: Insights from developing countries. *Machines, 10*(12), 1242. https://doi.org/10.3390/machines10121242

Khan, S., Khan, M. I., & Singh, R. (2023). Modeling the barriers of blockchain technology implementation in supply chain. *Journal of Industrial Integration and Management, 8*(3), 367–385. https://doi.org/10.1142/S2424862223500070

Khan, S., Singh, R., Alnahas, J., Abbate, S., & Centobelli, P. (2024). Navigating the smart circular economy: A framework for manufacturing firms. *Journal of Cleaner Production, 480*, 144007. https://doi.org/10.1016/j.jclepro.2024.144007

Kim, S.-K., & Huh, J. (2020). Blockchain of carbon trading for UN sustainable development goals. *Sustainability, 12*(10), 4021. https://doi.org/10.3390/su12104021

Kouhizadeh, M., & Sarkis, J. (2018). Blockchain practices, potentials, and perspectives in greening supply chains. *Sustainability, 10*(10), 3652. https://doi.org/10.3390/su10103652

Kumar, K., Singh, R., Sharma, A., & Lehri, B. P. S. (2024). Climate change adaptation strategies for achieving net-zero economy. In *Transition towards a sustainable future: Net zero policies and environmental sustainability* (pp. 163–183). Springer Nature. https://doi.org/10.1007/978-981-97-5756-5_8

Laarabi, M., & Maach, A. (2020). Understanding risk assessment in the context of fractional ownership using ethereum smart contract. *Advances in Science, Technology and Engineering Systems Journal, 5*(5), 1028. https://doi.org/10.25046/aj0505126

Lamba, A. (2022). Are carbon offsets the key to green cryptocurrencies? *PLOS Sustainability and Transformation, 1*(3), e0000002. https://doi.org/10.1371/journal.pstr.0000002

Manfrino, M. (2024). Beyond carbon: Navigating the path to biodiversity credits for climate action and ecosystem conservation. *SSRN Electronic Journal.* https://doi.org/10.2139/ssrn.4687255

Markheim, M., & Berentsen, A. (2021). Real estate trifft auf blockchain: Chancen und Herausforderungen der Tokenisierung von illiquiden Vermögenswerten. *Zeitschrift für Immobilienökonomie, 7*(1), 59. https://doi.org/10.1365/s41056-020-00051-3

McGrath, P., McCarthy, L., Marshall, D., & Rehme, J. (2021). Tools and technologies of transparency in sustainable global supply chains. *California Management Review, 64*(1), 67. https://doi.org/10.1177/00081256211045993

Nascimento, D. L. M., Alencastro, V., Quelhas, O. L. G., Caiado, R. G. G., Garza-Reyes, J. A., Rocha-Lona, L., & Tortorella, G. L. (2018). Exploring industry 4.0 technologies to enable circular economy practices in a manufacturing context. *Journal of Manufacturing Technology Management, 30*(3), 607. https://doi.org/10.1108/jmtm-03-2018-0071

Nikolakis, W., John, L., & Krishnan, H. (2018). How blockchain can shape sustainable global value chains: An evidence, verifiability, and enforceability (EVE) framework. *Sustainability, 10*(11), 3926. https://doi.org/10.3390/su10113926

Pandey, C. (2020). Scalability challenges and opportunities in blockchain-based systems: A systematic review. *Türk Bilgisayar ve Matematik Eğitimi Dergisi, 11*(3), 2022. https://doi.org/10.17762/turcomat.v11i3.13599

Ranta, V., & Saari, U. A. (2019). Circular economy: Enabling the transition towards sustainable consumption and production. In *Encyclopedia of the UN sustainable development goals* (p. 1). Springer International Publishing. https://doi.org/10.1007/978-3-319-71062-4_3-1

Richardson, N. (2022). How new sustainability typologies will reshape traditional approaches to loyalty. *Italian Journal of Marketing, 2022*(3), 289. https://doi.org/10.1007/s43039-022-00047-y

Rinaldi, F. R., Bernardino, C. D., Cram-Martos, V., & Pisani, M. T. (2022). Traceability and transparency: Enhancing sustainability and circularity in garment and footwear. *Sustainability Science Practice and Policy, 18*(1), 132. https://doi.org/10.1080/15487733.2022.2028454

Romero, D., Gaiardelli, P., Powell, D., Wuest, T., & Thürer, M. (2018). Digital lean cyber-physical production systems: The emergence of digital lean manufacturing and the significance of digital waste. In *IFIP advances in information and communication technology* (p. 11). Springer Science+Business Media. https://doi.org/10.1007/978-3-319-99704-9_2

Schär, F. (2021). Decentralized finance: On blockchain- and smart contract-based financial markets. *Review, 103*(2), 153–174. https://doi.org/10.20955/r.103.153-74

Schletz, M., Constant, A., Hsu, A., Schillebeeckx, S. J. D., Beck, R., & Wainstein, M. (2023). Blockchain and regenerative finance: Charting a path toward regeneration. *Frontiers in Blockchain, 6.* https://doi.org/10.3389/fbloc.2023.1165133

Sharif, M., & Ghodoosi, F. (2022). The ethics of blockchain in organizations. *Journal of Business Ethics, 178*(4), 1009. https://doi.org/10.1007/s10551-022-05058-5

Sheldon, M. D. (2022). Tracking tangible asset ownership and provenance with blockchain. *Journal of Information Systems, 36*(3), 153. https://doi.org/10.2308/isys-2020-042

Singh, R., Khan, S., & Centobelli, P. (2022). Investigating the interplay between social performance and organisational factors supporting circular economy practices. *Sustainability, 14*(24), 16781. https://doi.org/10.3390/su142416781

Singh, R., Khan, S., & Dsilva, J. (2023). A framework for assessment of critical factors for circular economy practice implementation. *Journal of Modelling in Management, 18*(5), 1476–1497. https://doi.org/10.1108/JM2-06-2021-0145

Singh, R., Crowther, D., & Ravesangar, K. (2024a). Harnessing human capital for a sustainable future: HR's pivotal role toward net zero policies. In *Transition towards a sustainable future: Net zero policies and environmental sustainability* (pp. 27–47). Springer. https://doi.org/10.1007/978-981-97-5756-5_2

Singh, R., Gupta, A., Bajpai, A., & Kandpal, V. (2024b). Exploring challenges of circular economy initiatives for smart cities. In V. Kandpal, E. D. Santibanez-Gonzalez, P. Chatterjee, & M. K. Nallapaneni (Eds.), *Smart cities and circular economy* (pp. 59–69). Emerald Publishing Limited. https://doi.org/10.1108/978-1-83797-957-820241005

Singh, R., Khan, S., Dsilva, J., et al. (2024c). Modelling the organisational factors for implementation of corporate social responsibility: A modified TISM approach. *Global Journal of Flexible Systems Management, 25*, 283–301. https://doi.org/10.1007/s40171-024-00388-x

Singh, R., Kumar, K., & Khan, S. (2024d). A comprehensive view of artificial intelligence (AI)–based technologies for sustainable development goals (SDGs). In *Artificial intelligence enabled management: An emerging economy perspective* (p. 183). Walter de Gruyter GmbH & Co KG.

Taylor, P., Steenmans, K., & Steenmans, I. (2020). Blockchain technology for sustainable waste management. *Frontiers in Political Science, 2*, 590923. https://doi.org/10.3389/fpos.2020.590923

Tian, Y., Minchin, R. E., Chung, K., Woo, J., & Adriaens, P. (2022). Towards inclusive and sustainable infrastructure development through blockchain-enabled asset tokenization: An exploratory case study. *IOP Conference Series: Materials Science and Engineering, 1218*(1), 012040. https://doi.org/10.1088/1757-899X/1218/1/012040

Yadav, N., Luthra, S., & Garg, D. (2023). Blockchain technology for sustainable supply chains: A network cluster analysis and future research propositions. *Environmental Science and Pollution Research, 30*(24), 64779. https://doi.org/10.1007/s11356-023-27049-3

Yeung, S. M. C. (2023). Transformative learning via integrated projects with sustainable development goals and innovations. *Corporate Governance and Sustainability Review, 7*(2), 29. https://doi.org/10.22495/cgsrv7i2p3

Index

R. Singh et al. (eds.), *Blockchain Technologies for Smart Circular Economy and Organisational Sustainability*, Environmental Footprints and Eco-design of Products and Processes, https://doi.org/10.1007/978-3-031-89183-0

The manufacturer's authorised representative in the EU is Springer Nature Customer Service Centre GmbH, Europaplatz 3, 69115 Heidelberg, Germany. If you have any concerns regarding our products, please contact ProductSafety@springernature.com

Printed and bound by CPI Group (UK) Ltd, Croydon, CR0 4YY
30/09/2026
02225563-0006